THEORY OF DISTRIBUTIONS
THE SEQUENTIAL APPROACH

Modern Analytic *and* Computational Methods *in* Science *and* Mathematics

A GROUP OF MONOGRAPHS AND ADVANCED TEXTBOOKS

Richard Bellman, EDITOR

Published

1. R. E. Bellman, R. E. Kalaba, and Marcia C. Prestrud, Invariant Imbedding and Radiative Transfer in Slabs of Finite Thickness, 1963
2. R. E. Bellman, Harriet H. Kagiwada, R. E. Kabala, and Marcia C. Prestrud, Invariant Imbedding and Time-Dependent Transport Processes, 1964
3. R. E. Bellman and R. E. Kalaba, Quasilinearization and Nonlinear Boundary-Value Problems, 1965
4. R. E. Bellman, R. E. Kalaba, and Jo Ann Lockett, Numerical Inversion of the Laplace Transform: Applications to Biology, Economics, Engineering, and Physics, 1966
5. S. G. Mikhlin and K. L. Smolitskiy, Approximate Methods for Solution of Differential and Integral Equations, 1967
6. R. N. Adams and E. D. Denman, Wave Propagation and Turbulent Media, 1966
7. R. L. Stratonovich, Conditional Markov Processes and Their Application to the Theory of Optimal Control, 1968
8. A. G. Ivakhnenko and V. G. Lapa, Cybernetics and Forecasting Techniques, 1967
9. G. A. Chebotarev, Analytical and Numerical Methods of Celestial Mechanics, 1967
10. S. F. Feshchenko, N. I. Shkil', and L. D. Nikolenko, Asymptotic Methods in the Theory of Linear Differential Equations, 1967
11. A. G. Butkovskiy, Distributed Control Systems, 1969
12. R. E. Larson, State Increment Dynamic Programming, 1968
13. J. Kowalik and M. R. Osborne, Methods for Unconstrained Optimization Problems, 1968
14. S. J. Yakowitz, Mathematics of Adaptive Control Processes, 1969
15. S. K. Srinivasan, Stochastic Theory and Cascade Processes, 1969
16. D. U. von Rosenberg, Methods for the Numerical Solution of Partial Differential Equations, 1969
17. R. B. Banerji, Theory of Problem Solving: An Approach to Artificial Intelligence, 1969
18. R. Lattès and J.-L. Lions, The Method of Quasi-Reversibility: Applications to Partial Differential Equations. Translated from the French edition and edited by Richard Bellman, 1969
19. D. G. B. Edelen, Nonlocal Variations and Local Invariance of Fields, 1969
20. J. R. Radbill and G. A. McCue, Quasilinearization and Nonlinear Problems Fluid and Orbital Mechanics, 1970
21. W. Squire, Integration for Engineers and Scientists, 1970
22. T. Parthasarathy and T. E. S. Raghavan, Some Topics in Two-Person Games, 1971
23. T. Hacker, Flight Stability and Control, 1970

24. D. H. Jacobson and D. Q. Mayne, Differential Dynamic Programming, 1970

25. H. Mine and S. Osaki, Markovian Decision Processes, 1970

26. W. Sierpiński, 250 Problems in Elementary Number Theory, 1970

27. E. D. Denman, Coupled Modes in Plasmas, Elastic Media, and Parametric Amplifiers, 1970

28. F. A. Northover, Applied Diffraction Theory, 1971

29. G. A. Phillipson, Identification of Distributed Systems, 1971

30. D. H. Moore, Heaviside Operational Calculus: An Elementary Foundation, 1971

31. S. M. Roberts and J. S. Shipman, Two-Point Boundary Value Problems: Shooting Methods, 1971

32. V. F. Demyanov and A. M. Rubinov, Approximate Methods in Optimization Problems, 1970

33. S. K. Srinivasan and R. Vasudevan, Introduction to Random Differential Equations and Their Applications, 1971

34. C. J. Mode, Multitype Branching Processes: Theory and Applications, 1971

35. R. Tomović and M. Vukobratović, General Sensitivity Theory, 1971

36. J. G. Krzyż, Problems in Complex Variable Theory, 1971

37. W. T. Tutte, Introduction to the Theory of Matroids, 1971

38. B. W. Rust and W. R. Burrus, Mathematical Programming and the Numerical Solution of Linear Equations, 1971

39. J. O. Mingle, The Invariant Imbedding Theory of Nuclear Transport, 1973

40. H. M. Lieberstein, Mathematical Physiology, 1973

THEORY
OF DISTRIBUTIONS
THE SEQUENTIAL APPROACH

by

PIOTR ANTOSIK

Special Research Centre of the Polish Academy of Sciences in Katowice

JAN MIKUSIŃSKI

Special Research Centre of the Polish Academy of Sciences in Katowice

ROMAN SIKORSKI

University of Warsaw

ELSEVIER SCIENTIFIC PUBLISHING COMPANY

AMSTERDAM

PWN—POLISH SCIENTIFIC PUBLISHERS

WARSZAWA

1973

Distribution of this book is being handled by the following publishers:

for the U.S.A. and Canada

American Elsevier Publishing Company, Inc.
52 Vanderbilt Avenue
New York, New York 10017

for Albania, Bulgaria, Chinese People's Republic, Czechoslovakia, Cuba, German Democratic Republic, Hungary, Korean People's Democratic Republic, Mongolia, Poland, Rumania, Democratic Republic of Vietnam, U.S.S.R., and Yugoslavia

PWN — Polish Scientific Publishers
00-251 Warszawa (Poland), Miodowa 10

for all remaining areas

Elsevier Scientific Publishing Company
335 Jan van Galenstraat
P.O. Box 1270, Amsterdam, The Netherlands

International Standard Book Number 0-444-41082-1
Library of Congress Catalog Card Number 73-78246

PRINTED IN POLAND (DRP)

Contents

Preface

In Classical Analysis, continuous functions are not necessarily differentiable. Distributions are, roughly speaking, a generalization of the concept of functions such that each continuous function becomes differentiable. Its derivative is a distribution. Moreover, each distribution is differentiable; its derivative is another distribution.

In order to introduce distributions one can proceed in various ways. One of the earliest methods is the method of functionals, founded by Sobolev [44] and Schwartz [33]. Actually there exists a large literature based on their description. A disadvantage of the functional approach is that it requires deep knowledge of functional analysis, which makes the theory difficult for average users. Consequently, any attempt to write a lucid and self-contained textbook results in a large introductory part, which largely exceeds the part devoted to the theory of distributions itself.

A more recent approach consists in considering distributions as limits of sequences of functions. The logical construction of such limits is based on Cantor's concept of equivalence classes, very familiar in mathematics. The sequential approach is not only simpler, but also more in agreement with the intuition of physicists: already Dirac, the precursor of the distribution theory, was aware of fact that his "function" delta could be approximated by sequences of ordinary functions.

Some advocates of the functional approach claim that their method is the only one that offers flexibility and power; this is of course not an accurate statement, but an expression of their belief. In fact, to each distribution in the functional approach there corresponds one distribution in the sequential approach, and conversely. Moreover, each theorem on distributions which has been proved by functional methods can also be proved by sequential methods. The point is that the proofs are easier or more difficult, depending on the case.

Besides the functional and sequential methods, there are also other possibilities of introducing distributions (see e.g., [9], [35], [37], [45]).

This book is based on the sequential method and was primarily intended as an elementary introduction to distributions. Its first part is concerned with the simplest case: distributions of finite order on the real line. The second part contains the generalization to arbitrary distributions in Euclidean spaces of any number of dimensions. Both parts are almost unaltered reprints of publications* which appeared a few years ago. This causes a certain inconsistency of the structure, form and notation, but shows the way in which the ideas were being gradually developed by two of the authors, namely Mikusiński and Sikorski. The third part has been written by Mikusiński in collaboration with Antosik. It includes some deeper and partly new results, and may be of interest also to specialists. More information about the contents is given in the separate introductions to each part.

* The first edition appeared in English under the title *The Elementary Theory of Distributions* (Warsaw, 1957 and 1961). Further editions appeared in Russian (1959 and 1963), in Chinese (1960), in French (1964) and in Polish (1964).

Part I

Elementary theory of distributions of a single real variable

Introduction to Part I

The purpose of Part I is to present the foundations of the theory of distributions in such a manner as to make it comprehensible to physicists and engineers as well as to mathematicians. To achieve this we shall not avail ourselves of the methods of functional analysis and we shall not define the distributions as functionals. In applied mathematics distributions are regarded as ordinary functions, e.g. the function $\delta(x)$ of Dirac. Essentially, however, distributions are not functions but in an intuitive sense, they may be approximated by functions. Approximation, strictly defined, is our starting point for the definition of distributions. This definition leads us in a natural way to denote distributions in the same way as functions, enabling us to retain the same form for analytical formulas and to make use of familiar methods of calculation.

Other authors have also felt the need for a simpler treatment of the theory of distributions by basing the definition of a distribution on simpler notions. This is to be seen in several works on the foundations of distributions (Halperin [8], König [9], Korevaar [10], Mikusiński [16], [17], Sikorski [37], Słowikowski [42], [43], Temple [45]).

Every distribution is a derivative of a continuous function. We establish this property early and develop the rest of the theory on the basis of both aspects of a distribution: as a limit of continuous functions and as a derivative of a function. This helps to make the proofs of all the theorems elementary and very simple.

In Part I we are mainly concerned with the theory of distributions of finite order, their role being fundamental. In Section 4.3 we show how to extend the basic definitions and theorems to the case of distributions of infinite order.

Some results included in Part I are the outcome of seminar work conducted by the authors at the Mathematical Institute of the Polish Academy of Sciences in the years 1954–55 and 1955–56. The contributions of S. Łojasiewicz, K. Urbanik, J. Wloka and Z. Zieleźny have been particularly useful in its preparation.

In Part I we restrict ourselves to distributions of one real variable. The theory of distributions of several variables will be the subject of Parts II and III.

1. Fundamental definitions

1.1. THE IDENTIFICATION PRINCIPLE

The identification principle consists in the identification of objects (mathematical entities) which have a common property. It is often applied in mathematics to construct new notions. We shall explain it by means of examples.

Oriented segments x and y are said to be *equivalent* if they are parallel and have the same length and orientation. We then write

$$x \sim y.$$

It is easy to see that the relation of equivalence defined above has the following properties:

($\mathscr{E}_1$) *$x \sim x$ (reflexivity);*

($\mathscr{E}_2$) *if $x \sim y$, then $y \sim x$ (symmetry);*

($\mathscr{E}_3$) *if $x \sim y$ and $y \sim z$, then $x \sim z$ (transitivity).*

Identifying equivalent oriented segments we obtain the notion of free vector. We now explain the mathematical sense of this identification. By means of the equivalence relation we obtain a decomposition of the set of all oriented segments into disjoint classes such that segments in the same class are equivalent and segments in different classes are not equivalent. Thus, from the logical point of view, each free vector is a class of equivalent oriented segments.

Another example of the identification principle is Cantor's definition of real numbers. Here the starting point is the notion of fundamental sequences of rational numbers. By a "fundamental sequence" we mean a sequence $\{a_n\}$ satisfying the Cauchy condition: for every (rational) number $\varepsilon > 0$ there exists an index n_0 such that

$$|a_m - a_n| < \varepsilon \quad \text{for} \quad m, n > n_0.$$

Fundamental sequences $\{a_n\}$ and $\{b_n\}$ of rational numbers are said to be equivalent if the sequence $\{a_n - b_n\}$ converges to zero, in which case we write $\{a_n\} \sim \{b_n\}$. It is easy to verify that the relation of equivalence defined above has the properties ($\mathscr{E}_1$), ($\mathscr{E}_2$) and ($\mathscr{E}_3$).

Identifying equivalent fundamental sequences we obtain the notion of real numbers. In the Cantor theory a real number is thus a class of equivalent fundamental sequences.

The identification principle can be applied to sets of arbitrary elements provided there exists an equivalence relation $\sim$, satisfying the conditions $(\mathscr{E}_1)$, $(\mathscr{E}_2)$ and $(\mathscr{E}_3)$.

For each element y, let us denote by $[y]$ the class of all elements x satisfying the relation $x \sim y$. The classes $[y]$ thus obtained will be called *equivalence classes.* It follows from $(\mathscr{E}_1)$, $(\mathscr{E}_2)$ and $(\mathscr{E}_3)$ that

(a) *y belongs to $[y]$;*

(b) *if $y \sim z$, then $[y] = [z]$, i.e., the classes $[y]$ and $[z]$ have the same elements;*

(c) *if the relation $y \sim z$ does not hold, then the classes $[y]$ and $[z]$ have no common element.*

Property (a) follows from $(\mathscr{E}_1)$.

To prove (b) suppose that $y \sim z$. If x belongs to $[y]$, then $x \sim y$. Thus $x \sim z$ on account of $(\mathscr{E}_3)$, i.e., x belongs to $[z]$. On the other hand, it follows from $(\mathscr{E}_2)$ that $z \sim y$. Therefore, if x belongs to $[z]$, i.e., $x \sim z$, then also $x \sim y$ by $(\mathscr{E}_3)$, i.e. x belongs to $[y]$.

To prove (c) suppose that the relation $y \sim z$ does not hold and that there exists an element x belonging to $[y]$ and $[z]$. Then $x \sim y$ and $x \sim z$ and, by $(\mathscr{E}_2)$ and $(\mathscr{E}_3)$, $y \sim z$ contrary to hypothesis.

It follows from (a), (b) and (c) that the whole set is decomposed into equivalence classes without common elements so that two elements are in the same equivalence class if and only if they are equivalent. The identification of equivalent elements consists in the passage from the elements of the set in question to the equivalence classes. The relation of equivalence is then transformed into simple equality.

1.2. FUNDAMENTAL SEQUENCES OF CONTINUOUS FUNCTIONS

Distributions are a generalization of functions. We shall introduce them in a way analogous to that used in the Cantor theory when generalizing rational to real numbers. The purpose of introducing real numbers is to make certain operations always possible, e.g., the evaluation of roots or logarithms. The purpose of introducing distributions is to make differentiation always feasible, which is not so for functions, even when continuity is assumed.

While rational numbers are the starting point in the Cantor theory, functions, continuous in a fixed interval $A < x < B$ $(-\infty \leqslant A < B \leqslant \infty)$, are the starting point for the theory to be developed here.

A sequence $\{f_n(x)\}$ of continuous functions defined for $A < x < B$ is said to be *fundamental* if there exists a sequence $\{F_n(x)\}$ of functions and an integer $k \geqslant 0$ such that

(F_1) $F_n^{(k)}(x) = f_n(x)$;

(F_2) *the sequence* $\{F_n(x)\}$ *converges almost uniformly.*

We say that a sequence $\{F_n(x)\}$ converges to a function $F(x)$ *almost uniformly* in the interval $A < x < B$ and we write

$$F_n(x) \rightrightarrows F(x)$$

if it converges to $F(x)$ uniformly on each finite closed interval contained in the interval $A < x < B$.

For instance, $x/n \rightrightarrows 0$ in the interval $-\infty < x < \infty$. It is slightly more difficult to see that $(1+x/n)^n \rightrightarrows e^x$ for $-\infty < x < \infty$. The sequence of partial sums of any power series converges almost uniformly in the interval of convergence.

Obviously, each uniformly convergent sequence also converges almost uniformly. The limit of an almost uniformly convergent sequence of continuous functions is itself continuous.

We write

$$F_n(x) \rightrightarrows$$

to denote that the sequence $\{F_n(x)\}$ converges almost uniformly to a function. We write

$$F_n(x) \rightrightarrows\leftleftarrows G_n(x)$$

to denote that the sequences $\{F_n(x)\}$ and $\{G_n(x)\}$ converge almost uniformly to the same function.

It follows directly from the definition, taking $k = 0$, that

1.2.1. *Each almost uniformly convergent sequence of continuous functions is fundamental.*

Some other examples of fundamental sequences will be given after the following lemmas.

1.2.2. *If* $\{f_n(x)\}$ *is a fundamental sequence of functions with continuous mth derivatives* $f_n^{(m)}(x)$, *then the sequence* $\{f_n^{(m)}(x)\}$ *is also fundamental.*

If the sequence $\{f_n(x)\}$ satisfies conditions (F_1) and (F_2), then the sequence $\{f_n^{(m)}(x)\}$ satisfies the condition $F_n^{(k+m)}(x) = f_n^{(m)}(x)$ and (F_2). This proves that $\{f_n^{(m)}(x)\}$ is fundamental.

1.2.3. *If a sequence* $\{f_n(x)\}$ *of continuous functions is bounded and* $f_n(x) \rightrightarrows f(x)$ *in the intervals* $A < x < x_0$ *and* $x_0 < x < B$, *then* $\int_{x_0}^{x} f_n(t)\,dt \rightrightarrows \int_{x_0}^{x} f(t)\,dt$ *in the interval* $A < x < B$. *Hence the sequence* $\{f_n(x)\}$ *is fundamental.*

Suppose that $|f_n(x)| \leqslant M$. Given any $\varepsilon > 0$ and an interval $a \leqslant x \leqslant b$ $(A < a < x_0 < b < B)$, we fix an index n_0 such that $|f_n(x)-f(x)| < \varepsilon/2(b-a)$ for $n > n_0$ in the intervals $a \leqslant x \leqslant x_0-\varepsilon/4M$ and $x_0+\varepsilon/4M \leqslant x \leqslant b$. Then for $n > n_0$, the integral of $|f_n(x)-f(x)|$ is $< \varepsilon/2$ over each of those intervals and also over each of the intervals $x_0-\varepsilon/4M \leqslant x \leqslant x_0$, $x_0 \leqslant x \leqslant x_0+\varepsilon/4M$. Consequently

$$\left|\int_{x_0}^{x} f_n(t)\,dt - \int_{x_0}^{x} f(t)\,dt\right| < \varepsilon \quad \text{for} \quad a \leqslant x \leqslant b \quad \text{and for} \quad n > n_0,$$

which proves the lemma.

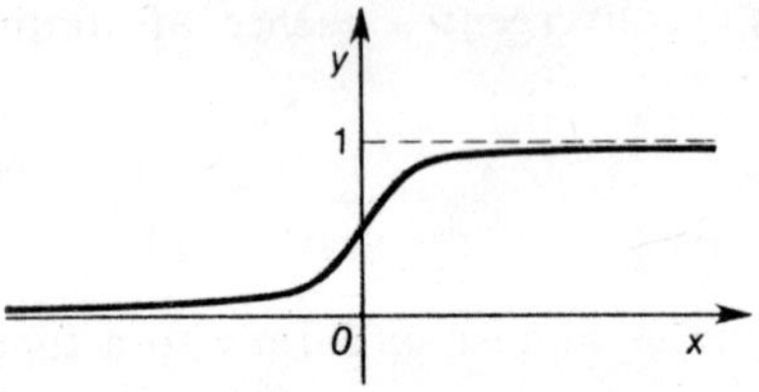

FIG. 1.1

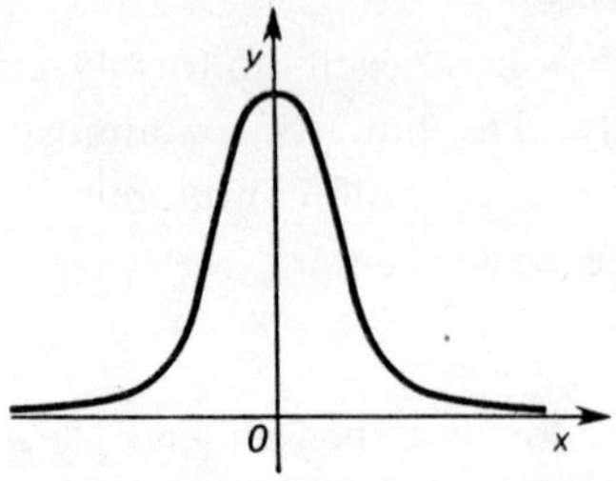

FIG. 1.2

EXAMPLES OF FUNDAMENTAL SEQUENCES. 1. The sequence

$$g_n(x) = 1/(1+e^{-nx})$$

is bounded by the number 1, $g_n(x) \rightrightarrows 0$ in $-\infty < x < 0$ and $g_n(x) \rightrightarrows 1$ in $0 < x < \infty$. In view of 1.2.3 it is fundamental.

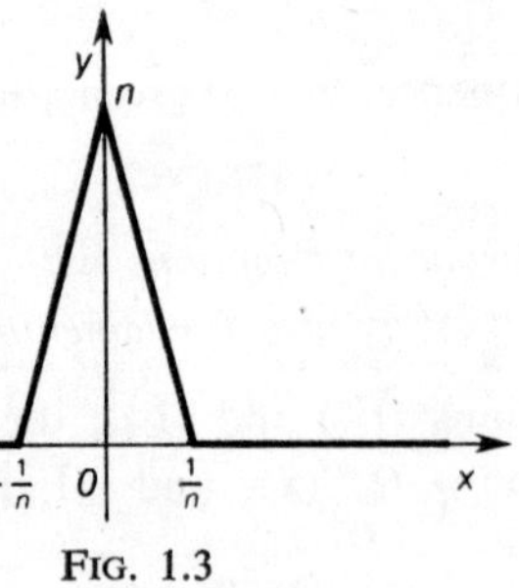

FIG. 1.3

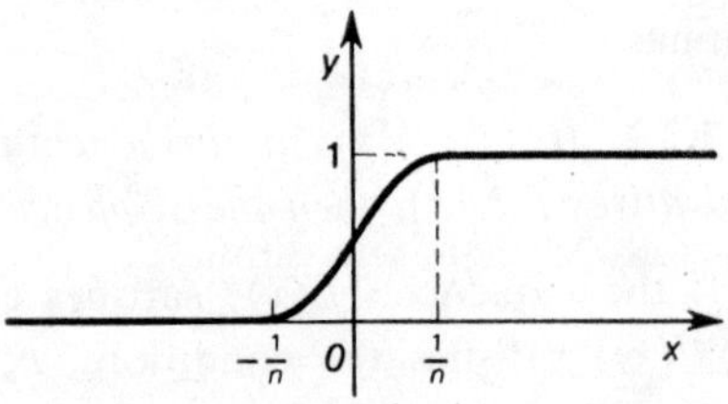

FIG. 1.4

2. The sequence

$$f_n(x) = (\sqrt{n/2\pi})\, e^{-nx^2/2}$$

is fundamental. In fact, the sequence

$$g_n(x) = \int_{-\infty}^{x} f_n(t)dt$$

is bounded by the number 1, $g_n(x) \rightrightarrows 0$ in $-\infty < x < 0$ and $g_n(x) \rightrightarrows 1$ in $0 < x < \infty$. By 1.2.3 the sequence $\{g_n(x)\}$ is fundamental. Hence, by 1.2.2, so is the sequence $\{f_n(x)\}$.

3. The sequence of functions $\{f_n(x)\}$ shown in Fig. 1.3 is fundamental since the functions $g_n(x) = \int_{-\infty}^{x} f_n(t)\,dt$ have the same properties as those in examples 1 and 2, which is easily seen from Fig. 1.4.

The following lemma is helpful when considering fundamental sequences of polynomials:

1.2.4. *If a sequence of polynomials of degree* $< k$

$$p_n(x) = a_{n0} + a_{n1}x + \dots + a_{n,k-1}x^{k-1} \qquad (n = 1, 2, \dots) \tag{1}$$

converges at k *points, then the limits*

$$a_j = \lim_{n\to\infty} a_{nj} \qquad (j = 0, 1, \dots, k-1) \tag{2}$$

exist. Conversely, if the limits (2) *exist, then* $p_n(x) \rightrightarrows p(x)$, *where*

$$p(x) = a_0 + a_1 x + \dots + a_{k-1}x^{k-1}.$$

Let $x_1, \dots, x_k$ be distinct numbers. Then, substituting these numbers into (1), we find that

$$u_{nj} = \frac{1}{A}(A_{1j}p_n(x_1) + \dots + A_{kj}p_n(x_k)),$$

where

$$A = \begin{vmatrix} 1 & x_1 & \dots & x_1^{k-1} \\ \dots & \dots & \dots & \dots \\ 1 & x_k & \dots & x_k^{k-1} \end{vmatrix} \neq 0$$

and A_{ij} are minors of A. Hence, if the limits $\lim_{n\to\infty} p_n(x_i)$ $(i = 1, 2, \dots, k)$ exist, then the limits (2) also exist. The uniform convergence in every interval $-c \leqslant x \leqslant c$ follows from the estimate

$$|p_n(x) - p(x)| \leqslant \sum_{j=0}^{n-1} |a_{nj} - a_j|\, c^j.$$

1.2.5. *A sequence* $\{p_n(x)\}$ *of polynomials of degree* $< m$ *is fundamental if and only if it converges almost uniformly.*

By 1.2.1 we have only to prove the necessity. If $\{p_n(x)\}$ is fundamental, then there exist an integer $k \geqslant 0$ and a sequence $\{P_n(x)\}$ of polynomials of degree $< m+k$ such that $P_n^{(k)}(x) = p_n(x)$ and $\{P_n(x)\}$ converges almost uniformly. By 1.2.4 the coefficients of $P_n(x)$ converge. Consequently the coefficients of $p_n(x)$ also converge. It follows from 1.2.4 that $\{p_n(x)\}$ converges almost uniformly.

It is useful to know that the integer k which occurs in the difinition of fundamental sequences can, if necessary, be replaced by any greater integer. This results from the following lemma:

1.2.6. *If* $\{F_n(x)\}$ *satisfies* (F_1) *and* (F_2), *then the sequence*

$$\tilde{F}_n(x) = \int_{x_0}^{x} dt_1 \int_{x_0}^{t_1} dt_2 \ldots \int_{x_0}^{t_{l-1}} F_n(t_l)\, dt_l,$$

where l is a positive integer, also satisfies (F_1) *and* (F_2) *where k is replaced by* $k+l$. *Moreover, if* $F_n(x) \rightrightarrows F(x)$, *then* $\tilde{F}_n(x) \rightrightarrows \tilde{F}(x)$ *where*

$$\tilde{F}(x) = \int_{x_0}^{x} dt_1 \int_{x_0}^{t_1} dt_2 \ldots \int_{x_0}^{t_{l-1}} F(t_l)\, dt_l.$$

1.3. THE DEFINITION OF DISTRIBUTIONS

We say that two fundamental sequences $\{f_n(x)\}$ and $\{g_n(x)\}$ are *equivalent* and we write

$$\{f_n(x)\} \sim \{g_n(x)\}$$

if there exist sequences $\{F_n(x)\}$ and $\{G_n(x)\}$ and an integer $k \geqslant 0$ such that

(E_1) $F_n^{(k)}(x) = f_n(x)$ *and* $G_n^{(k)}(x) = g_n(x)$;

(E_2) $F_n(x) \rightrightarrows \leftleftarrows G_n(x)$.

It follows from 1.2.6 that

1.3.1. *The integer k, appearing in the definition of equivalent sequences, can, if necessary, be replaced by any greater integer.*

For this purpose it suffices to replace the functions $F_n(x)$ by $\tilde{F}_n(x)$ and the functions $G_n(x)$ by the functions $\tilde{G}_n(x)$ defined analogously.

1 3.2. *Fundamental sequences* $\{f_n(x)\}$ *and* $\{g_n(x)\}$ *are equivalent if and only if the sequence*

$$f_1(x),\ g_1(x),\ f_2(x),\ g_2(x),\ \ldots \tag{1}$$

is fundamental.

If the sequence (1) is fundamental, then there exist an integer $k \geqslant 0$ and continuous functions $F_n(x)$ and $G_n(x)$ such that $F_n^{(k)}(x) = f_n(x)$, $G_n^{(k)}(x) = g_n(x)$ and the sequence

$$F_1(x),\ G_1(x),\ F_2(x),\ G_2(x),\ \ldots \tag{2}$$

converges almost uniformly. Consequently the conditions (E_1) and (E_2) are satisfied.

Conversely, suppose that (E_1) and (E_2) are satisfied. Then the sequence (2) converges almost uniformly, i.e., sequence (1) satisfies conditions (F_1) and (F_2).

It is easy to see that the relation $\sim$ satisfies conditions ($\mathscr{E}_1$) and ($\mathscr{E}_2$). We shall prove that the condition ($\mathscr{E}_3$) is also satisfied. If $\{f_n(x)\} \sim \{g_n(x)\}$ and $\{g_n(x)\} \sim \{h_n(x)\}$, then there exist an integer $k \geqslant 0$ and sequences $\{F_n(x)\}$ and $\{G_n(x)\}$ satisfying conditions (E_1) and (E_2), and there exist an integer $l \geqslant 0$ and sequences $\{\overline{G}_n(x)\}$, $\{H_n(x)\}$ satisfying the analogous conditions

$$\overline{G}_n^{(l)}(x) = g_n(x), \qquad H_n^{(l)}(x) = h_n(x), \qquad \overline{G}_n(x) \rightrightarrows\leftleftarrows H_n(x).$$

By 1.3.1 we may assume that $k = l$. Then, writing $\overline{H}_n(x) = G_n(x) - \overline{G}_n(x) + H_n(x)$, we have

$$F_n^{(k)}(x) = f_n(x), \qquad \overline{H}_n^{(k)}(x) = h_n(x), \qquad F_n(x) \rightrightarrows\leftleftarrows \overline{H}_n(x),$$

which implies that $\{f_n(x)\} \sim \{h_n(x)\}$.

Since the conditions ($\mathscr{E}_1$), ($\mathscr{E}_2$) and ($\mathscr{E}_3$) are satisfied, the set of all fundamental sequences $\{f_n(x)\}$ (defined in $A < x < B$) is partitioned into equivalence classes without common elements, such that two fundamental sequences are in the same equivalence class if and only if they are equivalent. These equivalence classes will be called *distributions* (defined in $A < x < B$). Thus the notion of distribution is obtained from the identification of equivalent fundamental sequences.

The distribution determined by the fundamental sequence $\{f_n(x)\}$, i.e., the class of all sequences equivalent to $\{f_n(x)\}$, will be denoted by the symbol $[f_n(x)]$. Two sequences $\{f_n(x)\}$ and $\{g_n(x)\}$ determine the same distribution if and only if they are equivalent. In other words,

$$[f_n(x)] = [g_n(x)] \quad \text{if and only if} \quad \{f_n(x)\} \sim \{g_n(x)\}.$$

The fundamental sequences given in examples 2 and 3 (Section 1.2) determine the same distribution, called the *Dirac delta distribution*.

In fact, if $\{f_n(x)\}$ denotes the sequence in example 2 or 3, then by 1.2.3 the sequence

$$\int_{-\infty}^{x} dt \int_{-\infty}^{t} f_n(\tau)\,d\tau$$

converges almost uniformly to the same function

$$F(x) = \begin{cases} 0 & \text{for} \quad x < 0, \\ x & \text{for} \quad x \geqslant 0. \end{cases}$$

Consequently the sequences $\{f_n(x)\}$ in examples 2 and 3 are equivalent.

1.4. DISTRIBUTIONS AS A GENERALIZATION OF THE NOTION OF FUNCTIONS

By 1.2.1 the constant sequence $\{f(x)\}$, i.e., the sequence each of whose terms is equal to the same continuous function $f(x)$, is fundamental. It thus determines a distribution $[f(x)]$.

Different functions $f(x)$ and $g(x)$ determine different distributions $[f(x)]$ and $[g(x)]$.

For, suppose that $[f(x)] = [g(x)]$, i.e., that there exist functions $F_n(x)$ and $G_n(x)$ and an integer $k \geqslant 0$ such that

$$F_n^{(k)}(x) = f(x), \quad G_n^{(k)}(x) = g(x), \quad F_n(x) \rightrightarrows\leftleftarrows G_n(x).$$

Then the functions $p_n(x) = \big(F_1(x)-G_1(x)\big)-\big(F_n(x)-G_n(x)\big)$ are polynomials of degree $< k$ since $p_n^{(k)}(x) \equiv 0$. Moreover, $p_n(x) \rightrightarrows \big(F_1(x)-G_1(x)\big)$. It follows from 1.2.4 that the function $F_1(x)-G_1(x)$ is a polynomial of degree $< k$. Hence its kth derivative, $f(x)-g(x)$, is equal to zero, i.e., the functions $f(x)$ and $g(x)$ are equal.

Consider the set of all distributions of the form $[f(x)]$. The one-to-one correspondence established above between functions $f(x)$ and distributions $[f(x)]$ makes it unnecessary to distinguish between the two notions. In the sequel we shall identify the distribution $[f(x)]$ with the function $f(x)$ and we shall write $[f(x)] = f(x)$.

However, not all distributions can be presented in the form $[f(x)]$, i.e., not all distributions can be identified with continuous functions. For instance, the Dirac delta distribution cannot be identified with any continuous function; this will be proved in Section 2.4. Therefore, the notion of distribution can be thought of as an essential generalization of the notion of continuous function. We shall show later on that this generalization also embraces a large class of non-continuous functions.

From the logical point of view, this generalization is of exactly the same type as the generalization of rational numbers by real numbers in the Cantor theory. In fact, a rational number a is identified there with the class $[a]$ of fundamental sequences equivalent to the constant sequence $\{a\}$.

1.4.1. *If* $f_n(x) \rightrightarrows f(x)$, *then* $[f_n(x)] = f(x)$.

To prove this, it suffices to remark that $f(x)$ is continuous and that the sequences $\{f_n(x)\}$ and $\{f(x)\}$ meet the conditions (E_1) and (E_2) with $k = 0$.

Since distributions are a generalization of the notion of functions, we shall retain for them the usual notation of functions and we shall denote distributions by symbols $f(x)$, $g(x)$, etc. It should be remarked that this notation is purely symbolic and, in general, it is not meaningful to substitute numbers for the variable x.

The Dirac delta distribution will be denoted by the symbol $\delta(x)$.

2. Operations on distributions

2.1. ALGEBRAIC OPERATIONS ON DISTRIBUTIONS

We now introduce the notion of the sum and the difference of two distributions and of the product of a distribution with a number. The definitions of these operations will generalize the same operations on functions, i.e., in the case where the distributions are functions, the operations introduced below will coincide with the usual operations on functions.

By the *sum* $f(x)+g(x)$ of the distributions $f(x) = [f_n(x)]$ and $g(x) = [g_n(x)]$ we mean the distribution $[f_n(x)+g_n(x)]$.

To verify the consistency of this definition we must prove that

(i) if the sequences $\{f_n(x)\}$ and $\{g_n(x)\}$ are fundamental, then so is $\{f_n(x)+ +g_n(x)\}$;

(ii) the distribution $[f_n(x)+g_n(x)]$ does not depend on the choice of sequences $\{f_n(x)\}$ and $\{g_n(x)\}$ representing the distributions $f(x)$ and $g(x)$, i.e., if $\{f_n(x)\} \sim \{\bar{f}_n(x)\}$ and $\{g_n(x)\} \sim \{\bar{g}_n(x)\}$, then $\{f_n(x)+g_n(x)\} \sim \{\bar{f}_n(x)+\bar{g}_n(x)\}$.

Property (i) ensures that addition is feasible, and property (ii) that the sum is unique.

To prove (i), suppose that there exist integers $k, k_1 \geqslant 0$ and functions $F_n(x)$ and $G_n(x)$ such that

$$F_n^{(k)}(x) = f_n(x), \qquad F_n(x) \rightrightarrows,$$
$$G_n^{(k_1)}(x) = g_n(x), \qquad G_n(x) \rightrightarrows.$$

By 1.2.6 we can assume that $k = k_1$. Since

$$(F_n(x)+G_n(x))^{(k)} = f_n(x)+g_n(x) \quad \text{and} \quad F_n(x)+G_n(x) \rightrightarrows,$$

the sequence $\{f_n(x)+g_n(x)\}$ is fundamental.

Suppose that the hypotheses of (ii) are satisfied. Then, by 1.3.2, the sequences

$$f_1(x),\ \bar{f}_1(x),\ f_2(x),\ \bar{f}_2(x),\ \ldots,$$
$$g_1(x),\ \bar{g}_1(x),\ g_2(x),\ \bar{g}_2(x),\ \ldots,$$

are fundamental. By (i) the sequence

$$f_1(x)+g_1(x),\ \bar{f}_1(x)+\bar{g}_1(x),\ f_2(x)+g_2(x),\ \bar{f}_2(x)+\bar{g}_2(x),\ \ldots$$

is also fundamental, and so $\{f_n(x)+g_n(x)\} \sim \{\bar{f}_n(x)+\bar{g}_n(x)\}$ by 1.3.2.

By the *difference* $f(x)-g(x)$ of the distributions $f(x) = [f_n(x)]$ and $g(x) = [g_n(x)]$ we mean the distribution $[f_n(x)-g_n(x)]$.

By the *product* $\lambda f(x)$ of a distribution $f(x) = [f_n(x)]$ with a number λ we mean the distribution $[\lambda f_n(x)]$.

The consistency of these definitions can be checked by a procedure similar to that already used for the sum.

It follows at once from the definitions of the operations introduced above that the following well-known properties of functions are retained for distributions:

(1) $f(x)+g(x) = g(x)+f(x)$,

(2) $\big(f(x)+g(x)\big)+h(x) = f(x)+\big(g(x)+h(x)\big)$,

(3) *the difference* $g(x) = h(x)-f(x)$ *is the only solution of the equation* $f(x)+g(x) = h(x)$,

(4) $\lambda\big(f(x)+g(x)\big) = \lambda f(x)+\lambda g(x)$,

(5) $(\lambda+\mu)f(x) = \lambda f(x)+\mu f(x)$,

(6) $\lambda\big(\mu f(x)\big) = (\lambda\mu)f(x)$,

(7) $1\cdot f(x) = f(x)$.

Denoting by 0 the *zero distribution*, i.e., the distribution which coincides with the function identically equal to zero, we have

$$0+f(x) = f(x) \quad \text{and} \quad 0\cdot f(x) = 0.$$

In the last formula the symbol 0 has two different meanings: on the left hand side it denotes the number zero, on the right hand side—the zero distribution. This ambiguity does not in practice lead to any confusion.

2.2. DERIVATION OF DISTRIBUTIONS

To define the derivative of a distribution we need the following lemmas:

2.2.1. *If the functions* $f_n(x)$ *and* $g_n(x)$ *have continuous mth derivatives and* $\{f_n(x)\} \sim \{g_n(x)\}$, *then*

$$\{f_n^{(m)}(x)\} \sim \{g_n^{(m)}(x)\}.$$

In fact, if the functions $f_n(x)$ and $g_n(x)$ satisfy conditions (E_1) and (E_2), then $f_n^{(m)}(x)$ and $g_n^{(m)}(x)$ satisfy condition (E_1), with k replaced by $k+m$, and condition (E_2).

2.2.2. *For every continuous function $F(x)$ there is a sequence $\{P_n(x)\}$ of polynomials such that $P_n(x) \rightrightarrows F(x)$.*

Let a_n be a decreasing sequence and b_n an increasing sequence of numbers such that $a_n \to A$ and $b_n \to B$.

By the well-known Weierstrass approximation theorem there exist polynomials $P_n(x)$ such that

$$|F(x) - P_n(x)| < 1/n \quad \text{for} \quad a_n \leqslant x \leqslant b_n.$$

Hence $P_n(x) \rightrightarrows F(x)$ in $A < x < B$.

2.2.3. *Each distribution can be represented in the form $[p_n(x)]$ where $p_n(x)$ are polynomials.*

In fact, if $[f_n(x)]$ is a distribution, then, for some integer $k \geqslant 0$, there is a sequence $F_n(x)$ of continuous functions and a continuous function $F(x)$ such that $F_n^{(k)}(x) = f_n(x)$ and $F_n(x) \rightrightarrows F(x)$. Suppose that $p_n(x) = P_n^{(k)}(x)$ where $\{P_n(x)\}$ is a sequence of polynomials convergent almost uniformly to $F(x)$. Then $\{p_n(x)\} \sim \{f_n(x)\}$, i.e., $[p_n(x)] = [f_n(x)]$.

By the *mth derivative* of a distribution represented in the form $f(x) = [p_n(x)]$ where $p_n(x)$ are polynomials we mean the distribution $[p_n^{(m)}(x)]$. This definition is consistent since the sequence $\{p_n^{(m)}(x)\}$ is fundamental by 1.2.2, and the distribution $[p_n^{(m)}(x)]$ does not depend on the representation of $f(x)$ in the form $[p_n(x)]$, by 2.2.1.

From 2.2.3 follows

2.2.4. THEOREM. *Each distribution has derivatives of all orders.*

In the definition of the mth derivative of a distribution we can replace the polynomials $p_n(x)$ by functions $f_n(x)$ with a continuous mth derivative. This is a consequence of the following lemma:

2.2.5. *If a fundamental sequence $\{f_n(x)\}$ consists of functions with continuous mth derivatives, then the distribution $[f_n^{(m)}(x)]$ is the mth derivative of the distribution $[f_n(x)]$.*

In fact, the sequence $\{f_n^{(m)}(x)\}$ is fundamental by 1.2.2. If $[f_n(x)] = [p_n(x)]$ where the $p_n(x)$ are polynomials, then $[f_n^{(m)}(x)] = [p_n^{(m)}(x)]$, by 2.2.1.

2.2.6. *If a distribution is a function with a continuous mth derivative, then its mth derivative in the distributional sense coincides with its mth derivative in the ordinary sense.*

In fact, by 2.2.5, the mth derivative of the distribution $f(x) = [f(x)]$ is the distribution $[f^{(m)}(x)] = f^{(m)}(x)$.

Thus the notion of the derivative of a distribution is a generalization of the notion of derivative in the domain of continuously derivable functions. Consequently we can apply the usual notation: the mth derivative of the distribution $f(x) = [f_n(x)]$ will be denoted by $f^{(m)}(x)$ or $[f_n(x)]^{(m)}$. In particular, the first derivative of the distribution $f(x) = [f_n(x)]$ will be denoted by $f'(x)$ or $[f_n(x)]'$. If the functions $f_n(x)$ have continuous mth derivatives, then by 2.2.5

$$[f_n(x)]^{(m)} = [f_n^{(m)}(x)].$$

The following formulas occurring in the ordinary differential calculus follow immediately, for distributions, from the definition of derivative:

$$\big(f(x)+g(x)\big)^{(m)} = f^{(m)}(x)+g^{(m)}(x),$$

$$\big(\lambda f(x)\big)^{(m)} = \lambda f^{(m)}(x) \quad (\lambda \text{ a number}),$$

$$\big(f^{(m)}(x)\big)^{(p)} = f^{(m+p)}(x).$$

2.2.7. THEOREM. *The equality $f^{(m)}(x) = 0$ holds if and only if the distribution $f(x)$ is a polynomial of degree $< m$.*

The sufficiency is obvious. To prove the necessity, suppose that $f^{(m)}(x) = 0$. We can represent $f(x)$ in the form $[f_n(x)]$ where $f_n^{(m)}(x)$ are continuous. Then $\{f_n^{(m)}(x)\} \sim \{0\}$. There exist an integer $k \geqslant m$ and sequences $\{F_n(x)\}$, $\{G_n(x)\}$ such that

$$f_n^{(m)}(x) = F_n^{(k)}(x), \quad G_n^{(k)}(x) = 0, \quad F_n(x) \rightrightarrows P(x), \quad G_n(x) \rightrightarrows P(x).$$

Hence $\{f_n(x) - F_n^{(k-m)}(x)\}$ is a fundamental sequence of polynomials of degree $< m$. By 1.2.5 this sequence converges almost uniformly to a polynomial $p(x)$. By 1.4.1 we have $[f_n(x) - F_n^{(k-m)}(x)] = p(x)$ and consequently

$$f(x) = [f_n(x)] = [F_n^{(k-m)}(x)] + p(x). \tag{1}$$

Since the functions $G_n(x)$ are polynomials of degree $< k$, so is the function $P(x)$, by 1.2.4. Since $[F_n(x)] = P(x)$ by 1.4.1, the distribution $[F_n^{(k-m)}(x)] = P^{(k-m)}(x)$ is a polynomial of degree $< m$. Hence, by (1), $f(x)$ is a polynomial of degree $< m$.

It follows from 2.2.7 that, in particular,

2.2.8. *The equality $f'(x) = 0$ holds if and only if the distribution $f(x)$ is a constant function.*

Replacing $f(x)$ by $f(x) - g(x)$ in 2.2.8 we get

2.2.9. *The equality $f'(x) = g'(x)$ holds if and only if the distributions $f(x)$ and $g(x)$ differ from each other by a constant function.*

The following lemma will be useful later on:

2.2.10. *If some derivative $f^{(m)}(x)$ of a distribution $f(x)$ is a continuous function, then $f(x)$ is a continuous function and $f^{(m)}(x)$ is its ordinary derivative.*

Let

$$g(x) = \int_0^x dt_1 \int_0^{t_1} dt_2 \dots \int_0^{t_{m-1}} f^{(m)}(t_m)\,dt_m.$$

It follows from 2.2.7 and 2.2.6 that $p(x) = f(x)-g(x)$ is a polynomial of degree $< m$. Thus $f(x)$ is the function $g(x)+p(x)$ and, by 2.2.6, $f^{(m)}(x)$ is its ordinary mth derivative.

By 2.2.4, each continuous function has a derivative. This derivative is, in general, not a function, but a distribution. For instance, the non-differentiable function of Weierstrass is differentiable in the sense of distributions, but its derivative is not a function.

2.2.11. THEOREM. *Each distribution is the derivative, of some order, of a continuous function.*

For, if $f(x) = [f_n(x)]$, $f_n(x) = F_n^{(k)}(x)$ and $F_n(x) \rightrightarrows F(x)$, then $F(x) = [F_n(x)]$ by 1.4.1 and $f(x) = [F_n^{(k)}(x)] = [F_n(x)]^{(k)} = F^{(k)}(x)$.

2.3. THE DEFINITION OF DISTRIBUTIONS BY DERIVATIVES

2.2.11 suggests another equivalent definition of distributions. This definition also makes use of the identification principle but in another way. As the starting point we take here not fundamental sequences but ordered pairs $(F(x), k)$ where $F(x)$ is a continuous function in $A < x < B$ and k is an integer $\geqslant 0$. We say that two pairs $(F(x),k)$ and $(G(x),l)$ are equivalent and we write

$$(F(x), k) \sim (G(x), l)$$

if either

(i) $k \leqslant l$ and the difference $\dfrac{d^{l-k}G(x)}{dx^{l-k}} - F(x)$ exists and is a polynomial of degree $< k$;

or

(ii) $l \leqslant k$ and the difference $\dfrac{d^{k-l}F(x)}{dx^{k-l}} - G(x)$ exists and is a polynomial of degree $< l$.

By a polynomial of degree < 0 we mean the function identically equal to 0.

It is easy to see that the relation $\sim$ has properties $(\mathscr{E}_1)$ and $(\mathscr{E}_2)$. We shall prove that it also has property $(\mathscr{E}_3)$. Let

$$(F(x), k) \sim (G(x), l) \quad \text{and} \quad (G(x), l) \sim (H(x), m).$$

Because of the symmetry in $(\mathscr{E}_2)$ we may assume that $k \leqslant l \leqslant m$. Then

$$\frac{d^{l-k}G(x)}{dx^{l-k}} - F(x) = p_1(x) \quad \text{and} \quad \frac{d^{m-l}H(x)}{dx^{m-l}} - G(x) = p_2(x),$$

where $p_1(x)$ is a polynomial of degree $< k$ and $p_2(x)$ is a polynomial of degree $< l$. Hence

$$\frac{d^{m-k}H(x)}{dx^{m-k}} - F(x) = \frac{d^{l-k}p_2(x)}{dx^{l-k}} + p_1(x)$$

which proves that $(F(x), k) \sim (H(x), m)$.

Since conditions $(\mathscr{E}_1)$, $(\mathscr{E}_2)$ and $(\mathscr{E}_3)$ are satisfied, the set of all pairs $(F(x), k)$ is decomposed into equivalence classes without common elements in such a way that two pairs are in the same class if and only if they are equivalent. These equivalence classes are called *distributions*. In this new formulation we obtain distributions from the identification of equivalent pairs.

The distribution determined by a pair $(F(x), k)$, i.e., the class of all pairs equivalent to $(F(x), k)$, will be denoted by the symbol $F^{(k)}(x)$ which, for the time being, should not be thought of as a derivative. We introduce this notation because, as we shall prove in the sequel, the distribution $F^{(k)}(x)$ can be interpreted as the kth derivative of the function $F(x)$.

It follows from the definition of the equivalence that $(F(x), 0) \sim (G(x), 0)$ if and only if $F(x) = G(x)$. Thus an equivalence class of the type $F^{(0)}(x)$ contains exactly one element of the form $(G(x), 0)$, viz., the element $(F(x), 0)$. Consequently we can identify the distribution $F^{(0)}(x)$ with the function $F(x)$ and write

$$F^{(0)}(x) = F(x).$$

By the mth derivative of the distribution $F^{(k)}(x)$ we mean the distribution $F^{(k+m)}(x)$. In particular, the kth derivative of the distribution $F^{(0)}(x)$ is the distribution $F^{(k)}(x)$.

If a distribution is a function with a continuous mth derivative, then its derivative in the sense defined above is the same as its derivative in the ordinary sense:

$$F^{(m)}(x) = \frac{d^m F(x)}{dx^m}.$$

In fact, the pair $(F(x), m)$ is equivalent to $(d^m F(x)/dx^m, 0)$.

Now 2.2.11 becomes obvious: the distribution $F^{(k)}(x)$ is the kth derivative of the function $F(x)$. This fact justifies the manner of denoting distributions, adopted in this paragraph.

Any distribution $F^{(k)}(x)$ considered in the new sense can be identified with the distribution $[F(x)]^{(k)}$ in the former sense:

$$F^{(k)}(x) = [F(x)]^{(k)}.$$

This identification is consistent because $F^{(k)}(x) = G^{(l)}(x)$ if and only if $[F(x)]^{(k)} = [G(x)]^{(l)}$. In fact, if $F^{(k)}(x) = G^{(l)}(x)$ and $k \leqslant l$, i.e., if $F(x) = d^{l-k}G(x)/dx^{l-k} + p(x)$ where $p(x)$ is a polynomial of degree $< k$, then, interpreting these functions as distributions in the former sense, we obtain

$$[F(x)]^{(k)} = \left[\frac{d^{l-k}G(x)}{dx^{l-k}}\right]^{(k)} + [p(x)]^{(k)} = [G(x)]^{(l)}$$

by 2.2.6. Conversely, if $[F(x)]^{(k)} = [G(x)]^{(l)}$ and $k \leqslant l$, then $([G(x)]^{(l-k)} - [F(x)])^{(k)} = 0$. Thus $[G(x)]^{(l-k)} = F(x) + p(x)$ where $p(x)$ is a polynomial of degree $< k$. Consequently, by 2.2.9,

$$\frac{d^{l-k}G(x)}{dx^{l-k}} = F(x) + p(x)$$

which proves the equality $F^{(k)}(x) = G^{(l)}(x)$.

After the identification of distributions in the former and the present sense it can be seen that differentiation is the same in both cases since

$$(F^{(k)}(x))^{(m)} = F^{(k+m)}(x) = [F(x)]^{(k+m)} = ([F(x)]^{(k)})^{(m)}.$$

The operations of addition, subtraction and multiplication by numbers and also further operations introduced in the following paragraphs can be defined in a simple way, using the definition of distribution given in this section.

2.4. LOCALLY INTEGRABLE FUNCTIONS

We have seen in Section 1.4 that distributions in $A < x < B$ are a generalization of continuous functions in $A < x < B$. We shall see now that they also include a larger class of functions.

If $f(x)$ is a continuous function, then, as is well known,

$$\left(\int_a^x f(t)\,dt\right)' = f(x). \tag{1}$$

If the function $f(x)$ is piecewise continuous, or — more generally — if it is Riemann-integrable, equation (1) still holds except at the points of discontinuity of $f(x)$. The reader acquainted with the Lebesgue integral can interpret formula (1) in an even wider sense, viz. as equality almost everywhere.

In the case discussed above the left hand side of (1) can also be interpreted as a distribution which is the distributional derivative of the continuous function $\int_a^x f(t)\,dt$. Equality (1) then allows the identification of the distribution on the left hand side of (1) with the function $f(x)$. With this convention we include in the set of distributions in $A < x < B$ a class of non-continuous functions. The extent of this class depends on the adopted definition of integrability. Viz. this class is the set of all locally integrable functions, i.e., of functions integrable over each interval $a \leqslant x \leqslant b$ such that $A < a < b < B$.

The interpretation of these functions as distributions makes the following definition necessary: locally integrable functions $f(x)$ and $g(x)$ are equal if and only if they are equal as distributions, i.e., if $\int_a^x f(t)\,dt = \int_a^x g(t)\,dt$ for every x. In particular, if the functions are piecewise continuous, or — more generally — if they are locally Riemann-integrable, then they are equal provided their values are equal at all common points of continuity. Lebesgue-integrable functions are equal if and only if they have the same values almost everywhere. It should be observed here that this definition of equality is that usually adopted in the theory of the Lebesgue integral.

The class of indefinite Lebesgue integrals of locally integrable functions is the class of absolutely continuous functions. The ordinary derivative of an absolutely continuous function exists almost everywhere and coincides with the distributional derivative.

If $f(x)$ and $g(x)$ are locally integrable functions, then the symbol $f(x)+g(x)$ has the same meaning whether we interpret them as functions or as distributions. The same remark is true for subtraction and multiplication by numbers.

The incorporation of locally integrable functions allows us to strengthen 2.2.10 without changing its proof:

2.4.1. *If a derivative $f^{(m)}(x)$ $(m > 0)$ of a distribution is a locally integrable function, then $f(x)$ is a continuous function and $f^{(m)}(x)$ is its ordinary mth derivative.*

In applications we often come across the so called *Heaviside function*

$$H(x) = \begin{cases} 0 & \text{for} \quad x < 0, \\ 1 & \text{for} \quad x \geqslant 0. \end{cases}$$

Its integral

$$G(x) = \int_0^x H(t)\,dt = \begin{cases} 0 & \text{for} \quad x < 0, \\ x & \text{for} \quad x \geqslant 0 \end{cases}$$

is continuous. The Heaviside function $H(x)$ is the distributional derivative of $G(x)$; it is also its ordinary derivative except at the point $x = 0$.

Since $G(x)$ is the limit of the integrals

$$\int_{-\infty}^{x} g_n(t)\,dt$$

where $g_n(x)$ are the functions from example 1 of Section 1.2, we have

$$[g_n(x)] = G'(x) = H(x),$$

i.e., the fundamental sequence $\{g_n(x)\}$ represents the Heaviside function.

Similarly, the fundamental sequences $\{g_n(x)\}$ from examples 2 and 3 (Section 1.2) represent the Heaviside function. Hence, for the functions $f_n(x)$ from examples 2 or 3 (1.2), we have

$$\delta(x) = [f_n(x)] = [g_n'(x)] = H'(x).$$

Thus the Dirac distribution $\delta(x)$ is the distributional derivative of the Heaviside function.

The Dirac distribution $\delta(x)$ is an example of a distribution in $-\infty < x < \infty$ which is not a locally integrable function. Indeed, suppose $\delta(x)$ to be an integrable function. From $H'(x) = \delta(x)$ it follows, by 2.4.1, that the function $H(x)$ is continuous. This is not true.

It is worth remarking that the ordinary derivative of the Heaviside function $H(x)$ vanishes everywhere except at the point $x = 0$ at which it does not exist. This example shows that the ordinary derivative does not always coincide with the distributional derivative, even when the ordinary derivative exists everywhere except at one point.

2.5. SEQUENCES AND SERIES OF DISTRIBUTIONS

We say that a sequence $\{f_n(x)\}$ of distributions *converges* to a distribution $f(x)$ and we write

$$f_n(x) \to f(x) \quad \text{or} \quad \lim_{n\to\infty} f_n(x) = f(x)$$

if there exist an integer $k \geqslant 0$, a continuous function $F(x)$ and a sequence $\{F_n(x)\}$ of continuous functions such that

$$F_n(x) \rightrightarrows F(x), \quad F_n^{(k)}(x) = f_n(x) \quad \text{and} \quad F^{(k)}(x) = f(x). \tag{1}$$

The limit $f(x)$, if it exists, is unique.

In fact, if

$$f_n(x) \to f(x) \text{ and } f_n(x) \to g(x),$$

then there exist sequences of continuous functions

$$F_n(x) \rightrightarrows F(x) \text{ and } G_n(x) \rightrightarrows G(x)$$

and integers k, l such that

$$F_n^{(k)}(x) = G_n^{(l)}(x) = f_n(x), \quad F^{(k)}(x) = f(x) \quad \text{and} \quad G^{(l)}(x) = g(x).$$

We may assume that $k \leqslant l$. By 1.2.6 there exist functions $\tilde{F}_n(x)$ and $\tilde{F}(x)$ such that

$$\tilde{F}_n^{(l)}(x) = f_n(x), \quad \tilde{F}_n(x) \rightrightarrows \tilde{F}(x) \quad \text{and} \quad \tilde{F}^{(l)}(x) = f(x).$$

Since $\tilde{F}_n^{(l)}(x) - G_n^{(l)}(x) = 0$, the differences $\tilde{F}_n(x) - G_n(x)$ form a sequence of polynomials of degree $< l$. By 1.2.4, its limit $\tilde{F}(x) - G(x)$ is also a polynomial of degree $< l$. Consequently

$$\tilde{F}^{(l)}(x) - G^{(l)}(x) = 0, \quad \text{i.e.,} \quad f(x) = g(x).$$

2.5.1. THEOREM. *For every sequence of distributions $\{f_n(x)\}$ and for every integer $m \geqslant 0$,*

$$f_n(x) \to f(x) \quad \textit{implies} \quad f_n^{(m)}(x) \to f^{(m)}(x).$$

In fact, if conditions (1) are satisfied, then

$$F_n^{(k+m)}(x) = f_n^{(m)}(x), \quad F_n(x) \rightrightarrows F(x) \quad \text{and} \quad F^{(k+m)}(x) = f^{(m)}(x).$$

It follows immediately from the definition of convergence (in the case $k = 0$) that

2.5.2. *If a sequence $\{f_n(x)\}$ of continuous functions converges almost uniformly to a function $f(x)$, then it also converges to $f(x)$ in the distributional sense.*

More generally,

2.5.3. *If a sequence $\{f_n(x)\}$ of locally integrable functions converges almost everywhere to a function $f(x)$ and is bounded by a locally integrable function, then it also converges to $f(x)$ in the distributional sense.*

This follows from the almost uniform convergence of the sequence of integrals $\int_a^x f_n(t)\,dt$ to the function $\int_a^x f(t)\,dt$.

2.5.4. THEOREM. *A sequence $\{f_n(x)\}$ of continuous functions converges to a distribution $f(x)$ if and only if it is a fundamental sequence of $f(x)$.*

In other words:

$$f_n(x) \to f(x) \quad \textit{if and only if} \quad [f_n(x)] = f(x).$$

If $f_n(x) \to f(x)$, then $\{f_n(x)\}$ is fundamental by (1). Moreover, it follows from the second of conditions (1) and from 1.4.1 that $[F_n(x)] = F(x)$. Hence, by 2.2.5, we obtain

$$[F_n^{(k)}(x)] = F^{(k)}(x), \quad \text{i.e.,} \quad [f_n(x)] = f(x).$$

On the other hand, if $[f_n(x)] = f(x)$, then the first two conditions (1) are satisfied and $F(x) = [F_n(x)]$ by 1.4.1. Differentiating k times we obtain

$$F^{(k)}(x) = [F_n^{(k)}(x)] = f(x).$$

Thus the third of the conditions (1) is also satisfied, which proves that $f_n(x) \to f(x)$.

An analogous theorem is in the Cantor theory of real numbers, where it is proved that a sequence of rational numbers converges to a real number a if and only if it is a fundamental sequence of a.

In view of 2.5.4 the sequences $\{f_n(x)\}$ from examples 2 and 3 (1.2) converge distributionally to $\delta(x)$. In classical analysis various sequences of functions which are distributionally convergent to $\delta(x)$ have been studied, e.g.

$$\frac{\sin nx}{\pi x} \qquad \text{(Dirichlet)},$$

$$\tfrac{1}{2} n e^{-n|x|} \qquad \text{(Picard)},$$

$$\frac{1}{\pi} \cdot \frac{n}{e^{nx} + e^{-nx}} \qquad \text{(Stieltjes)}.$$

It follows from 2.5.4 and 1.2.5 that

2.5.5. *A sequence of polynomials of degree $< k$ converges in the distributional sense if and only if it converges almost uniformly.*

In particular:

2.5.6. *A sequence of constant functions converges in the distributional sense if and only if it converges in the usual sense.*

It follows immediately from the definition of convergence that arithmetical operations on limits of sequences of distributions can be performed in the same way as on limits of sequences of functions:

2.5.7. *If* $f_n(x) \to f(x)$ *and* $g_n(x) \to g(x)$, *then* $f_n(x)+g_n(x) \to f(x)+g(x)$ *and* $f_n(x)-g_n(x) \to f(x)-g(x)$.

If $f_n(x) \to f(x)$ *and* $\lambda_n \to \lambda$, *then* $\lambda_n f_n(x) \to \lambda f(x)$.

A series of distributions $\sum_{n=1}^{\infty} g_n(x)$ is said to be *convergent* if the sequence of its partial sums $f_n(x) = g_1(x)+ \ldots +g_n(x)$ converges. The limit $g(x) = \lim_{n\to\infty} f_n(x)$ is said to be the *sum* of this series. We then write

$$g(x) = \sum_{n=1}^{\infty} g_n(x).$$

From 2.5.1 follows

2.5.8. THEOREM. *For every convergent series of distributions,*

$$\Big(\sum_{n=1}^{\infty} g_n(x)\Big)' = \sum_{n=1}^{\infty} g_n'(x).$$

Roughly speaking, each convergent series of distributions can be differentiated term by term.

Theorems 2.2.4, 2.5.1 and 2.5.8 are of considerable importance for the practical use of the theory of distributions. They make it possible to differentiate each function without any restriction and to commute the differentiation with the passage to a limit.

Theorems 2.5.1 and 2.5.8 are considerably simpler than the analogous theorems in the differential calculus, where some additional hypotheses are necessary. Thus the introduction of distributions, of generalized derivatives and of distributional convergence makes the calculations easier and routine, which reaffirms the adequency of these concepts.

The definition of convergence can be generalized to sequences of distributions defined in different intervals.

We say that a sequence of functions $F_n(x)$ defined in $A_n < x < B_n$ converges to a function $F(x)$ *almost uniformly* in an interval $A < x < B$ if each finite closed interval $a \leqslant x \leqslant b$ contained in $A < x < B$ is also contained in all the intervals $A_n < x < B_n$ for n sufficiently large, and the functions $F_n(x)$ converge to $F(x)$ uniformly in $a \leqslant x \leqslant b$. We then write $F_n(x) \rightrightarrows F(x)$ in $A < x < B$.

The above definition is a generalization of the almost uniform convergence in Section 1.2.

We now say that a sequence $f_n(x)$ of distributions defined in $A_n < x < B_n$ *converges* to a distribution $f(x)$ in $A < x < B$ if the conditions at the beginning of this paragraph are satisfied provided the almost uniform convergence $F_n(x) \rightrightarrows F(x)$ is understood in the above generalized sense. We then write $f_n(x) \to f(x)$ in $A < x < B$.

Theorem 2.5.1, 2.5.2, and 2.5.5–2.5.7 remain true in this generalized setting.

2.6. DISTRIBUTIONS DEPENDING ON A CONTINUOUS PARAMETER

It is convenient to define here convergence of distributions depending on a continuous parameter in full generality, i.e., when the distributions are defined in intervals which also depend on the parameter.

A function $F_\alpha(x)$ defined in $A_\alpha < x < B_\alpha$ is said to converge, for $\alpha \to \alpha_0$, to a function $F(x)$ *almost uniformly* in an interval $A < x < B$ if each finite closed interval $a \leqslant x \leqslant b$ contained in $A < x < B$ is also contained in the intervals $A_\alpha < x < B_\alpha$ for all α sufficiently near to α_0, and $F_\alpha(x)$ converges to $F(x)$ uniformly in $a \leqslant x \leqslant b$ as $\alpha \to \alpha_0$. We then write

$$F_\alpha(x) \rightrightarrows F(x) \quad \text{for} \quad \alpha \to \alpha_0. \tag{1}$$

We say that a distribution $f_\alpha(x)$ defined in $A_\alpha < x < B_\alpha$ *converges* for $\alpha \to \alpha_0$ to a distribution $f(x)$ defined in $A < x < B$ and we write

$$f_\alpha(x) \to f(x) \quad \text{for} \quad \alpha \to \alpha_0$$

or

$$\lim_{\alpha \to \alpha_0} f_\alpha(x) = f(x) \tag{2}$$

if there exist an integer $k \geqslant 0$, a continuous function $F(x)$ defined in $A < x < B$ and a continuous function $F_\alpha(x)$ defined (for α in some neighbourhood of α_0) in $A_\alpha < x < B_\alpha$, such that

$$F_\alpha(x) \rightrightarrows F(x) \text{ for } \alpha \to \alpha_0, \quad F_\alpha^{(k)}(x) = f_\alpha(x) \quad \text{and} \quad F^{(k)}(x) = f(x).$$

If the limit $f(x)$ exists, it is unique. The proof is similar to that for sequences.

As it is well known, (1) holds if and only if

$$F_{\alpha_n}(x) \rightrightarrows F(x) \quad \text{for every sequence } \alpha_n \to \alpha_0.$$

We are going to prove that, analogously,

2.6.1. *Equality* (2) *holds if and only if*

$$f_{\alpha_n}(x) \to f(x) \quad \textit{for every sequence } \alpha_n \to \alpha_0. \tag{3}$$

The necessity of condition (3) is obvious.

In the proof of sufficiency we shall write $f_n(x) \stackrel{k}{\to} f(x)$ if there exist an index n_0 and functions $\tilde{F}_n(x)$ and $F(x)$ such that

$$\tilde{F}_n^{(k)}(x) = f_n(x) \text{ for } n > n_0, \quad \tilde{F}_n(x) \rightrightarrows F(x) \quad \text{and} \quad F^{(k)}(x) = f(x). \tag{4}$$

We shall prove that, assuming (3), there is an integer $k \geqslant 0$ such that

$$f_{\alpha_n}(x) \stackrel{k}{\to} f(x) \quad \text{for every sequence } \alpha_n \to \alpha_0. \tag{5}$$

Suppose the contrary. Then there is an increasing sequence of integers k_m, and moreover there is, for any fixed m, a sequence $\alpha_{mn} \to \alpha_0$ such that the relation

$$f_{\alpha_{mn}}(x) \stackrel{l}{\to} f(x) \quad (n \to \infty) \tag{6}$$

holds for $l = k_m$, but does not hold for $l < k_m$. We can assume, without any loss of generality, that all the numbers α_{mn} can be arranged in a simple sequence $\{\alpha_n\}$ convergent to α_0 (if necessary we can omit a finite number of initial terms in each of the sequences $\{\alpha_{1n}\}, \{\alpha_{2n}\}, \ldots$). It follows from (3) that there is an integer $k_0 \geqslant 0$ such that $f_{\alpha_n}(x) \stackrel{k_0}{\to} f(x)$ and, consequently, $f_{\alpha_{mn}}(x) \stackrel{k_0}{\to} f(x)$ for every m. This contradicts the hypotheses that $k_m \to \infty$ and that (6) does not hold for $l < k_m$.

In what follows, let k be a fixed integer with property (5). The function $F(x)$ occurring in (4) is determined up to a polynomial of degree $< k$. Including this polynomial among the functions $\tilde{F}_n(x)$ we may assume that $F(x)$ is the same for all sequences $\{\tilde{F}_n(x)\}$.

Let $\{a_n\}$ be a decreasing sequence and $\{b_n\}$ an increasing sequence of numbers, such tha $a_n \to A$ and $b_n \to B$. We shall prove that, for any integer $m > 0$, there exists a number $\eta_m > 0$ with the following property: If $|\alpha - \alpha_0| < \eta_m$, then there exists a function $\tilde{F}_\alpha(x)$ such that

$$\tilde{F}_\alpha^{(k)}(x) = f_\alpha(x) \quad \text{in} \quad A_\alpha < x < B_\alpha, \tag{7}$$

$$|\tilde{F}_\alpha(x) - F(x)| < 1/m \quad \text{in} \quad a_m \leqslant x \leqslant b_m. \tag{8}$$

For otherwise there would be a sequence $\alpha_n \to \alpha_0$ such that, for $\alpha = \alpha_n$ $(n = 1, 2, \ldots)$, every function satisfying (7) would not satisfy (8). On the other hand, by (5), there exist functions $\tilde{F}_{\alpha_m}(x)$ such that, for n sufficiently large,

$$\tilde{F}_{\alpha_n}^{(k)}(x) = f_{\alpha_n}(x) \quad \text{and} \quad \tilde{F}_{\alpha_n}(x) \rightrightarrows F(x).$$

These functions satisfy (7) and (8) for n sufficiently large. Contradiction!

We may assume that $\eta_{m+1} < \eta_m$ $(m = 1, 2, \ldots)$ and $\eta_m \to 0$. By the above argument we can define $F_\alpha(x)$ in such a manner that

$$F_\alpha^{(k)}(x) = f_\alpha(x) \quad \text{in} \quad A_\alpha < x < B_\alpha \tag{9}$$

and

$$|F_\alpha(x) - F(x)| < 1/m \quad \text{in} \quad a_m \leqslant x \leqslant b_m$$

for $\eta_{m+1} \leqslant |\alpha - \alpha_0| < \eta_m$ $(m = 1, 2, \ldots)$. In this way, the functions $F_\alpha(x)$ are defined for $0 < |\alpha - \alpha_0| < \eta_1$. Given any fixed integer $q > 0$, we have, for $m > q$,

$$|F_\alpha(x) - F(x)| < 1/m \quad \text{in} \quad a_q \leqslant x \leqslant b_q$$

if $0 < |\alpha - \alpha_0| < \eta_m$. Thus $F_\alpha(x)$ converges to $F(x)$ uniformly in $a_q \leqslant x \leqslant b_q$ $(q = 1, 2, \ldots)$, i.e., $F_\alpha(x) \rightrightarrows F(x)$ as $\alpha \to \alpha_0$. This, together with (9), means that $f_\alpha(x) \to f(x)$ as $\alpha \to \alpha_0$, and completes the proof in the case where α_0 is finite.

When $\alpha_0 = \infty$ or $-\infty$ the proof needs obvious modifications only.

The following theorems on distributions depending on a parameter can be proved by methods analogous to the case of sequences of distributions.

2.6.2. THEOREM. *If $f_\alpha(x) \to f(x)$ for $\alpha \to \alpha_0$, then $f_\alpha^{(m)}(x) \to f^{(m)}(x)$ for $\alpha \to \alpha_0$ and $m \geqslant 0$.*

2.6.3. *If $F_\alpha(x) \rightrightarrows F(x)$ for $\alpha \to \alpha_0$, then also $F_\alpha(x) \to F(x)$ for $\alpha \to \alpha_0$, i.e., the almost uniform convergence of a function with a parameter implies its distributional convergence.*

2.6.4. *A polynomial of degree $< k$ with coefficients depending on a parameter α converges for $\alpha \to \alpha_0$ in the distributional sense if and only if it converges almost uniformly.*

2.6.5. *If $f_\alpha(x) \to f(x)$ and $g_\alpha(x) \to g(x)$ for $\alpha \to \alpha_0$, then*

$$f_\alpha(x)+g_\alpha(x) \to f(x)+g(x) \quad \text{and} \quad f_\alpha(x)-g_\alpha(x) \to f(x)-g(x)$$

for $\alpha \to \alpha_0$.

If $\lambda_\alpha \to \lambda$ and $f_\alpha(x) \to f(x)$ for $\alpha \to \alpha_0$, then

$$\lambda_\alpha f_\alpha(x) \to \lambda f(x)$$

for $\alpha \to \alpha_0$.

2.7. MULTIPLICATION OF DISTRIBUTIONS BY FUNCTIONS

We shall restrict ourselves, in this paper, to the multiplication of a distribution $f(x)$ by a function $\omega(x)$ which is indefinitely derivable. By the *product* $\omega(x)f(x)$ we mean the distribution $[\omega(x)f_n(x)]$, where $[f_n(x)] = f(x)$.

To check that this definition is consistent it is enough to show that

(i) the sequence $\{\omega(x)f_n(x)\}$ is fundamental;

(ii) $\{f_n(x)\} \sim \{g_n(x)\}$ implies $\{\omega(x)f_n(x)\} \sim \{\omega(x)g_n(x)\}$.

There exist an integer $k \geqslant 0$ and functions $F_n(x)$ such that $F_n(x) \rightrightarrows$ and $F_n^{(k)}(x) = f_n(x)$. The sequence

$$\omega(x)F_n'(x) = \big(\omega(x)F_n(x)\big)' - \omega'(x)F_n(x)$$

is fundamental as it is the difference of two sequences which are fundamental by 1.2.2 and 1.2.1. For the same reason the sequence $\{\omega'(x)F_n'(x)\}$ is fundamental. Thus the sequence

$$\omega(x)F_n''(x) = \big(\omega(x)F_n'(x)\big)' - \omega'(x)F_n'(x)$$

is also fundamental. By the same argument the sequences $\{\omega(x)F_n'''(x)\}, \ldots,$ $\{\omega(x)F_n^{(k)}(x)\}$ are fundamental. Thus the sequence $\{\omega(x)f_n(x)\}$ is fundamental.

To prove (ii), observe that if the fundamental sequences

$$f_1(x), f_2(x), \dots \quad \text{and} \quad g_1(x), g_2(x), \dots$$

are equivalent, then the sequence

$$f_1(x), g_1(x), f_2(x), g_2(x), \dots$$

is fundamental by 1.3.2. Consequently, by (i), the sequence

$$\omega(x)f_1(x), \omega(x)g_1(x), \omega(x)f_2(x), \omega(x)g_2(x), \dots$$

is fundamental. By 1.3.2 the sequences

$$\omega(x)f_1(x), \omega(x)f_2(x), \dots \quad \text{and} \quad \omega(x)g_1(x), \omega(x)g_2(x), \dots$$

are equivalent.

The following usual properties of multiplication follow directly from the definition:

$$\omega_1(x)\big(\omega_2(x)f(x)\big) = \big(\omega_1(x)\omega_2(x)\big)f(x),$$

$$\big(\omega_1(x)+\omega_2(x)\big)f(x) = \omega_1(x)f(x)+\omega_2(x)f(x),$$

$$\omega(x)\big(f(x)+g(x)\big) = \omega(x)f(x)+\omega(x)g(x).$$

If $f(x)$ is a function, then the product defined above is the ordinary product of functions. Moreover if $\omega(x)$ is a constant function and $f(x)$ is an arbitrary distribution, this product coincides with the product defined in Section 2.1.

If we assume that the functions $f_n(x)$ in the definition of multiplication are polynomials, we can easily prove the formula

$$\big(\omega(x)f(x)\big)' = \omega'(x)f(x)+\omega(x)f'(x). \tag{1}$$

This formula may be considered as a particular case (where $k = 1$) of the formula

$$\omega(x)f^{(k)}(x) = \sum_{j=0}^{k}(-1)^j\binom{k}{j}\big(\omega^{(j)}(x)f(x)\big)^{(k-j)}, \tag{2}$$

which can be proved by induction in the same way as for functions.

Replacing $f(x)$ by a continuous function $F(x)$ we get from the last formula

$$\omega(x)f(x) = \sum_{j=0}^{k}(-1)^j\binom{k}{j}\big(\omega^{(j)}(x)F(x)\big)^{(k-j)} \quad \text{for} \quad f(x) = F^{(k)}(x). \tag{3}$$

Since the products under the sign $\sum$ are products of continuous functions, the sense of the right hand side is well defined even before the notion of product of a distribution by a function is introduced. Formula (3) could therefore be used as an alternative definition of the product $\omega(x)f(x)$.

As an application of (1) we shall prove that

$$\omega(x)\,\delta(x) = \omega(0)\,\delta(x). \tag{4}$$

In fact, we can easily verify that

$$\int_0^x \omega'(t)H(t)\,dt = \big(\omega(x)-\omega(0)\big)H(x).$$

Differentiating distributionally we get

$$\omega'(x)H(x) = \omega'(x)H(x)+\big(\omega(x)-\omega(0)\big)\delta(x),$$

which implies (4).

Substituting $f(x) = \delta(x)$ in (2) we obtain, by (4),

$$\omega(x)\,\delta^{(k)}(x) = \sum_{j=0}^{k} (-1)^j \binom{k}{j} \omega^{(j)}(0)\,\delta^{(k-j)}(x).$$

2.7.1. *If* $f_n(x) \to f(x)$, *then*

$$\omega(x)f_n(x) \to \omega(x)f(x).$$

More generally, if $\omega_n^{(m)}(x) \rightrightarrows \omega^{(m)}(x)$ *for* $m = 0, 1, 2, \ldots$ *and* $f_n(x) \to f(x)$, *then*

$$\omega_n(x)f_n(x) \to \omega(x)f(x).$$

In fact, in view of (3) we have

$$\omega_n(x)f_n(x) = \sum_{j=0}^{k} (-1)^j \binom{k}{j} \big(\omega_n^{(j)}(x)F_n(x)\big)^{(k-j)},$$

where $f_n(x) = F_n^{(k)}(x)$, $F_n(x) \rightrightarrows F(x)$ and $f(x) = F^{(k)}(x)$. The expression on the right hand side of this formula converges to the expression on the right hand side of (3), and consequently the same is true of the expressions on the left hand sides. This proves the second part of 2.7.1. The first part is a particular case of the second part.

2.7.2. *The formula*

$$\omega(x)\sum_{n=1}^{\infty} f_n(x) = \sum_{n=1}^{\infty} \omega(x)f_n(x)$$

holds whenever the series of distributions on the left hand side converges.

This follows from 2.7.1.

The following lemma is the continuous analogue of 2.7.1.

2.7.3. *If* $f_\alpha(x) \to f(x)$ *for* $\alpha \to \alpha_0$ $(-\infty \leqslant \alpha_0 \leqslant \infty)$, *then*

$$\omega(x)f_\alpha(x) \to \omega(x)f(x).$$

More generally, if $\omega_\alpha^{(m)}(x) \rightrightarrows \omega^{(m)}(x)$ $(m = 0, 1, 2, \ldots)$ *and* $f_\alpha(x) \to f(x)$ *for* $\alpha \to \alpha_0$, *then*

$$\omega_\alpha(x) f_\alpha(x) \to \omega(x) f(x).$$

2.8. COMPOSITIONS

Let $\varphi(x)$ be an infinitely derivable function defined in $A_0 < x < B_0$, and suppose that $A < \varphi(x) < B$ and $\varphi'(x) \neq 0$ for every x. Finally let $f(x) = [f_n(x)]$ be a distribution in $A < x < B$. By the *composition* $f(\varphi(x))$ of the distribution $f(x)$ with the function $\varphi(x)$ we shall mean the distribution $[f_n(\varphi(x))]$ defined in $A_0 < x < B_0$.

To verify the consistency of this definition we have to show that

(i) if a sequence $\{f_n(x)\}$ is fundamental, so is $\{f_n(\varphi(x))\}$;

(ii) from $\{f_n(x)\} \sim \{g_n(x)\}$ follows $\{f_n(\varphi(x))\} \sim \{g_n(\varphi(x))\}$.

Observe first that if the functions $g_n(x)$ are continuously derivable and the sequence $g_n(\varphi(x))$ is fundamental, then the sequence

$$g_n'(\varphi(x)) = \frac{1}{\varphi'(x)} \big(g_n(\varphi(x))\big)'$$

is fundamental, in view of 1.2.2 and property (i) from Section 2.7.

Let $\{F_n(x)\}$ be an almost uniformly convergent sequence of continuous functions, such that $F_n^{(k)}(x) = f_n(x)$. Then $\{F_n(\varphi(x))\}$ also converges almost uniformly, and is therefore fundamental. Consequently the sequences $\{F_n'(\varphi(x))\}, \ldots,$ $\{F_n^{(k)}(\varphi(x))\}$ are also fundamental. The last of these sequences coincides with $f_n(\varphi(x))$, and (i) is proved.

Property (ii) can be proved by the same argument as property (ii) in Section 2.7.

Calculations with compositions of distributions can be carried out in the same way as those with compositions of functions. In particular, we have the formula

$$f(\varphi(x))' = f'(\varphi(x))\varphi'(x) \tag{1}$$

since

$$f(\varphi(x))' = [f_n(\varphi(x))]' = [f_n'(\varphi(x))\varphi'(x)] = [f_n'(\varphi(x))]\varphi'(x) = f'(\varphi(x))\varphi'(x).$$

From (1) it follows in particular that

$$\delta(\varphi(x)) = \frac{1}{\varphi'(x)} \big(H(\varphi(x))\big)'.$$

If $\varphi(x) \neq 0$ everywhere, then the function $H(\varphi(x))$ is everywhere equal to 0 or everywhere equal to 1. Hence

$$\delta(\varphi(x)) = 0.$$

If $\varphi(x)$ admits the value 0 at some point x_0, which must be unique since $\varphi'(x)$ does not vanish, then $H(\varphi(x)) = H(x-x_0)$ as $\varphi(x)$ increases, or $H(\varphi(x)) = 1-H(x-x_0)$ as $\varphi(x)$ decreases. Consequently

$$\delta(\varphi(x)) = \frac{1}{|\varphi'(x_0)|}\delta(x-x_0). \tag{2}$$

2.8.1. *If a sequence of distributions $f_n(x)$ converges to $f(x)$, then the sequence $f_n(\varphi(x))$ converges to $f(\varphi(x))$. More generally, if $\varphi_n^{(m)}(x) \rightrightarrows \varphi^{(m)}(x)$ for $m = 0, 1, 2, \ldots$ and $f_n(x) \to f(x)$, then $f_n(\varphi_n(x)) \to f(\varphi(x))$.*

In fact, there are functions $F_n(x)$ and $F(x)$ such that $F_n^{(k)}(x) = f_n(x)$, $F^{(k)}(x) = f(x)$ and $F_n(x) \rightrightarrows F(x)$. Thus $F_n(\varphi_n(x)) \to F(\varphi(x))$. Hence by differentiation

$$F_n'(\varphi_n(x))\,\varphi_n'(x) \to F'(\varphi(x))\,\varphi'(x). \tag{3}$$

It is easy to verify that

$$\left(\frac{1}{\varphi_n'(x)}\right)^{(m)} \rightrightarrows \left(\frac{1}{\varphi'(x)}\right)^{(m)} \qquad (m = 0, 1, 2, \ldots). \tag{4}$$

It follows from (3) and (4) by 2.7.1 that

$$F_n'(\varphi_n(x)) \to F'(\varphi(x)),$$

and after k similar steps we obtain

$$F_n^{(k)}(\varphi_n(x)) \to F^{(k)}(\varphi(x)),$$

i.e., $f_n(\varphi_n(x)) \to f(\varphi(x))$.

The following lemma is the continuous analogue of 2.8.1.

2.8.2. *If $f_\alpha(x) \to f(x)$ for $\alpha \to \alpha_0$ $(-\infty \leqslant \alpha_0 \leqslant \infty)$, then*

$$f_\alpha(\varphi(x)) \to f(\varphi(x)).$$

More generally, if $\varphi_\alpha^{(m)}(x) \rightrightarrows \varphi^{(m)}(x)$ $(m = 0, 1, 2, \ldots)$ and $f_\alpha(x) \to f(x)$ for $\alpha \to \alpha_0$, then

$$f_\alpha(\varphi_\alpha(x)) \to f(\varphi(x)).$$

The proof is the same as that of 2.8.1.

The case of a linear substitution is especially important.

2.8.3. *For every distribution $f(x)$ and every integer $k \geqslant 0$ we have*

$$(f(\alpha x+\beta))^{(k)} = \alpha^k f^{(k)}(\alpha x+\beta) \qquad (\alpha \neq 0).$$

In fact, if the polynomials $p_n(x)$ form a fundamental sequence for the distribution $f(x)$, then

$$(f(\alpha x+\beta))^{(k)} = [p_n(\alpha x+\beta)]^{(k)} = [\alpha^k p_n^{(k)}(\alpha x+\beta)] = \alpha^k f^{(k)}(\alpha x+\beta).$$

By applying 2.8.3 to the trivial equality

$$\frac{1}{\alpha}\left(H(\alpha x+\beta)-\frac{1}{2}\right)=\frac{1}{|\alpha|}\left(H\left(x+\frac{\beta}{\alpha}\right)-\frac{1}{2}\right)\qquad(\alpha\neq 0)$$

we get

$$\delta(\alpha x+\beta)=\frac{1}{|\alpha|}\,\delta\left(x+\frac{\beta}{\alpha}\right),\tag{5}$$

which is a particular case of (2). Differentiating (5) further we obtain

$$\delta^{(m)}(\alpha x+\beta)=\frac{1}{|\alpha|\alpha^m}\,\delta^{(m)}\left(x+\frac{\beta}{\alpha}\right).$$

It follows at once from 2.8.3 that

2.8.4. *If a distribution $f(x)$ is the kth derivative of a continuous function $F(x)$, then $f(\alpha x+\beta)$ is the kth derivative of the function $(1/\alpha^k)\,F(\alpha x+\beta)$.*

3. Local properties

3.1. EQUALITY OF DISTRIBUTIONS IN INTERVALS

Any distribution $f(x)$ in $A < x < B$ can be interpreted, if necessary, as a distribution in a subinterval $a < x < b$, since the functions of any fundamental sequence representing $f(x)$ can be interpreted (by restriction) as functions in the subinterval.

For instance, in the difference $f(x+\alpha)-f(x)$ where $f(x)$ is a distribution in $A < x < B$ both the distributions $f(x+\alpha)$ and $f(x)$ should be interpreted as distributions in the overlap of the intervals $A-\alpha < x < B-\alpha$ and $A < x < B$. The difference is then a distribution in that overlap. If the overlap is empty, the difference has no meaning.

Generally, the sum $f(x)+g(x)$ and the difference $f(x)-g(x)$ are distributions in the overlap of the intervals where $f(x)$ and $g(x)$ are defined. An similar remark holds for the product $\omega(x)f(x)$.

If we simply write the equality

$$f(x) = g(x)$$

we always mean that the distributions on both sides are defined in the same interval and are equal. This is what we have done hitherto and what we shall continue to do.

If we write

$$f(x) = g(x) \quad \text{in} \quad a < x < b$$

we understand that the interval $a < x < b$ is contained in each of the intervals where $f(x)$ and $g(x)$ are defined and that $f(x)$ and $g(x)$ interpreted as distributions in $a < x < b$ are equal.

For instance, we have

$$\delta(x) = 0 \quad \text{in} \quad -\infty < x < 0$$

and

$$\delta(x) = 0 \quad \text{in} \quad 0 < x < \infty .$$

In fact, the fundamental sequences $\{f_n(x)\}$ from examples 2 and 3 (Section 1.2), when restricted to one of the intervals $-\infty < x < 0$, $0 < x < \infty$, converge to zero almost uniformly and, therefore, determine the null distribution.

More generally,

$$\alpha_0 \delta(x-x_0)+\alpha_1 \delta'(x-x_0)+ \dots +\alpha_k \delta^{(k)}(x-x_0) = 0 \quad \text{for} \quad x \neq x_0; \tag{1}$$

this means that the above equality holds in $-\infty < x < x_0$ and in $x_0 < x < \infty$.

3.1.1. THEOREM. *If $f(x) = 0$ for $x \neq x_0$, then the distribution $f(x)$ is of the form*

$$\alpha_0 \delta(x-x_0)+\alpha_1 \delta'(x-x_0)+ \dots +\alpha_k \delta^{(k)}(x-x_0). \tag{2}$$

In fact, there exists a continuous function $F(x)$ such that $F^{(k)}(x) = f(x)$. By 2.2.7, $F(x)$ is a polynomial of degree $< k$ in each of the intervals $-\infty<x<x_0$ and $x_0 < x < \infty$, i.e.,

$$F(x) = \begin{cases} F(x_0)+a_1(x-x_0)+ \dots +a_{k-1}(x-x_0)^{k-1} & \text{for} \quad -\infty < x < x_0, \\ F(x_0)+b_1(x-x_0)+ \dots +b_{k-1}(x-x_0)^{k-1} & \text{for} \quad x_0 < x < \infty. \end{cases}$$

The function $F(x)$ can be written in the form

$$F(x) = F(x_0)+\psi_1(x)+ \dots +\psi_{k-1}(x),$$

where

$$\psi_i(x) = \begin{cases} a_i(x-x_0)^i & \text{for} \quad -\infty < x < x_0, \\ b_i(x-x_0)^i & \text{for} \quad x_0 < x < \infty. \end{cases}$$

It is easy to see that

$$\psi_i^{(i)}(x) = i!\big(a_i+(b_i-a_i)H(x-x_0)\big)$$

and

$$\psi_i^{(k)}(x) = i!(b_i-a_i)\delta^{(k-i-1)}(x-x_0),$$

which proves the theorem.

If $f(x) = 0$ for $x \neq x_0$, then the representation of the distribution $f(x)$ in the form (2) is unique. This is a consequence of the following lemma:

3.1.2. *If $g(x)$ is a function and*

$$g(x)+\alpha_0 \delta(x-x_0)+ \dots +\alpha_k \delta^{(k)}(x-x_0) = 0 \tag{3}$$

on the whole axis $-\infty < x < \infty$, then $g(x) = 0$ and $\alpha_0 = \dots = \alpha_k = 0$.

We argue by induction. The case $k = 0$ is obvious since the distribution $\delta(x-x_0)$ is not a function. Suppose that the theorem is true for $k-1$.

From (3) it follows that $g(x) = 0$ for $x \neq 0$. Thus the function $g(x)$ is the distribution equal to 0. Applying 2.2.8 to (3) we obtain

$$c+\alpha_0 H(x-x_0)+\alpha_1 \delta(x-x_0)+ \dots +\alpha_k \delta^{(k-1)}(x-x_0) = 0.$$

Hence, from the induction hypothesis, we infer that $\alpha_1 = \ldots = \alpha_k = 0$ and $c+\alpha_0 H(x-x_0) = 0$; hence also $\alpha_0 = 0$.

It follows from (1) that if distributions are equal in intervals $-\infty < x < x_0$ and $x_0 < x < \infty$, i.e., if they differ only at one point, then we cannot infer that they are equal. However, it follows from 3.1.1 that their difference is then a finite linear combination of the distribution $\delta(x-x_0)$ and its derivatives. Nevertheless, if the distributions differ at one point at most and are functions, then they are equal.

3.2. FUNCTIONS WITH POLES

The function $1/x$ is a distribution in $-\infty < x < 0$ and in $0 < x < \infty$. However it is not identified with any distribution in $-\infty < x < \infty$ since it is not integrable in any neighbourhood of $x = 0$.

On the other hand, there exist distributions $f(x)$ in $-\infty < x < \infty$ such that

$$f(x) = 1/x \quad \text{for} \quad x \neq 0. \tag{1}$$

For instance,

$$(\ln|x|)' = 1/x \quad \text{for} \quad x \neq 0 \tag{2}$$

where the derivative is understood in the distributional sense. If we add an arbitrary linear combination of $\delta(x)$ and its derivatives to the left hand side of (2), equation (2) remains true.

We can also interpret equality (2) as the identification of the function $1/x$ with the distribution $(\ln|x|)'$. Such an identification can be extended to a wider class of functions which have poles at some points and are locally integrable elsewhere. More precisely, we shall consider functions $f(x)$ in $A < x < B$ which, in a neighbourhood of any point x_0, are of the form

$$f(x) = f_0(x) + \sum_{\nu=1}^{k} \frac{c_\nu}{(x-x_0)^\nu}, \tag{3}$$

where $f_0(x)$ is an integrable function. The decomposition into the integrable function $f_0(x)$ and the remaining singular part is unique. The singular part can vanish. The points x_0, for which at least one of the coefficients c_ν differs from zero are called *poles* of $f(x)$. In every finite closed subinterval with end points a and b there is at most a finite number of poles, and we can write

$$f(x) = f_1(x) + \sum_{\mu=1}^{m} \sum_{\nu=1}^{k} \frac{c_{\mu\nu}}{(x-x_\mu)^\nu} \tag{4}$$

where $f_1(x)$ is an integrable function and $x_1, \ldots, x_m$ are points of this interval.

This decomposition is unique. If a and b are not poles, we define the integral from a to b by the formula

$$\int_a^b f(t)\,dt = \left(\int_a^b f_1(t)\,dt + \sum_{\mu=1}^{m} c_{\mu 1} \ln|x-x_\mu|\Big|_a^b\right) + \sum_{\mu=1}^{m}\sum_{\nu=2}^{k} \frac{-c_{\mu\nu}}{(\nu-1)(x-x_\mu)^{\nu-1}}\Big|_a^b,$$

which is obtained from (4) by formal integration. By an indefinite integral of $f(x)$ we mean any function of the form

$$\int_a^x f(t)\,dt + C,$$

where a is not a pole and C is an arbitrary constant. The indefinite integral just defined, determined up to the constant, is also a function with poles, but the orders of the poles are decreased by 1.

In the sequel we shall deal only with functions which can be represented, at every point x_0, in the form (3) with the same integer k. Integrating $f(x)$ k times in the above sense, we obtain a locally integrable function $F(x)$ which is determined up to a polynomial of degree $< k$. Its distributional derivative $F^{(k)}(x)$ is uniquely determined by the function $f(x)$. We identify the distribution $F^{(k)}(x)$ with the function $f(x)$.

By these methods we have thus included all rational functions in the calculus of distributions; in particular, the function $1/(x-x_0)^k$ is to be identified with the kth distributional derivative of the function

$$\frac{(-1)^{k-1}}{(k-1)!} \ln|x-x_0|.$$

Moreover, we have included all rational expressions of sine and cosine; in particular we have the formulas

$$\tan x = (-\ln|\cos x|)', \qquad \cot x = (\ln|\sin x|)'.$$

Several other functions, very useful from the practical point of view, also enter into the calculus, e.g., the elliptic functions, the Euler function, etc.

3.3. DERIVATIVE AS THE LIMIT OF A DIFFERENCE QUOTIENT

We can define the derivative of a distribution in the same way as in the case of functions. In fact,

3.3.1. *For every distribution $f(x)$,*

$$f'(x) = \lim_{\alpha\to 0} \frac{f(x+\alpha)-f(x)}{\alpha}.$$

There exists a function $F(x)$ with a continuous derivative, such that $F^{(k)}(x) = f(x)$. Since

$$\frac{F(x+\alpha)-F(x)}{\alpha} \rightrightarrows F'(x) \quad \text{as} \quad \alpha \to 0,$$

we have

$$\frac{f(x+\alpha)-f(x)}{\alpha} = \left(\frac{F(x+\alpha)-F(x)}{\alpha}\right)^{(k)} \to F'(x)^{(k)} = f'(x)$$

in view of 2.6.2.

3.3.2. *For every distribution* $f(x)$,

$$xf'(x) = \lim_{\alpha\to 0} \frac{f(x+\alpha x)-f(x)}{\alpha}.$$

There exists a function $F(x)$ with a continuous derivative such that $F^{(k)}(x) = f(x)$. Since

$$\frac{F(x+\alpha x)-F(x)}{\alpha} \rightrightarrows xF'(x) \quad \text{as} \quad \alpha \to 0,$$

we have, differentiating

$$\frac{F'(x+\alpha x)-F'(x)}{\alpha} + F'(x+\alpha x) \to xF''(x)+F'(x).$$

In view of 2.8.2 we have $F'(x+\alpha x) \to F'(x)$ and therefore

$$\frac{F'(x+\alpha x)-F'(x)}{\alpha} \to xF''(x).$$

After k such steps we obtain

$$\frac{F^{(k)}(x+\alpha x)-F^{(k)}(x)}{\alpha} \to xF^{(k+1)}(x),$$

i.e., the formula in 3.3.2 holds.

Multiplying by $1/x$, we obtain from 3.3.2

$$f'(x) = \lim_{\alpha\to 0} \frac{f(x+\alpha x)-f(x)}{\alpha x} \quad \text{for} \quad x \neq 0 \tag{1}$$

by 2.7.3. According to Section 3.1, if $f(x)$ is a distribution in $A < x < B$ and $A < 0 < B$, then equality (1) should be understood as an equality in the intervals $A < x < 0$ and $0 < x < B$.

We can also generalize the well known theorem of Peano to the case of distributions.

3.3.3. *For every distribution* $f(x)$ *in* $A < x < B$ *and for every integer* $m > 0$ *we have*

$$f(x+\alpha) = f(x)+\frac{\alpha}{1!}f'(x)+ \ldots +\frac{\alpha^m}{m!}f^{(m)}(x)+\alpha^m r_\alpha(x)$$

in the common part of intervals $A-\alpha < x < B-\alpha$ *and* $A < x < B$, *where the distribution* $r_\alpha(x)$ *is defined in this common part and, as* $\alpha \to 0$,

$$r_\alpha(x) \to 0 \quad in \quad A < x < B.$$

There exists a function $F(x)$ with a continuous mth derivative, such that $F^{(k)}(x) = f(x)$ for some integer $k \geqslant 0$. We have

$$F(x+\alpha) = F(x)+\frac{\alpha}{1!}F'(x)+\ \dots\ +\frac{\alpha^m}{m!}F^{(m)}(x)+\alpha^m R_\alpha(x)$$

where $R_\alpha(x) \rightrightarrows 0$ as $\alpha \to 0$. Differentiating k times, we obtain the required equality where $r_\alpha(x) = R_\alpha^{(k)}(x)$.

3.4. THE VALUE OF A DISTRIBUTION AT A POINT

We shall first prove the following lemma:

3.4.1 (Zieleźny [46]). *If a distribution* $g(x)$ *in* $-\infty < x < \infty$ *satisfies the equation*

$$g(\lambda x) = g(x) \quad \textit{for each} \quad \lambda \neq 0, \tag{1}$$

then it is a constant function.

Equality (1) implies, by formula (1) in 3.3, that $g'(x) = 0$ for $x \neq 0$. Hence, by 3.1.1, we have

$$g'(x) = \alpha_0\delta(x)+\alpha_1\delta'(x)+\ \dots\ +\alpha_k\delta^{(k)}(x).$$

Consequently by 2.2.9

$$g(x) = c+\alpha_0 H(x)+\alpha_1\delta(x)+\ \dots\ +\alpha_k\delta^{(k-1)}(x)$$

where c is a constant. From (1) it follows, by formula (5) in 2.8, that

$$\alpha_0\big(H(\lambda x)-H(x)\big)+\alpha_1\left(\frac{1}{|\lambda|}-1\right)\delta(x)+\ \dots\ +\alpha_k\left(\frac{1}{|\lambda|\lambda^{k-1}}-1\right)\delta^{(k-1)}(x) = 0.$$

Hence we infer, by 3.1.2, that $\alpha_0 = \alpha_1 = \dots = \alpha_k = 0$. This proves that $g(x) = c$.

3.4.2. *If the limit*

$$g(x) = \lim_{\alpha\to 0} f(\alpha x+x_0) \tag{2}$$

exists, it is a constant function.

In fact, the distribution $g(x)$ satisfies the hypothesis of 3.4.1.

The value of the function defined by formula (2) will be called the *value* of the distribution $f(x)$ at the point x_0. If the value of the distribution at x_0 exists, i.e., if the limit (2) exists, the point x_0 is said to be *regular*. Otherwise it is said to be *singular*.

3.4.3. *If a distribution $f(x)$ is a locally integrable function continuous at x_0, then $f(\alpha x+x_0) \rightrightarrows f(x_0)$ as $\alpha \to 0$. Consequently the point x_0 is regular and the value of $f(x)$ at x_0 in the distributional sense is equal to $f(x_0)$.*

3.4.4. *If a locally integrable function $F(x)$ possesses an (ordinary) derivative at x_0, then this derivative is the value of the distribution $F'(x)$ at that point.*

By hypothesis there exists a limit c such that

$$\frac{F(\alpha x+x_0)-F(x_0)}{\alpha} \to c \quad \text{as} \quad \alpha \to 0$$

where all the symbols are interpreted in the classical sense. Consequently

$$\frac{F(\alpha x+x_0)-F(x_0)}{\alpha} \rightrightarrows cx \quad \text{in} \quad -\infty < x < \infty$$

as $\alpha \to 0$. Differentiating this formula distributionally we obtain

$$F'(\alpha x+x_0) \to c \quad \text{for} \quad \alpha \to 0,$$

which proves the theorem.

The converse theorem is not true. It is possible for the ordinary derivative not to exist at some point although the distributional derivative has a value at this point. For instance, the function $F(x) = 3x^2\sin(1/x)-x\cos(1/x)$ does not have an ordinary derivative at the point 0. However, the distributional derivative $F'(x)$ has the value 0 at the point 0. In fact, we have

$$(\alpha x)^3 \sin\frac{1}{\alpha x} \rightrightarrows 0 \quad \text{as} \quad \alpha \to 0.$$

Hence, by successive differentiation,

$$\frac{F(\alpha x)}{\alpha} \to 0 \quad \text{and} \quad F'(\alpha x) \to 0 \quad \text{as} \quad \alpha \to 0.$$

The following lemma is a particular case of 3.4.4.

3.4.5. *If $f(x)$ is a locally integrable function and the function $F(x) = \int_a^x f(t)dt$ has an ordinary derivative at x_0, then this derivative is the value of the distribution $f(x)$ at x_0.*

The value of a distribution $f(x)$ at a point x_0 will be denoted by $f(x_0)$ as in the case of functions. This notation does not give rise to any misunderstanding. In fact, if the distribution $f(x)$ is a continuous function, both meanings of $f(x_0)$ coincide by 3.4.3. If $f(x)$ is only locally integrable (or, more generally, a function with poles), then, by 3.4.5, the values of the distribution $f(x)$ exist almost everywhere. Both meanings of $f(x_0)$ then coincide almost everywhere but, in general, not everywhere. When the two values differ we adopt the convention of denoting by $f(x_0)$ the value in the distributional sense.

Example 1. By 3.4.3 each point $x_0 \neq 0$ is a regular point of the Heaviside function $H(x)$, and the value at x_0 in the distributional sense is the same as the value in the usual sense. The point $x_0 = 0$ is singular since the limit of

$$H(\alpha x) = \frac{\alpha}{|\alpha|}\left(H(x)-\tfrac{1}{2}\right)+\tfrac{1}{2} \tag{3}$$

does not exist as $\alpha \to 0$.

Example 2. The Dirac distribution $\delta(x)$ has the value 0 at each point $x_0 \neq 0$ and has no value at the point $x_0 = 0$. In fact, differentiating (3) we obtain

$$|\alpha|\delta(\alpha x) = \delta(x).$$

The existence of the limit $\lim\limits_{\alpha\to 0}\delta(\alpha x)$ would imply that $\delta(x)$ is the function identically equal to 0.

Example 3. It follows from 3.4.5 that the value (in the distributional sense) of the function $\sin(1/x)$ at the point 0 is equal to 0.

In fact, by the substitution $t = 1/\tau$ and the second mean value theorem,

$$\frac{1}{x}\int_0^x \sin\frac{1}{t}\,dt = \frac{1}{x}\int_{1/x}^{\infty}\frac{1}{\tau}\cdot\frac{\sin\tau}{\tau}\,d\tau = \int_{\xi}^{\infty}\frac{\sin\tau}{\tau}\,d\tau \qquad (1/x < \xi < \infty)$$

and hence

$$\left(\int_0^x \sin\frac{1}{t}\,dt\right)'_{x=0} = \lim_{x\to 0}\frac{1}{x}\int_0^x \sin\frac{1}{t}\,dt = 0.$$

It can be proved that if the value of a distribution $f(x)$ is 0 everywhere, then $f(x)$ is the null function (Łojasiewicz [14]). Thus a distribution is uniquely determined by its values provided they exist everywhere.

3.5. EXISTENCE THEOREMS FOR VALUES OF DISTRIBUTIONS

In applications the following theorem is of great importance.

3.5.1. Theorem (Łojasiewicz [13] and [14]). *A distribution $f(x)$ has the value c at a point x_0 if and only if there exist an integer $k \geqslant 0$ and a continuous function $F(x)$ such that $F^{(k)}(x) = f(x)$ and*

$$\lim_{x\to x_0}\frac{F(x)}{(x-x_0)^k} = \frac{c}{k!}.$$

It suffices to prove the theorem in the case $x_0 = 0$.

Suppose that $F^{(k)}(x) = f(x)$ and

$$\lim_{x\to 0} \frac{F(x)}{x^k} = \frac{c}{k!}. \tag{1}$$

Then $\alpha^{-k}F(\alpha x) \rightrightarrows cx^k/k!$ in $-\infty < x < \infty$ as $\alpha \to 0$. Since $f(\alpha x) = (\alpha^{-k}F(\alpha x))^{(k)}$, we have $\lim_{\alpha\to 0} f(\alpha x) = c$, i.e., $f(x)$ has the value c at the point 0.

Suppose now that $f(x)$ has the value c at the point 0. Then there exist an integer $k > 0$ and a function $F_\alpha(x)$ with a parameter α such that

$$F_\alpha^{(k)}(x) = f(\alpha x) \quad \text{and} \quad F_\alpha(x) \rightrightarrows cx^k/k! \quad \text{as} \quad \alpha \to 0.$$

In particular

$$F_1^{(k)}(x) = f(x).$$

Since $(\alpha^k F_\alpha(x/\alpha) - F_1(x))^{(k)} = 0$, the functions $\alpha^k F_\alpha(x/\alpha)$ and $F_1(x)$ only differ from each other by a polynomial of degree $< k$:

$$\alpha^k F_\alpha(x/\alpha) = F_1(x) + b_0(\alpha) + b_1(\alpha)x + \ldots + b_{k-1}(\alpha)x^{k-1}.$$

We are going to prove that the limits $b_i = \lim_{\alpha\to 0} b_i(\alpha)$ exist ($i = 0, 1, \ldots, k-1$), and that the function

$$F(x) = F_1(x) + b_0 + b_1 x + \ldots + b_{k-1}x^{k-1}$$

has the required property.

Since

$$F_\alpha(x) = \alpha^{-k}\big(F_1(\alpha x) + b_0(\alpha) + \alpha b_1(\alpha)x + \ldots + \alpha^{k-1}b_{k-1}(\alpha)x^{k-1}\big),$$

for the function

$$G(x) = F_1(x) - cx^k/k!$$

we have

$$\alpha^{-k}\big(G(\alpha x) + b_0(\alpha) + \alpha b_1(\alpha)x + \ldots + \alpha^{k-1}b_{k-1}(\alpha)x^{k-1}\big) \rightrightarrows 0 \quad \text{as} \quad \alpha \to 0.$$

Consequently there exists an increasing function $\varepsilon(\alpha)$ of the parameter $\alpha > 0$ such that $\varepsilon(\alpha) \to 0$ as $\alpha \to 0$ and

$$|G(\alpha x) + b_0(\alpha) + \alpha b_1(\alpha)x + \ldots + \alpha^{k-1}b_{k-1}(\alpha)x^{k-1}| < \alpha^k \varepsilon(\alpha) \tag{2}$$

for $-1 \leqslant x \leqslant 1$.

Fix k points $-1 \leqslant x_1 < \ldots < x_k \leqslant 1$. We have, by (2),

$$|G(\alpha x_i) + b_0(\alpha) + \alpha b_1(\alpha)x_i + \ldots + \alpha^{k-1}b_{k-1}(\alpha)x_i^{k-1}| < \alpha^k \varepsilon(\alpha).$$

Replacing α by β ($\beta > \alpha$) in (2) and then x by $x_i\alpha/\beta$ we obtain

$$|G(\alpha x_i) + b_0(\beta) + \alpha b_1(\beta)x_i + \ldots + \alpha^{k-1}b_{k-1}(\beta)x_i^{k-1}| < \beta^k \varepsilon(\beta).$$

From the last two inequalities it follows that

$$\left|\big(b_0(\alpha)-b_0(\beta)\big)+\alpha\big(b_1(\alpha)-b_1(\beta)\big)x+\dots+\alpha^{k-1}\big(b_{k-1}(\alpha)-b_{k-1}(\beta)\big)x^{k-1}\right| < 2\beta^k\varepsilon(\beta) \qquad (3)$$

for $x = x_1, \dots, x_k$.

Denoting by $p(x)$ the polynomial within the modulus signs (| |) on the left hand side of (3), we can write

$$\alpha^i\big(b_i(\alpha)-b_i(\beta)\big) = \frac{1}{A}\big(A_{1i}p(x_1)+\dots+A_{ki}p(x_k)\big)$$

where A_{ji} are minors of the determinant

$$A = \begin{vmatrix} 1 & x_1 & \dots & x_1^{k-1} \\ \dots & \dots & \dots & \dots \\ 1 & x_k & \dots & x_k^{k-1} \end{vmatrix},$$

just as in the proof of 1.2.4. Thus there exists a constant K such that

$$\alpha^i|b_i(\alpha)-b_i(\beta)| < 2K\varepsilon(\beta)\beta^k.$$

If $\alpha < \beta \leqslant 2\alpha$, then

$$|b_i(\alpha)-b_i(\beta)| < 2^{i+1}K\varepsilon(\beta)\beta^{k-i}. \qquad (4)$$

For $\alpha = 2^{-n-1}$ and $\beta = 2^{-n}$ we have, by (4),

$$|b_i(1/2^{n+1})-b_i(1/2^n)| < 2^{i+1}K\varepsilon(1/2^{n(k-i)}) \qquad (n = 1, 2, \dots).$$

Writing down these inequalities for $n, n+1, \dots, m-1$ and adding them we obtain the estimate

$$|b_i(1/2^m)-b_i(1/2^n)| \leqslant 2^{i+2}K\varepsilon(1/2^n)(1/2^{n(k-i)}) \qquad (m \geqslant n). \qquad (5)$$

For arbitrary α, β such that $0 < \alpha < \beta \leqslant 1/2$ there exist positive integers m, n such that

$$\alpha < 1/2^m \leqslant 2\alpha \quad \text{and} \quad 1/2^n < \beta \leqslant 2/2^n.$$

Replacing β by $1/2^m$ in (4) we have

$$|b_i(\alpha)-b_i(1/2^m)| < 2^{i+1}K\varepsilon(1/2^m)(1/2^{m(k-i)}). \qquad (6)$$

Replacing α by $1/2^n$ in (4) we have

$$|b_i(1/2^n)-b_i(\beta)| < 2^{i+1}K\varepsilon(\beta)\beta^{k-i}. \qquad (7)$$

From (5), (6) and (7) we find

$$|b_i(\alpha)-b_i(\beta)| < 2^{i+3}K\varepsilon(\beta)\beta^{k-i} \qquad (0 < \alpha < \beta < 1/2).$$

This inequality proves that the limits

$$b_i = \lim_{\alpha\to 0+} b_i(\alpha) \qquad (i = 0, 1, \dots, k-1)$$

exist and

$$|b_i - b_i(\alpha)| \leqslant 2^{i+3} K\varepsilon(\alpha)\alpha^{k-i} \qquad (i = 0, 1, \ldots, k-1). \tag{8}$$

It follows from (2) and (8) that for $-1 \leqslant x \leqslant 1$

$$|G(\alpha x) + b_0 + \alpha b_1 x + \ldots + \alpha^{k-1} b_{k-1} x^{k-1}| < K_1 \varepsilon(\alpha)\alpha^k \tag{9}$$

where $K_1 = 1 + 2^{k+3}K$. We shall show that the function

$$F(x) = G(x) + b_0 + b_1 x + \ldots + b_{k-1} x^{k-1} + \frac{c}{k!} x^k$$

satisfies the equation (1). In fact, by (9), we have for $x = 1$

$$|F(\alpha)/\alpha^k - c/k!| \leqslant K_1 \varepsilon(\alpha).$$

Hence the right hand limit of $F(\alpha)/\alpha^k$ is equal to $c/k!$. In the same way, for $x = -1$, we verify that the left hand limit also has this value.

To complete the proof it suffices to observe that $F^{(k)}(x) = f(x)$.

3.5.2. THEOREM (Łojasiewicz [13] and [14]). *If the distribution $f'(x)$ has a value at x_0, then the distribution $f(x)$ also has a value at x_0.*

In fact, there exist an integer $k \geqslant 0$ and a continuous function $F(x)$ such that $F^{(k)}(x) = f'(x)$ and the limit

$$\lim_{x \to x_0} \frac{F(x)}{(x - x_0)^k}$$

exists. If $k = 0$, then $f'(x)$ is a continuous function and so is $f(x)$; therefore the theorem is true.

If $k > 0$, then

$$\lim_{x \to x_0} \frac{F(x)}{(x - x_0)^{k-1}} = 0.$$

It follows from 3.5.1 that the distribution $F^{(k-1)}(x)$ has a value at x_0. The distribution $f(x)$ differs from $F^{(k-1)}(x)$ by a constant, and so $f(x)$ also has a value at x_0.

Let $\omega(x)$ satisfy the hypothesis of Section 2.7.

3.5.3. *If a distribution $f(x)$ has a value $f(x_0)$ at x_0, then the distribution $\omega(x)f(x)$ has the value $\omega(x_0)f(x_0)$ at x_0.*

If fact, we have $f(\alpha x + x_0) \to f(x_0)$, $\omega(\alpha x + x_0) \rightrightarrows \omega(x_0)$ and $\omega(\alpha x + x_0)^{(m)} \rightrightarrows 0$ $(m = 1, 2, \ldots)$ as $\alpha \to 0$. Hence we infer from 2.7.3 that

$$\omega(\alpha x + x_0) f(\alpha x + x_0) \to \omega(x_0) f(x_0).$$

Let $\varphi(x)$ satisfy the hypothesis of Section 2.8.

3.5.4. *If a distribution $f(x)$ has a value at the point $x = \varphi(x_0)$, then the distribution $f(\varphi(x))$ also has this value at x_0.*

In fact, by 3.5.1 there is an integer $k \geqslant 0$ and a continuous function $F(x)$ such that

$$\lim_{x\to\varphi(x_0)} \frac{F(x)}{(x-\varphi(x_0))^k} = \frac{c}{k!}$$

where c is the value of $f(x)$ at $\varphi(x_0)$. Since

$$\frac{F(\varphi(x))}{(x-x_0)^k} = \frac{F(\varphi(x))}{(\varphi(x)-\varphi(x_0))^k}\left(\frac{\varphi(x)-\varphi(x_0)}{x-x_0}\right)^k$$

we have

$$\lim_{x\to x_0} \frac{F(\varphi(x))}{(x-x_0)^k} = \frac{c}{k!}(\varphi'(x_0))^k.$$

Consequently

$$\frac{F(\varphi(\alpha x+x_0))}{\alpha^k} \rightrightarrows \frac{c}{k!}(\varphi'(x_0))^k x^k.$$

Differentiating distributionally we obtain

$$\frac{F'(\varphi(\alpha x+x_0))}{\alpha^{k-1}}\varphi'(\alpha x+x_0) \to \frac{c}{(k-1)!}(\varphi'(x_0))^k x^{k-1}.$$

Since

$$\frac{1}{\varphi'(\alpha x+x_0)} \rightrightarrows \frac{1}{\varphi'(x_0)} \quad \text{and} \quad \left(\frac{1}{\varphi'(\alpha x+x_0)}\right)^{(m)} \rightrightarrows 0 \quad (m = 1, 2, \ldots),$$

we have by 2.7.3

$$\frac{F'(\varphi(\alpha x+x_0))}{\alpha^{k-1}} \to \frac{c}{(k-1)!}(\varphi'(x_0))^{k-1}x^{k-1}.$$

After k similar steps we finally obtain

$$F^{(k)}(\varphi(\alpha x+x_0)) \to c, \quad \text{i.e.,} \quad f(\varphi(\alpha x+x_0)) \to c \quad \text{as} \quad \alpha\to 0.$$

3.6. THE VALUE OF A DISTRIBUTION AT INFINITY

We shall prove the lemma

3.6.1. *For any distribution $f(x)$, if the limit*

$$\lim_{\beta\to\infty} f(x+\beta) \tag{1}$$

exists, then it is a constant function.

Denote the limit (1) by $g(x)$. For every α we have

$$g(x+\alpha) = \lim_{\beta\to\infty} f(x+\alpha+\beta) = \lim_{\beta\to\infty} f(x+\beta) = g(x).$$

Hence

$$g'(x) = \lim_{\alpha\to 0} \frac{g(x+\alpha)-g(x)}{\alpha} = 0$$

and, consequently, $g(x)$ is a constant function.

The value of the function defined by (1) is said to be the *value* of the distribution $f(x)$ at ∞ and is denoted by $f(\infty)$.

The value $f(-\infty)$ of the distribution $f(x)$ at $-\infty$ is defined similarly.

Obviously the symbols $f(\infty), f(-\infty)$ have a meaning if and only if the corresponding limits exist.

3.6.2. *If a distribution $f(x)$ is a continuous function and has the ordinary limit c at ∞, resp. $-\infty$, then $f(x+\beta) \rightrightarrows c$ as $\beta\to\infty$, resp. $\beta\to-\infty$. Consequently $f(\infty) = c$, resp. $f(-\infty) = c$.*

The next lemma follows immediately from the definition of $f(\infty)$ and 2.6.2.

3.6.3. *If $f(\infty)$ exists, then $f'(\infty) = 0$. If $f(-\infty)$ exists, then $f'(-\infty) = 0$.*

Hence we obtain, in particular, for $f(x) = \int_a^x g(t)\,dt$:

3.6.4. *If $g(x)$ is a locally integrable function and the improper integral $\int_a^\infty g(t)\,dt$ exists, then $g(\infty) = 0$.*

From this remark it follows that there exist continuous functions $g(x)$ such that the distributional value $g(\infty)$ exists but the ordinary limit at ∞ does not exist.

Let $\omega(x)$ satisfy the hypothesis of Section 2.7.

3.6.5. *If a distribution $f(x)$ has the value $f(\infty)$ at ∞ and if $\omega(x+\beta) \rightrightarrows \omega(\infty)$ and $\omega^{(m)}(x+\beta) \rightrightarrows 0$ $(m = 1, 2, \ldots)$ as $\beta\to\infty$, then the distribution $\omega(x)f(x)$ has the value $\omega(\infty)f(\infty)$ at ∞.*

This follows immediately from 2.7.3.

4. Extension of the theory

4.1. THE INTEGRAL OF A DISTRIBUTION

By an *indefinite integral* of a distribution $f(x)$ in $A < x < B$ we mean any distribution $\psi(x)$ in $A < x < B$ such that $\psi'(x) = f(x)$. It follows from 2.2.9 and 2.2.11 that, for every distribution, the indefinite integral exists and is uniquely determined up to a constant (just as for functions). In the case where the distribution is a function, this definition coincides with the usual one.

It follows from 3.5.2 that

4.1.1. *Each regular point of a distribution $f(x)$ is also a regular point of the indefinite integral of $f(x)$.*

We introduce the following notation

$$\int_a^b f(x+t)\,dt = \psi(x+b) - \psi(x+a), \tag{1}$$

where $\psi'(x) = f(x)$ and a and b belong to the interval $A < x < B$. The distribution (1) is defined in the overlap of the intervals $A-a < x < B-a$, $A-b < x < B-b$, and does not depend on the choice of indefinite integral $\psi(x)$. If $f(x)$ is a locally integrable function, then the meaning of the integral (1) coincides with its usual meaning.

The following formulas hold for these integrals:

$$\int_a^b f(x+t)\,dt = -\int_b^a f(x+t)\,dt,$$

$$\int_a^b \lambda f(x+t)\,dt = \lambda \int_a^b f(x+t)\,dt,$$

$$\int_a^b (f(x+t)+g(x+t))\,dt = \int_a^b f(x+t)\,dt + \int_a^b g(x+t)\,dt,$$

$$\int_a^b (f(x+t)-g(x+t))\,dt = \int_a^b f(x+t)\,dt - \int_a^b g(x+t)\,dt,$$

$$\int_a^b f(x+t)\,dt + \int_b^c f(x+t)\,dt = \int_a^c f(x+t)\,dt,$$

$$\left(\int_a^b f(x+t)\,dt\right)' = \int_a^b f'(x+t)\,dt = f(x+b)-f(x+a).$$

4.1.2. THEOREM. *If* $f_n(x) \to f(x)$, *then*

$$\int_a^b f_n(x+t)\,dt \to \int_a^b f(x+t)\,dt. \tag{2}$$

In fact, there exist an integer $k > 0$ and continuous functions $F_n(x) \rightrightarrows F(x)$ such that

$$F_n^{(k)}(x) = f_n(x) \quad \text{and} \quad F^{(k)}(x) = f(x).$$

Consequently

$$F_n(x+b)-F_n(x+a) \rightrightarrows F(x+b)-F(x+a).$$

Differentiating $k-1$ times we obtain (2) by 2.5.1.

The following is an immediate corollary:

4.1.3. THEOREM. *The equality*

$$\int_a^b \sum_{n=1}^{\infty} g_n(x+t)\,dt = \sum_{n=1}^{\infty} \int_a^b g_n(x+t)\,dt$$

holds for every convergent series of distributions.

It is possible, to give an equivalent definition of the integral (1) as the limit of the Riemann sums

$$\sum_{j=1}^{m} f(x+\tau_j)(t_j-t_{j-1})$$

(where $a = t_0 < t_1 < \ldots < t_m = b$ and $t_{j-1} \leqslant \tau_j \leqslant t_j$). To prove this equivalence, it suffices to present the integral $\int_a^b F(x+t)dt$, where $F(x)$ is a continuous function and $F^{(k)}(x) = f(x)$, as a limit of such sums and to differentiate them k times.

The value at $x=0$ of the distribution $\int_a^b f(x+t)\,dt$, if it exists, will be denoted by

$$\int_a^b f(t)\,dt; \tag{3}$$

this symbol, when meaningful, always stands for a number. If the distribution is a locally integrable function, (3) is its ordinary definite integral.

It follows directly from the definition that if a and b are regular points of the indefinite integral $\psi(x)$ of the distribution $f(x)$, then

$$\int_a^b f(t)\,dt = \psi(b)-\psi(a). \tag{4}$$

Equality (4) will also be used to define the *definite integral* in cases where a or b is infinite. In particular we have the formulas

$$\int_a^\infty f(t)\,dt = \psi(\infty)-\psi(a), \qquad \int_{-\infty}^b f(t)\,dt = \psi(b)-\psi(-\infty),$$
$$\int_{-\infty}^\infty f(t)\,dt = \psi(\infty)-\psi(-\infty). \tag{5}$$

Replacing x by $x+t$ in 2.7 (1), we have, after integration,

$$\int_a^b \omega(x+t)f'(x+t)\,dt+\int_a^b \omega'(x+t)f(x+t)\,dt$$
$$= \omega(x+b)f(x+b)-\omega(x+a)f(x+a). \tag{6}$$

This corresponds to the classical integration by parts.

Hence, substituting $x = 0$, we get by 3.5.3

$$\int_a^b \omega(t)f'(t)\,dt+ \int_a^b \omega'(t)f(t)\,dt = \omega(b)f(b)-\omega(a)f(a) \tag{7}$$

under the hypotheses that the values exist and that at least one of the integrals on the left hand side is meaningful. Formula (7) is also true when either one or both of the numbers a, b are infinite, provided $\omega(x)$ satisfies the hypotheses of 3.6.5 or the analogous hypotheses at $x = -\infty$.

Applying (7) we can easily prove that

$$\int_a^b \omega(t)\,\delta(t-x_0)\,dt = \omega(x_0) \quad \text{for} \quad a < x_0 < b. \tag{8}$$

In fact

$$\int_a^b \omega(t)H'(t-x_0)\,dt = \omega(b)\,H(b-x_0)-\omega(a)\,H(a-x_0)-\int_a^b \omega'(t)\,H(t-x_0)\,dt$$
$$= \omega(b)-\int_{x_0}^b \omega'(t)\,dt = \omega(x_0).$$

Similarly it can be proved that

$$\int_a^b \omega(t)\,\delta(t-x_0)\,dt = 0 \quad \text{if} \quad x_0< a< b \quad \text{or} \quad a < b< x_0. \tag{9}$$

If $\varphi(x)$ satisfies the hypotheses of Section 2.8, the ordinary formula for substitution

$$\int_{\varphi(a)}^{\varphi(b)} f(t)\,dt = \int_a^b f(\varphi(t))\,\varphi'(t)\,dt \tag{10}$$

holds provided the indefinite integral of $f(x)$ has a value at the points $x = \varphi(a)$ and $x = \varphi(b)$.

In fact, let $F(x)$ be an indefinite integral of $f(x)$, i.e., $F'(x) = f(x)$. Setting

$$\psi_1(x) = F(\varphi(b)+x) - F(\varphi(a)+x),$$

the left hand side of (10) is equal to $\psi_1(0)$. On the other hand, with

$$\psi_2(x) = F(\varphi(b+x)) - F(\varphi(a+x)),$$

the right hand side of (10) becomes

$$\int_a^b F'(\varphi(t))\varphi'(t)\,dt = \int_a^b (F(\varphi(t)))'\,dt = \psi_2(0).$$

Since the values of $F(x)$ exist at $\varphi(a)$ and $\varphi(b)$, we have $\psi_1(0) = \psi_2(0)$ in view of 3.5.4. Thus the two sides of (10) are equal.

Observe that if the hypothesis on the existence of values of $F(x)$ at $\varphi(a)$ and $\varphi(b)$ is not satisfied, the equality may fail. E.g. let $f(x) = (\ln|x^2-2x|)'$, $\varphi(x) = x^2+x$ $(\frac{1}{2}<x<\infty)$ and $a = 0$, $b = 1$. We then have $F(x) = \ln|x^2-2x|$ and, it easily follows that

$$\psi_1(x) = \ln|(2+x)/(2-x)|, \qquad \psi_2(x) = \ln|(3+x)/(1-x)|.$$

Hence $\psi_1(0) = 0$, $\psi_2(0) = \ln 3$ and the two sides od (10) are not equal.

4.2. PERIODIC DISTRIBUTIONS

We say that a distribution $f(x)$ in $-\infty < x < \infty$ is *periodic* (with the period 2π) if

$$f(x+2\pi) = f(x).$$

We shall prove the following theorems:

4.2.1. *The integral*

$$\int_{-\pi}^{\pi} f(t)\,dt \tag{1}$$

exists for any periodic distribution $f(x)$.

4.2.2. *For every convergent series of periodic distributions the following equality holds*:

$$\int_{-\pi}^{\pi}\sum_{n=1}^{\infty} g_n(t)\,dt = \sum_{n=1}^{\infty}\int_{-\pi}^{\pi} g_n(t)\,dt.$$

4.2.3. *If, for a periodic distribution $f(x)$, the integral* (1) *vanishes, then there exists a periodic distribution $\psi(x)$ such that $\psi'(x) = f(x)$ and*

$$\int_{-\pi}^{\pi}\psi(t)\,dt = 0.$$

If $\psi_0'(x) = f(x)$, then

$$\int_{-\pi}^{\pi} f(x+t)\,dt = \psi_0(x+\pi)-\psi_0(x-\pi). \tag{2}$$

Since the derivative of this distribution is $f(x+\pi)-f(x-\pi) = 0$, the integral (1) is a constant function. Thus at the point 0 it has the value

$$\int_{-\pi}^{\pi} f(t)\,dt = \int_{-\pi}^{\pi} f(x+t)\,dt, \tag{3}$$

which proves 4.2.1.

4.2.2 follows from (3) and 4.1.3.

If the integral (1) vanishes and $\psi_0'(x) = f(x)$, then, by (2), we have

$$\psi_0(x+\pi)-\psi_0(x-\pi) = 0.$$

This means that the distribution $\psi_0(x)$ is periodic. The distribution

$$\psi(x) = \psi_0(x)-c \quad \text{where} \quad c = \frac{1}{2\pi}\int_{-\pi}^{\pi}\psi_0(t)\,dt$$

has all the properties stipulated in 4.2.3.

4.2.4. Theorem (Schwartz [33]; see also Łojasiewicz, Wloka and Zieleźny [15]) *Every periodic distribution $f(x)$ is the sum of a trigonometric series*

$$f(x) = \frac{a_0}{2}+\sum_{n=1}^{\infty}(a_n\cos nx+b_n\sin nx), \tag{4}$$

where

$$a_n = \frac{1}{\pi}\int_{-\pi}^{\pi} f(t)\cos nt\,dt \quad (n = 0, 1, 2, \ldots),$$
$$b_n = \frac{1}{\pi}\int_{-\pi}^{\pi} f(t)\sin nt\,dt \quad (n = 1, 2, \ldots). \tag{5}$$

The expansion (4) *is unique.*

4.2.5. THEOREM (Schwartz [33]). *A series*

$$\sum_{n=1}^{\infty} (a_n \cos nx + b_n \sin nx) \tag{6}$$

converges distributionally if and only if there is an integer k such that

$$a_n/n^k \to 0 \quad \text{and} \quad b_n/n^k \to 0 \quad (n \to \infty). \tag{7}$$

Suppose that the series converges and that equality (4) holds. Multiplying both sides of (4) by $\cos mx$ and integrating, we obtain by 4.2.2

$$\int_{-\pi}^{\pi} f(t) \cos mt\, dt$$

$$= \tfrac{1}{2} a_0 \int_{-\pi}^{\pi} \cos mt\, dt + \sum_{n=1}^{\infty} \left(a_n \int_{-\pi}^{\pi} \cos nt \cos mt\, dt + b_n \int_{-\pi}^{\pi} \sin nt \cos mt\, dt \right) = \pi a_m$$

since all the trigonometric integrals vanish, except the integral at a_m which is equal to π. We thus obtain the first of the formulas (5), and we obtain the second one in a similar way.

It follows from this that the representation of $f(x)$ in the form (4), if it exists, is unique.

Now let $f(x)$ be any periodic distribution. For the distribution $g(x) = f(x) - \frac{1}{2} a_0$ we have $\int_{-\pi}^{\pi} g(t)\, dt = 0$. Applying 4.2.3 k times we deduce that there exists a periodic distribution $G_0(x)$ with $G_0^{(k)}(x) = g(x)$. On the other hand, for sufficiently large k, there exists a continuous function $G(x)$ such that $G^{(k)}(x) = g(x)$. We may assume that k is a multiple of the number 4 and that $G(x)$ has a continuous derivative. The distributions $G_0(x)$ and $G(x)$ differ from each other only be a polynomial of degree $< k$, and so $G_0(x)$ is a periodic function with a continuous derivative.

By an elementary theorem on Fourier series, $G_0(x)$ is the sum of a uniformly convergent trigonometric series

$$G_0(x) = \tfrac{1}{2} \alpha_0 + \sum_{n=1}^{\infty} (\alpha_n \cos nx + \beta_n \sin nx) \tag{8}$$

where

$$\alpha_n \to 0, \quad \beta_n \to 0. \tag{9}$$

Differentiating formula (8) distributionally k times and adding $\frac{1}{2} a_0$ to both sides, we obtain the equality

$$f(x) = \tfrac{1}{2} a_0 + \sum_{n=1}^{\infty} (n^k \alpha_n \cos nx + n^k \beta_n \sin nx),$$

where the convergence is to be understood distributionally.

Since the representation of $f(x)$ in the form (4) is unique, it follows that

$$a_n = n^k \alpha_n \quad \text{and} \quad b_n = n^k \beta_n \quad (n = 1, 2, \ldots), \tag{10}$$

where a_n and b_n are given by (5). This completes the proof of 4.2.4.

Property (7) follows from (9) and (10). In order to prove 4.2.5 it is enough to show that (7) implies the distributional convergence of (6). Let k_0 be a multiple of the number 4 and let $k_0 > k+1$. Then the series

$$\sum_{n=1}^{\infty} (a_n/n^{k_0}) \quad \text{and} \quad \sum_{n=1}^{\infty} (b_n/n^{k_0})$$

converge absolutely. Thus the series

$$\sum_{n=1}^{\infty} \left(\frac{a_n}{n^{k_0}} \cos nx + \frac{b_n}{n^{k_0}} \sin nx \right)$$

converges uniformly. Differentiating the last series k_0 times we obtain the series (6) which, by 2.5.8, converges distributionally.

In Classical Analysis it is usual to distinguish between trigonometric series and Fourier series. In the theory of distributions such a distinction is unnecessary since each convergent trigonometric series is an expansion of a periodic distribution. Note that if a trigonometric series is the expansion of a periodic function in the classical sense, then is also its expansion in the distributional sense. This follows from the fact that the coefficients of the expansion are calculated from the same formulas (5).

As an example of a periodic distribution which is not a function, let us mention

$$\delta_{2\pi}(x) = \sum_{n=-\infty}^{\infty} \delta(x+2n\pi). \tag{11}$$

The distribution $\delta_{2\pi}(x)$ is the derivative of the function $\mathrm{E}(x/2\pi)$, where the symbol $\mathrm{E}(a)$ denotes the greatest integer not exceeding a. In fact, the function $\mathrm{E}(x/2\pi)$ can be expressed in terms of the Heaviside function:

$$\mathrm{E}\left(\frac{x}{2\pi}\right) = \sum_{n=1}^{\infty} H(x-2n\pi) + \sum_{n=-\infty}^{0} (H(x-2n\pi)-1). \tag{12}$$

Hence the right hand side of (12) is transformed under differentiation into the right hand side of (11).

To find the coefficients of the expansion for $\delta_{2\pi}(x)$ we use formulas (8) and (9) from Section 4.1:

$$a_n = \frac{1}{\pi} \int_{-\pi}^{\pi} \delta_{2\pi}(t) \cos nt\, dt = \frac{1}{\pi} \int_{-\pi}^{\pi} \delta(t) \cos nt\, dt = \frac{1}{\pi} \cos 0 = \frac{1}{\pi}$$

and, similarly,

$$b_n = 0.$$

Thus, by 4.2.4

$$\delta_{2\pi}(x) = \frac{1}{\pi}\left(\tfrac{1}{2}+\cos x+\cos 2x+ \ldots\right). \tag{13}$$

Hence we obtain the following formula for the sum of cosines:

$$\cos x+\cos 2x+ \ldots = \pi\delta_{2\pi}(x)-\tfrac{1}{2}.$$

This series does not converge in the usual sense at any point. The distributional convergence can be proved by double integration, while it can also be deduced from 4.2.5.

Similarly the series of sines

$$\sin x+\sin 2x+ \ldots$$

converges distributionally. To find its sum, we start from the well known formula

$$\frac{\cos x}{1}+\frac{\cos 2x}{2}+ \ldots = -\ln|\sin\tfrac{1}{2}x| - \ln 2. \tag{14}$$

The function on the right hand side is locally integrable, and is therefore a distribution. Differentiating (13) distributionally and changing the sign, we have

$$\sin x+\sin 2x+ \ldots = \tfrac{1}{2}\cot\tfrac{1}{2}x$$

according to Section 3.2. Hence we obtain the expansion of $\cot x$,

$$\cot x = 2(\sin 2x+\sin 4x+\sin 6x+ \ldots).$$

Replacing x by $\frac{1}{2}\pi -x$ we get the expansion of $\tan x$,

$$\tan x = 2(\sin 2x-\sin 4x+\sin 6x- \ldots).$$

Since $1/\sin x = \tan\frac{1}{2}x+\cot x$, we obtain

$$1/\sin x = 2(\sin x+\sin 3x+\sin 5x+ \ldots).$$

Replacing x by $\frac{1}{2}\pi-x$, we have

$$1/\cos x = 2(\cos x-\cos 3x+\cos 5x- \ldots).$$

It is interesting to observe that the above formulas continue to hold if the summation is understood not in the distributional sense but in the sense of arithmetical mean, for example.

As is well-known, the nth partial sum of the series (13) is just the Dirichlet kernel

$$\frac{1}{\pi}\cdot\frac{\sin(n+\frac{1}{2})x}{2\sin\frac{1}{2}x}.$$

By (13), the Dirichlet kernel converges to $\delta_{2\pi}(x)$ distributionally. Other kernels considered in the theory of trigonometric series, e.g.

$$\frac{1}{2n\pi}\left(\frac{\sin\frac{1}{2}nx}{\sin\frac{1}{2}x}\right)^2 \qquad \text{(Fejér)},$$

$$\tfrac{1}{2}\sqrt{n/\pi}\cos^{2n}\tfrac{1}{2}x \qquad \text{(de la Vallée-Poussin)},$$

share this property.

In some cases the following complex form of trigonometric series

$$\sum_{n=-\infty}^{\infty} c_n e^{inx} \tag{15}$$

is useful. It follows from 4.2.4 that each periodic distribution can be represented as the sum of a series (15) where

$$c_n = \frac{1}{2\pi}\int_{-\pi}^{\pi} f(t)e^{-int}\,dt.$$

By 4.2.5, the series (15) converges if and only if there exists an integer k such that $c_n/n^k \to 0$ as $n \to \infty$.

4.3. DISTRIBUTIONS OF INFINITE ORDER

The notion of distribution was introduced to mathematics by S. Soboleff [44] in 1936. The name "distribution" is due to L. Schwartz ([32], [33], [34]) who developed the theory in 1945 and subsequently. The definition of Soboleff differs from the one given here. The distributions defined as yet are called *distributions of finite order*, since they are derivatives of functions of some finite order in the whole considered interval. The distributions which were considered by Schwartz are derivatives of continuous functions locally only. In this paragraph we modify our definition to obtain a notion equivalent to that of Soboleff and Schwartz.

A sequence of continuous functions $f_n(x)$ defined in a fixed interval $A < x < B$ (finite or infinite) is called *fundamental* if for every closed interval $a \leqslant x \leqslant b$ $(A < a < b < B)$ there exist an integer $k \geqslant 0$ and a sequence of continuous functions $F_n(x)$ defined in $a \leqslant x \leqslant b$ such that

(i) $F_n^{(k)}(x) = f_n(x)$ for $a \leqslant x \leqslant b$;

(ii) $F_n(x)$ converges uniformly.

We say that two fundamental sequences $\{f_n(x)\}$ and $\{g_n(x)\}$ of continuous functions defined in $A < x < B$ are *equivalent* if for every closed interval $a \leqslant x \leqslant b$ $(A < a < b < B)$ there exist an integer $k \geqslant 0$ and sequences of continuous functions $\{F_n(x)\}$ and $\{G_n(x)\}$ such that

(i) $F_n^{(k)}(x) = f_n(x)$ and $G_n^{(k)}(x) = g_n(x)$;

(ii) the sequences $F_n(x)$ and $G_n(x)$ converge to the same limit uniformly in the interval $a \leqslant x \leqslant b$.

In this new definition the fact that the choice of the integer k depends on the interval $a \leqslant x \leqslant b$ is crucial.

Classes of equivalent sequences are called *distributions*.

A sequence which is fundamental in the earlier sense is clearly fundamental in the new sense. Conversely, if a sequence is fundamental in the new sense and if it is possible to choose the same integer k for all the intervals $a \leqslant x \leqslant b$ $(A < a < b < B)$, then the functions $F_n(x)$ can also be taken to be the same for all these intervals; thus the sequence is fundamental in the earlier sense. Distributions in the new sense which are defined by such sequences are called *distributions of finite order*.

Sequences which are fundamental in the earlier sense are equivalent in the new sense if and only if they are equivalent in the earlier sense; the proof is omitted here.

By the foregoing remarks, the distributions in the earlier sense can be identified with distributions of finite order.

For the new distributions one can define addition, subtraction, multiplication by a number or by a function, composition, differentiation, integration, etc. in such a way that the basic properties are preserved.

For any interval $a \leqslant x \leqslant b$ $(A < a < b < B)$ there exists an integer $k \geqslant 0$ such that the distribution is the kth derivative of a continuous function in this interval. The integer k can be chosen to be the same for all intervals $a \leqslant x \leqslant b$ if and only if the distribution is of finite order. In such a case the distribution is the kth derivative of a function continuous in the whole interval $A < x < B$. Otherwise the distribution is *of infinite order*.

Since the new distributions are not, in general, derivatives of any order of functions, continuous in $A < x < B$, the definition of convergence must be modified.

Namely, we say that a sequence of distributions $\{f_n(x)\}$ *converges* to a distribution $f(x)$ if for every closed interval $a \leqslant x \leqslant b$ $(A < a < b < B)$ there exist an integer $k \geqslant 0$ and continuous functions $F(x)$ and $F_n(x)$ such that $F_n^{(k)}(x) = f_n(x)$, $F^{(k)}(x) = f(x)$ and the sequence $\{F_n(x)\}$ converges uniformly to $F(x)$.

All the previous theorems relating to convergence remain valid. Convergence in the earlier sense implies convergence in the new sense, but not conversely. For instance the series

$$\delta(x)+\delta'(x-1)+\delta''(x-2)+\ \ldots$$

converges in the new sense, but does not converge in the earlier sense. It represents a distribution of infinite order.

Part II

Elementary theory of distributions of several real variables

Introduction to Part II

Part II contains an introduction to the theory of distributions of several variables. The case of a single variable was the subject of Part I. For teaching reasons, and particularly for beginners, it is advisable to read Part I before Part II. However, the exposition in Part II is complete and does not presuppose any knowledge of Part I. Most of important theorems in Part I are particular cases of results obtained in Part II.

The main idea is the same, but some modifications introduced in the case of several variables lead to striking improvements. In Part I, distributions are defined, roughly speaking, as limits of sequences of continuous functions. This could also be done for several variables, but would cause complications due to the necessity of introducing the superflous auxiliary notion of generalized derivatives of continuous functions. This notion can be avoided by using infinitely derivable functions or polynomials. But polynomials have no local properties, which — as we have verified experimentally — spoils the elegance of the theory. Thus we have decided on infinitely derivable functions as the starting point of the theory.

Another modification is that in Part I (except for the last Section) we dealt with distributions of finite order, while in Part II this restriction is dropped.

Part II contains a theory of elementary operations on distributions, such as addition, multiplication, derivation, substitution. Other operations, such as the integral, convolution and the Fourier transform, will be introduced in Part III.

1. Fundamental definitions

1.1. TERMINOLOGY AND NOTATION

Given two systems of finite or infinite numbers

$$a = (\alpha_1, \ldots, \alpha_q), \quad b = (\beta_1, \ldots, \beta_q),$$

we write

$$a < b,$$

iff

$$\alpha_j < \beta_j \quad \text{for} \quad j = 1, \ldots, q.$$

Similarly we write

$$a \leqslant b,$$

iff

$$\alpha_j \leqslant \beta_j \quad \text{for} \quad j = 1, \ldots, q.$$

If the real numbers $\xi_1, \ldots, \xi_q$ are finite, then

$$x = (\xi_1, \ldots, \xi_q)$$

can be thought of as a point of q-dimensional Euclidean space.

With the above convention, we can denote the q-dimensional open interval

$$\alpha_j < \xi_j < \beta_j \quad (j = 1, \ldots, q)$$

by

$$a < x < b,$$

just as in the one-dimensional case. Similarly, if α_j and β_j are finite, then the q-dimensional closed interval

$$\alpha_j \leqslant \xi_j \leqslant \beta_j \quad (j = 1, \ldots, q)$$

will be denoted by

$$a \leqslant x \leqslant b,$$

Unless the contrary is explicitly stated, "interval" always means "bounded interval". Usually the word "interval" means "open interval". An interval $a < x < b$ is said to be *inside* an open set O iff the closed interval $a \leqslant x \leqslant b$ is contained in O.

We adopt the usual notation

$$x+y = (\xi_1+\eta_1, \ldots, \xi_q+\eta_q), \quad x-y = (\xi_1-\eta_1, \ldots, \xi_q-\eta_q),$$

$$\lambda x = (\lambda\xi_1, \ldots, \lambda\xi_q), \quad |x| = \sqrt{\xi_1^2+ \ldots +\xi_q^2},$$

where $y = (\eta_1, \ldots, \eta_q)$ and λ is a number.

Functions defined on subsets of q-dimensional space will usually be denoted by $\varphi(x), f(x), F(x), \ldots$ rather than $\varphi(\xi_1, \ldots, \xi_q)$, $f(\xi_1, \ldots, \xi_q)$, $F(\xi_1, \ldots, \xi_q)$ etc. All functions under consideration are defined on open subsets of q-dimensional Euclidean space, unless the contrary is explicitly stated.

Let $F(x)$ be a continuous function in an interval I, let $x_0 = (\xi_{01}, \ldots, \xi_{0q})$ be a fixed point of this interval, and let $k = (\varkappa_1, \ldots, \varkappa_q)$ be a system of non-negative integers. Integrating $F(x)$ $\varkappa_1$ times in ξ_1, then $\varkappa_2$ times in ξ_2, and so on, we obtain the iterated integral of order k

$$\int_{\xi_{0q}}^{\xi_q} d\tau_{q\varkappa_q} \ldots \int_{\xi_{0q}}^{\tau_{q2}} d\tau_{q1} \ldots \int_{\xi_{01}}^{\xi_1} d\tau_{1\varkappa_1} \ldots \int_{\xi_{01}}^{\tau_{12}} d\tau_{11} F(\tau_{11}, \ldots, \tau_{q\varkappa_q}).$$

This integral will be denoted in abbreviated form by

$$\int_{x_0}^{x} F(t)\, dt^k,$$

or, in the particular case $k = (1, \ldots, 1)$, by

$$\int_{x_0}^{x} F(t)\, dt.$$

The following simple facts should be noted

$$\int_{x_0}^{x} \lambda f(t)\, dt^k = \lambda \int_{x_0}^{x} f(t)\, dt^k \qquad (\lambda \text{ a number}),$$

$$\int_{x_0}^{x} \big(f(t)+g(t)\big)\, dt^k = \int_{x_0}^{x} f(t)\, dt^k + \int_{x_0}^{x} g(t)\, dt^k,$$

$$\frac{\partial^{\varkappa_1+\cdots+\varkappa_q}}{\partial\xi_1^{\varkappa_1} \ldots \partial\xi_q^{\varkappa_q}} \int_{x_0}^{x} f(t)\, dt^k = f(x).$$

Infinitely continuously derivable functions will be called *smooth functions.* If $\varphi(x)$ is a smooth function and $k = (\varkappa_1, \ldots, \varkappa_q)$ is a system of non-negative integers, then by its *derivative of order k* we mean the function

$$\varphi^{(k)}(x) = \frac{\partial^{\varkappa_1+\cdots+\varkappa_q}}{\partial\xi_1^{\varkappa_1} \ldots \partial\xi_q^{\varkappa_q}} \varphi(\xi_1, \ldots, \xi_q).$$

Generally the term *order* means any sequence $k = (\varkappa_1, \ldots, \varkappa_q)$ of non-negative integers. The following notation will also be useful:

$$e_1 = (1, 0, \ldots, 0),$$
$$e_2 = (0, 1, \ldots, 0),$$
$$\ldots\ldots\ldots\ldots\ldots\ldots$$
$$e_q = (0, 0, \ldots, 1).$$

Instead of $(0, \ldots, 0)$, we may write 0. This does not usually lead to any misunderstanding.

1.2. UNIFORM AND ALMOST UNIFORM CONVERGENCE

Given any set I, we say that a sequence of function $f_n(x)$ *converges uniformly* in I to $f(x)$ and we write

$$f_n(x) \rightrightarrows f(x) \quad \text{in } I,$$

iff the function $f(x)$ is defined on I and, for any given number $\varepsilon > 0$, there is an index n_0 such that for every $n > n_0$ the function $f_n(x)$ is defined on the whole set I and satisfies the inequality $|f_n(x) - f(x)| < \varepsilon$ troughout I. Thus for initial indices n the functions $f_n(x)$ need not be defined in I.

We write

$$f_n(x) \rightrightarrows \quad \text{in } I,$$

iff there exists a function $f(x)$ such that $f_n(x) \rightrightarrows f(x)$ in I. We shall use this symbol when there is no need to give the limit function explicitly.

We write

$$f_n(x) \rightrightarrows\leftleftarrows g_n(x) \quad \text{in } I,$$

iff both sequences $f_n(x)$ and $g_n(x)$ converge uniformly on I to the same limit.

A sequence $f_n(x)$ is said to converge to $f(x)$ *almost uniformly* in an open set O iff $f_n(x) \rightrightarrows f(x)$ on every interval I inside O. The limit function is defined in the whole set O, but according to the definition none of the functions $f_n(x)$ need be defined in the whole set O. If O_n is the open set where $f_n(x)$ is defined, then for every interval I inside O there exists an integer n_0 such that I is inside O_n for $n > n_0$.

1.3. FUNDAMENTAL SEQUENCES OF SMOOTH FUNCTIONS

Let O be an open set in q-dimensional space.

A sequence $\varphi_n(x)$ of smooth functions is said to be *fundamental* in O iff for every interval I inside O there exists an order k and smooth functions $\Phi_n(x)$ such that

(F_1) $\Phi_n^{(k)}(x) = \varphi_n(x)$,

(F_2) $\Phi_n(x) \rightrightarrows$ in I.

The order k and the sequence $\Phi_n(x)$ depend, in general, on I. According to the definition, none of the functions $\varphi_n(x)$ need be defined in the whole set O. If O_n is the open set where $\varphi_n(x)$ is defined, then for every interval I inside O there exists an index n_0 such that I is inside O_n for $n > n_0$.

The functions $\Phi_n(x)$ are defined in I for $n > n_0$ and satisfy (F_1) and (F_2) there.

It follows immediately from the definition (for $k = 0$) that:

1.3.1. *Every sequence of smooth functions almost uniformly convergent in O is fundamental.*

Differentiating (F_1) m times, we find that:

1.3.2. *If $\varphi_n(x)$ is a fundamental sequence, so is $\varphi_n^{(m)}(x)$.*

It is useful to observe that the order k which occurs in the condition (F_1) can, if necessary, be replaced by any greater order. This is a consequence of the following statement:

1.3.3. *If $\Phi_n(x)$ satisfies (F_1) and (F_2) and if $l \geqslant k$, then the sequence of smooth functions*

$$\tilde{\Phi}_n(x) = \int_{x_0}^{x} \Phi_n(t)\,dt^{l-k} \qquad (x_0 \text{ in } I)$$

also satisfies (F_1) and (F_2), with k replaced by l.

We also note that:

1.3.4. *If a sequence $\varphi_n(x)$ is fundamental in every interval I inside O, then it is fundamental in O.*

For, if I is any interval inside O, there is an interval I' inside O such that I is inside I'. Since the sequence $\varphi_n(x)$ is fundamental in I', there are smooth functions $\Phi_n(x)$ and an order k such that (F_1) and (F_2) hold in I.

1.4. THE DEFINITION OF DISTRIBUTIONS

We say that two sequences $\varphi_n(x)$ and $\psi_n(x)$ fundamental in O are *equivalent* in O and we write

$$\varphi_n(x) \sim \psi_n(x),$$

iff the interlaced sequence

$$\varphi_1(x),\ \psi_1(x),\ \varphi_2(x),\ \psi_2(x),\ \ldots$$

is fundamental.

The following condition is plainly necessary and sufficient that $\varphi_n(x)$ and $\psi_n(x)$ be equivalent: For each interval I inside O there exist sequences of smooth functions $\Phi_n(x)$ and $\Psi_n(x)$ and an order k such that

(E_1) $\Phi_n^{(k)}(x) = \varphi_n(x)$ and $\Psi_n^{(k)}(x) = \psi_n(x)$,

(E_2) $\Phi_n(x) \rightrightarrows\leftleftarrows \Psi_n(x)$ in I.

The sequences $\Phi_n(x)$ and $\Psi_n(x)$ and the order k, in general, depend on I.

From 1.3.3 it follows that

1.4.1. *The order k in the condition* (E_1) *can, if necessary, be replaced by any greater order l.*

It is easy to see that the relation $\sim$ is reflexive and symmetric, i.e., that

(i) $\varphi_n(x) \sim \varphi_n(x)$,

(ii) $\varphi_n(x) \sim \psi_n(x)$ implies $\psi_n(x) \sim \varphi_n(x)$.

It is also transitive, i.e.,

(iii) $\varphi_n(x) \sim \psi_n(x)$ and $\psi_n(x) \sim \vartheta_n(x)$ imply $\varphi_n(x) \sim \vartheta_n(x)$.

In fact, the hypothesis of (iii) means that, for each interval I inside O, there exist an order k and smooth functions $\Phi_n(x)$ and $\Psi_n(x)$ satisfying (E_1) and (E_2), and there exist an order l and smooth functions $\tilde{\Psi}_n(x)$, $\Theta_n(x)$ such that

$$\tilde{\Psi}_n^{(l)}(x) = \psi_n(x), \qquad \Theta_n^{(l)}(x) = \vartheta_n(x), \qquad \tilde{\Psi}_n(x) \rightrightarrows\leftleftarrows \Theta_n(x).$$

By 1.4.1 we may assume that $k = l$. The sequences $\Phi_n(x)$ and $\tilde{\Theta}_n(x) = \Psi_n(x) - \tilde{\Psi}_n(x) + \Theta_n(x)$ converge uniformly in I to the same limit, and $\Phi_n^{(k)}(x) = \varphi_n(x)$, $\tilde{\Theta}_n^{(k)}(x) = \vartheta_n(x)$, which proves that $\varphi_n(x) \sim \vartheta_n(x)$. Since the relation $\sim$ is reflexive, symmetric and transitive, the set of all sequences fundamental in O decomposes into disjoint classes (equivalence classes of the relation $\sim$) such that two fundamental sequences are in the same class iff they are equivalent. These equivalence classes are called *distributions* (defined in O). Thus the notion of distribution is obtained from the identification of equivalent fundamental sequences.

The distribution determined by a fundamental sequence $\varphi_n(x)$, i.e., the class of all fundamental sequences equivalent to $\varphi_n(x)$, will be denoted by the symbol $[\varphi_n(x)]$. Two sequences $\varphi_n(x)$ and $\psi_n(x)$ fundamental in O determine the same distribution iff they are equivalent. Thus

$$[\varphi_n(x)] = [\psi_n(x)] \quad \text{iff} \quad \varphi_n(x) \sim \psi_n(x).$$

Distributions will be denoted by $f(x)$, $g(x)$, etc., in the same way as functions. It should be emphasized that this notation is purely symbolic and, in general, it does not allow us to substitute points for the variable x.

2. Operations on distributions

2.1. MULTIPLICATION BY A NUMBER

The operation $\lambda\varphi(x)$ of multiplication of a function $\varphi(x)$ by a number λ has the following property:

(i) If $\varphi_n(x)$ is a fundamental sequence, so is $\lambda\varphi_n(x)$.

This property enables us to extend the operation to arbitrary distributions $f(x) = [\varphi_n(x)]$ by setting

$$\lambda f(x) = [\lambda\varphi_n(x)].$$

To establish the uniqueness of the *product* $\lambda\varphi(x)$ we have to show that the product does not depend on the choice of the fundamental sequence $\varphi_n(x)$. In other words:

(ii) If $\varphi_n(x) \sim \bar{\varphi}_n(x)$, then $\lambda\varphi_n(x) \sim \lambda\bar{\varphi}_n(x)$.

In fact, the sequence

$$\varphi_1(x),\ \bar{\varphi}_1(x),\ \varphi_2(x),\ \bar{\varphi}_2(x),\ \ldots$$

is fundamental. Thus by (i) so is the sequence

$$\lambda\varphi_1(x),\ \lambda\bar{\varphi}_1(x),\ \lambda\varphi_2(x),\ \lambda\bar{\varphi}_2(x),\ \ldots,$$

which implies the assertion.

2.2. ADDITION

The operation $\varphi(x)+\psi(x)$ of addition of two functions $\varphi(x)$ and $\psi(x)$ has the following property:

(i) If $\varphi_n(x)$ and $\psi_n(x)$ are fundamental sequences, so is $\varphi_n(x)+\psi_n(x)$.

In order to prove (i), suppose that, for any interval I inside O, there are orders k and l and functions $\Phi_n(x)$, $\Psi_n(x)$ such that

$$\Phi_n^{(k)}(x) = \varphi_n(x), \qquad \Phi_n(x) \rightrightarrows,$$

$$\Psi_n^{(l)}(x) = \psi_n(x), \qquad \Psi_n(x) \rightrightarrows.$$

We can assume that $k = l$ since each of the orders k and l can be increased if necessary (see 1.3.3). Since

$$(\Phi_n(x)+\Psi_n(x))^{(k)} = \varphi_n(x)+\psi_n(x), \quad \Phi_n(x)+\Psi_n(x) \rightrightarrows,$$

the sequence $\varphi_n(x)+\psi_n(x)$ is fundamental.

Property (i) enables us to extend addition to arbitrary distributions $f(x) = [\varphi_n(x)]$ and $g(x) = [\psi_n(x)]$ by setting

$$f(x)+g(x) = [\varphi_n(x)+\psi_n(x)].$$

The sum so defined is unique, for it does not depend on the choice of fundamental sequences $\varphi_n(x)$ and $\psi_n(x)$. In other words:

(ii) If $\varphi_n(x) \sim \bar{\varphi}_n(x)$ and $\psi_n(x) \sim \bar{\psi}_n(x)$, then

$$\varphi_n(x)+\psi_n(x) \sim \bar{\varphi}_n(x)+\bar{\psi}_n(x).$$

In fact, the sequences

$$\varphi_1(x),\ \bar{\varphi}_1(x),\ \varphi_2(x),\ \bar{\varphi}_2(x),\ \ldots,$$

$$\psi_1(x),\ \bar{\psi}_1(x),\ \psi_2(x),\ \bar{\psi}_2(x),\ \ldots$$

are fundamental. By (i) so is the sequence

$$\varphi_1(x)+\psi_1(x),\ \bar{\varphi}_1(x)+\bar{\psi}_1(x),\ \varphi_2(x)+\psi_2(x),\ \bar{\varphi}_2(x)+\bar{\psi}_2(x),\ \ldots,$$

which implies the assertion.

2.3. REGULAR OPERATIONS

Multiplication by a given number λ is an operation on a single function (or distribution). Addition is an operation on two functions (or distributions). More generally, we may consider operations on an arbitrary number of functions and extend them to distributions. The method of extension is similar. It would be tedious and unnecessary to repeat the argument in each particular case. In this section it will be shown generally that the extension is feasible for a large class of operations.

Denote by

$$A(\varphi(x),\ \psi(x),\ \ldots)$$

an operation on a finite number of functions $\varphi(x)$, $\psi(x)$, ... We assume that the functions $\varphi(x)$, $\psi(x)$, ... are defined in open sets $P, Q, \ldots$ respectively; $P, Q, \ldots$ are subsets of some Euclidean spaces, not necessarily all of the same dimension. We assume that the operation A is feasible on all smooth functions, defined in $P, Q, \ldots$ and that the result of the operation is another smooth function defined in a fixed open set O, a subset of some Euclidean space.

Suppose that the operation A has the following property:

(i) If $\varphi_n(x)$, $\psi_n(x)$, ... are fundamental sequences in P, Q, ... respectively, then $A(\varphi_n(x), \psi_n(x), \ldots)$ is a fundamental sequence in O.

Such an operation extends to distributions $f(x) = [\varphi_n(x)]$, $g(x) = [\psi_n(x)]$, ..., by setting

$$A(f(x), g(x), \ldots) = [A(\varphi_n(x), \psi_n(x), \ldots)].$$

The extension is unique, i.e., it does not depend on the choice of the fundamental sequences $\varphi_n(x)$, $\psi_n(x)$, ... In other words:

(ii) If

$$\varphi_n(x) \sim \bar{\varphi}_n(x), \quad \psi_n(x) \sim \bar{\psi}_n(x), \quad \ldots, \tag{1}$$

then

$$A(\varphi_n(x), \psi_n(x), \ldots) \sim A(\bar{\varphi}_n(x), \bar{\psi}_n(x), \ldots). \tag{2}$$

In fact, by hypothesis, the sequences

$$\begin{array}{l} \varphi_1(x),\ \bar{\varphi}_1(x),\ \varphi_2(x),\ \bar{\varphi}_2(x),\ \ldots, \\ \psi_1(x),\ \bar{\psi}_1(x),\ \psi_2(x),\ \bar{\psi}_2(x),\ \ldots, \\ \cdots\cdots\cdots\cdots\cdots \end{array}$$

are fundamental. By (i) so is the sequence

$$A(\varphi_1(x), \psi_1(x), \ldots), \quad A(\bar{\varphi}_1(x), \bar{\psi}_1(x), \ldots), \quad A(\varphi_2(x), \psi_2(x), \ldots), \quad \ldots,$$

which proves the assertion.

All operations $A(\varphi(x), \psi(x), \ldots)$ with property (i) will be called *regular operations*. Every regular operation defined on smooth functions can be extended automatically to distributions, and this extension is always unique.

Multiplication by a number and addition are regular operations, as we have seen in Sections 2.1 and 2.2.

2.4. SUBTRACTION, TRANSLATION, DERIVATION

We are now going to give some further examples of regular operations.

SUBTRACTION. The subtraction $\varphi(x)-\psi(x)$ is a regular operation. In fact, if $\varphi_n(x)$ and $\psi_n(x)$ are fundamental sequences, so is $\varphi_n(x)-\psi_n(x)$. The proof is similar to that given in Section 2.2. We thus define the difference of two distributions $f(x) = [\varphi_n(x)]$ and $g(x) = [\psi_n(x)]$ by the formula

$$f(x)-g(x) = [\varphi_n(x)-\psi_n(x)].$$

TRANSLATION. The translation $\varphi(x+h)$ is a regular operation. More precisely, if $\varphi_n(x)$ is a fundamental sequence in the open set O, then $\varphi_n(x+h)$ is a fundamental sequence in the translated set O_h, consisting of all points x such that $x+h$ is in O. Thus, if $f(x) = [\varphi_n(x)]$ is a distribution defined in O, then

$$f(x+h) = [\varphi_n(x+h)]$$

is a distribution defined in O_h.

DERIVATION. The derivation $\varphi^{(m)}(x)$ of an arbitrary order m is a regular operation. In fact, by 1.3.2, if $\varphi_n(x)$ is a fundamental, so is $\varphi_n^{(m)}(x)$. Thus we define the derivative of order m of any distribution $f(x) = [\varphi_n(x)]$ by setting

$$f^{(m)}(x) = [\varphi_n^{(m)}(x)].$$

Evidently:

2.4.1. *Each distribution has derivatives of all orders.*

Property 2.4.1 is a great help in calculations with distributions, and makes them easier and more elegant than the calculations in Classical Analysis.

2.5. MULTIPLICATION OF A DISTRIBUTION BY A SMOOTH FUNCTION

The multiplication $\varphi(x)\psi(x)$, when considered as an operation on two functions $\varphi(x)$ and $\psi(x)$, is not regular, for if the sequences $\varphi_n(x)$ and $\psi_n(x)$ are fundamental, their product $\varphi_n(x)\psi_n(x)$ need not be fundamental.

However, multiplication may also be thought of as an operation on a single function, the other factor being kept fixed. Denote by $\omega(x)$ this fixed factor. We shall prove that, if $\omega(x)$ is a smooth function, the multiplication $\omega(x)\varphi(x)$ is a regular operation on $\varphi(x)$. In other words, if the sequence $\varphi_n(x)$ is fundamental, so is $\omega(x)\varphi_n(x)$.

In fact, since $\varphi_n(x)$ is fundamental, for every interval I inside O there exist an order k and smooth functions $\Phi_n(x)$ such that

$$\Phi_n^{(k)}(x) = \varphi_n(x) \quad \text{and} \quad \Phi_n(x) \rightrightarrows \text{ in } I.$$

For every order m and every smooth function $\omega(x)$ the sequence $\omega(x)\Phi_n^{(m)}(x)$ is fundamental in I. The proof is by induction. The case $m = 0$ follows from 1.3.1. If the sequence is fundamental for some m, then the sequence is also fundamental for $m+e_j$, since

$$\omega(x)\Phi^{(m+e_j)}(x) = \left(\omega(x)\Phi_n^{(m)}(x)\right)^{(e_j)} - \omega^{(e_j)}(x)\Phi_n^{(m)}(x)$$

and the right hand side is the difference of two sequences which are fundamental by 1.3.2 and the induction hypothesis. For $m = k$ we find that $\omega(x)\varphi_n(x)$ is fundamental in I. Since the interval I is arbitrary, the sequence $\omega(x)\varphi_n(x)$ is fundamental in the whole set O, by 1.3.4. We have just proved that multiplication by a smooth function $\omega(x)$ is a regular operation.

According to the general method, we define the *product* of an arbitrary distribution $f(x) = [\varphi_n(x)]$ by a smooth function $\omega(x)$ by means of the formula

$$\omega(x) f(x) = [\omega(x)\,\varphi_n(x)].$$

We shall occasionally write $f(x)\omega(x)$ instead of $\omega(x)f(x)$.

Note that if $\omega(x)$ is a constant function, then the multiplication just defined coincides with that given in Section 2.1.

2.6. SUBSTITUTION

Let $\sigma(x)$ be a fixed smooth function in a q-dimensional open set O such that

$$\left(\frac{\partial\sigma(x)}{\partial\xi_1}\right)^2 + \ldots + \left(\frac{\partial\sigma(x)}{\partial\xi_q}\right)^2 \neq 0 \quad \text{in } O; \tag{1}$$

suppose that the values of $\sigma(x)$ belong to an open set O' of real numbers y. The substitution

$$\varphi(\sigma(x))$$

is a regular operation on $\varphi(y)$ ($\sigma(x)$ being fixed). More precisely, we shall show that if $\varphi_n(y)$ is fundamental in O', then $\varphi_n(\sigma(x))$ is fundamental in O.

Let I be any interval inside O. The function $\sigma(x)$ maps I onto an interval I' inside O'.

Observe first that if, for some smooth functions $\Phi_n(y)$, the sequence $\Phi_n(\sigma(x))$ is fundamental in I, so is the sequence $\Phi_n'(\sigma(x))$. In fact, from

$$\frac{\partial}{\partial\xi_j}\Phi_n(\sigma(x)) = \Phi_n'(\sigma(x))\frac{\partial}{\partial\xi_j}\sigma(x) \qquad (j = 1, \ldots, q)$$

we find by calculation that

$$\Phi_n'(\sigma(x)) = \frac{\dfrac{\partial}{\partial\xi_1}\Phi_n(\sigma(x))\cdot\dfrac{\partial}{\partial\xi_1}\sigma(x) + \ldots + \dfrac{\partial}{\partial\xi_q}\Phi_n(\sigma(x))\cdot\dfrac{\partial}{\partial\xi_q}\sigma(x)}{\left(\dfrac{\partial\sigma(x)}{\partial\xi_1}\right)^2 + \ldots + \left(\dfrac{\partial\sigma(x)}{\partial\xi_q}\right)^2}. \tag{2}$$

Here the derivatives $\dfrac{\partial}{\partial\xi_i}\Phi_n(\sigma(x))$ form fundamental sequences, by 1.3.2. Also the products of those derivatives with the smooth functions $\dfrac{\partial}{\partial\xi_j}\sigma(x)$ are fundamental sequences, since multiplication by smooth functions is a regular operation. Thus the numerator in (2) is a fundamental sequence, as it is the sum of fundamental sequences. Finally, the complete fraction on the right hand side represents a fundamental sequence, for it can be represented as the product of the numerator with the inverse of the denominator.

By induction, if $\Phi_n(\sigma(x))$ is fundamental, so is $\Phi_n^{(k)}(\sigma(x))$ for every non-negative integer k.

Now, let $\Phi_n(y)$ be a sequence of smooth functions such that, for an integer $k \geqslant 0$,

$$\Phi_n^{(k)}(y) = \varphi_n(y) \quad \text{and} \quad \Phi_n(y) \rightrightarrows \quad \text{in } I'.$$

Then $\Phi_n(\sigma(x)) \rightrightarrows$ in I. Thus $\Phi_n(\sigma(x))$ is a fundamental sequence and so is $\Phi_n^{(k)}(\sigma(x))$, i.e., $\varphi_n(\sigma(x))$. Since the interval I is arbitrary, the sequence $\varphi_n(\sigma(x))$ is fundamental in the whole set O, by 1.3.4.

We have thus proved that the substitution of a given smooth function $\sigma(x)$ satisfying (1) is a regular operation. Following the general method we define the substitution of $\sigma(x)$ into an arbitrary distribution $f(y) = [\varphi_n(y)]$ in O' by the formula

$$f(\sigma(x)) = [\varphi_n(\sigma(x))].$$

The distribution $f(y)$ is one-dimensional, i.e., is defined in a one-dimensional set, while the distribution $f(\sigma(x))$ is q-dimensional, i.e., is defined in an open subset of q-dimensional Euclidean space.

In Section 4.2 we shall also consider the more general case where the outer distribution $f(y)$ is p-dimensional, $1 \leqslant p \leqslant q$.

2.7. PRODUCT OF DISTRIBUTIONS WITH SEPARATED VARIABLES

The product of two smooth functions $\varphi(\xi_1, \ldots, \xi_q)$, $\psi(\eta_1, \ldots, \eta_r)$ can be written in the form $\varphi(x)\psi(y)$, where $x = (\xi_1, \ldots, \xi_q)$, $y = (\eta_1, \ldots, \eta_r)$. If $\varphi(x)$ is defined in an open subset O' of q-dimensional space, and $\psi(y)$ is defined in an open subset O'' of r-dimensional space, then the product $\varphi(x)\psi(y)$ is defined in the open set O consisting of all points $(\xi_1, \ldots, \xi_q, \eta_1, \ldots, \eta_r)$ such that $(\xi_1, \ldots, \xi_q)$ is in O' and $(\eta_1, \ldots, \eta_r)$ is in O''.

It is plain that the product $\varphi(x)\psi(y)$ is a regular operation on two functions $\varphi(x)$ and $\psi(y)$. It can therefore be extended to arbitrary distributions $f(x) = [\varphi_n(x)]$, $g(y) = [\psi_n(y)]$ by setting

$$f(x)g(y) = [\varphi_n(x)\psi_n(y)].$$

Since the distributions $f(x)$ and $g(y)$ are defined in O' and O'' respectively, their product is defined in O. The distributions $f(x)$ and $g(y)$ are q-dimensional and r-dimensional respectively, while their product is $(q+r)$-dimensional.

2.8. CONVOLUTION WITH A SMOOTH FUNCTION VANISHING OUTSIDE AN INTERVAL

First we shall show that there exist smooth functions vanishing outside a given interval I, but not vanishing everywhere.

The function

$$\Omega(\xi) = \begin{cases} 0 & \text{for} \quad \xi \leqslant 0, \\ e^{-1/\xi} & \text{for} \quad \xi > 0 \end{cases}$$

is smooth, positive for $\xi > 0$, and vanishes for $\xi \leqslant 0$. The product

$$\Omega(\xi - \alpha)\,\Omega(\beta - \xi)$$

is also smooth, positive for $\alpha < \xi < \beta$ and vanishes elsewhere.

For any interval I:

$$a < x < b, \qquad a = (\alpha_1, \ldots, \alpha_q), \qquad b = (\beta_1, \ldots, \beta_q),$$

we define the function $\Omega_I(x)$ by the following formula:

$$\Omega_I(x) = \prod_{j=1}^{q} \Omega(\xi_j - \alpha_j)\,\Omega(\beta_j - \xi_j).$$

This function has the required properties: it is smooth, positive in I, and vanishes elsewhere.

If $f(x)$ is continuous or locally integrable in an open set O and $\omega(x)$ is continuous everywhere and vanishes outside an interval $a < x < b$, then by the *convolution* of $f(x)$ with $\omega(x)$ we mean the function

$$f(x) * \omega(x) = \int_a^b f(x-t)\,\omega(t)\,dt, \tag{1}$$

defined in the open set O' of all points x such that the interval

$$x - b < t < x - a$$

is inside O. The convolution (1) can be written in the form

$$\int_{-\infty}^{\infty} f(x-t)\,\omega(t)\,dt \quad \text{or} \quad \int_{-\infty}^{\infty} f(t)\,\omega(x-t)\,dt,$$

if we adopt the convention that the product is defined and equal to zero if one of the factor is equal to zero, whether or not the second factor is defined.

The convolution of a continuous or locally integrable function $f(x)$ with a smooth function $\omega(x)$ (vanishing outside $a < x < b$) is a smooth function and

$$\big(f(x) * \omega(x)\big)^{(m)} = f(x) * \omega^{(m)}(x) \tag{2}$$

for every order m.

If $f(x)$ is smooth, then also

$$\big(f(x)*\omega(x)\big)^{(m)} = f^{(m)}(x)*\omega(x) \tag{3}$$

for every order m.

If $f_n(x) \rightrightarrows f(x)$ in an interval $a_0+a < x < b_0+b$, then

$$f_n(x)*\omega(x) \rightrightarrows f(x)*\omega(x) \tag{4}$$

in $a_0 < x < b_0$. Hence it follows that

2.8.1. *If $\varphi_n(x)$ is fundamental in O and $\omega(x)$ is smooth, then $\varphi_n(x)*\omega(x)$ converges almost uniformly in O'.*

In fact, let I': $a_0 < x < b_0$ be an interval inside O'; then the interval I: $a_0-b < x < b_0-a$ is inside O. Since $\varphi_n(x)$ is fundamental in O, we have

$$\varphi_n(x) = \Phi_n^{(k)}(x) \quad \text{and} \quad \Phi_n(x) \rightrightarrows \quad \text{in } I.$$

Hence, by (3), (2) and (4),

$$\varphi_n(x)*\omega(x) = \big(\Phi_n(x)*\omega(x)\big)^{(k)} = \Phi_n(x)*\omega^{(k)}(x) \rightrightarrows \quad \text{in } I'.$$

It follows from 2.8.1 and 1.3.1 that convolution with a smooth function vanishing outside an interval is a regular operation.

Following the general method, we define the convolution of an arbitrary distribution $f(x) = [\varphi_n(x)]$ with a smooth function $\omega(x)$, vanishing outside an interval I, by means of the formula

$$f(x)*\omega(x) = [\varphi_n(x)*\omega(x)].$$

We shall occasionally write $\omega(x)*f(x)$ instead of $f(x)*\omega(x)$.

2.9. CALCULATIONS WITH DISTRIBUTIONS

In calculations, various identities are useful, e.g.

$$\begin{aligned}
\big(\varphi(x)-\psi(x)\big)+\psi(x) &= \varphi(x),\\
\lambda\big(\varphi(x)+\psi(x)\big) &= \lambda\varphi(x)+\lambda\psi(x),\\
\big(\omega(x)\varphi(x)\big)^{(e_j)} &= \omega^{(e_j)}(x)\varphi(x)+\omega(x)\varphi^{(e_j)}(x),\\
\varphi\big(\sigma(x)\big)^{(e_j)} &= \varphi'\big(\sigma(x)\big)\sigma^{(e_j)}(x),\\
\big(\omega(x)*\varphi(x)\big)^{(m)} &= \omega^{(m)}(x)*\varphi(x) = \omega(x)*\varphi^{(m)}(x).
\end{aligned} \tag{1}$$

All these formulae and many others can be extended to distributions. We need not justify the feasibility of this extension for each formula separately. We shall give a simple rule which enables us to describe a large class of formulas, valid both for smooth functions and for distributions. This rule is based on the concept of *iterations of operations.* For instance, the expression $\lambda\big(f(x)+g(x)\big)$ is an iteration of addition and multiplication (by the number λ).

Generally, by iteration of operations we mean an expression of the form

$$A\big(B(\varphi(x), \psi(x), \ldots), \; C(\chi(x), \vartheta(x), \ldots), \ldots\big)$$

where $A, B, C, \ldots$ are given operations. Each of the five examples quoted at the beginning of this section display equalities between iterations of operations, provided the identity operation $\mathscr{I}\big(\varphi(x)\big) = \varphi(x)$ is admitted. The identity operation is trivially a regular operation. In examples (1) there appear iterations of regular operations only. It follows immediately from the definition of regular operations that iterations of regular operations of regular operations are again regular operations.

The meaning of formulas (1) is that the left and the right hand sides of each equality represent the same operation. All these operations are regular, and so their extensions to distributions are unique. This implies that the same formulas hold if we replace smooth functions $\varphi(x)$, $\psi(x)$ by distributions.

Further, second order iterations, i.e. iterations of iterations of regular operations, are regular operations, and so are iterations of any arbitrary finite order, i.e. any finite iterations of regular operations. We thus have the following general rule:

2.9.1. *If an equality both of whose sides are finite iterations of regular operations holds for smooth functions, then it also holds for arbitrary distributions.*

This rule is not merely of theoretical but above all of practical importance, for it allows us to perform calculations on distributions in the same way as on smooth functions, provided all the operations occurring in those calculations are regular.

3. Local properties

3.1. DELTA-SEQUENCES AND THE DELTA-DISTRIBUTION

If $\Omega_I(x)$ is the function defined in Section 2.8, then the function

$$\omega_I(x) = \gamma^{-1}\Omega_I(x),$$

where

$$\gamma = \int_{-\infty}^{\infty} \Omega_I(x)\,dx$$

is smooth, positive in I, and vanishes elsewhere. Moreover, it has the property

$$\int_{-\infty}^{\infty} \omega_I(x)\,dx = 1.$$

Let α_n be positive numbers such that $\alpha_n \to 0$. There exist smooth functions $\delta_n(x)$, non-negative for $|x| < \alpha_n$ and vanishing elsewhere, such that

$$\int_{-\infty}^{\infty} \delta_n(x)\,dx = 1.$$

The existence of such sequences is ensured by the preceding example. Any sequence $\delta_n(x)$ with the above properties will be called a *δ-sequence.*

Every δ-sequence is fundamental. In fact, the sequence

$$\Delta_n(x) = \int_{-\infty}^{x} \delta_n(t)\,dt^k,$$

where $k = (2, \ldots, 2)$, converges uniformly everywhere and $\Delta_n^{(k)}(x) = \delta_n(x)$.,

All δ-sequences are equivalent, for the interlaced sequence formed from two δ-sequences is again a δ-sequence.

Thus the δ-sequences determine a distribution

$$\delta(x) = [\delta_n(x)];$$

this is called the *q-dimensional Dirac delta-distribution.* The dimension of $\delta(x)$ is given by the dimension of the variable $x = (\xi_1, \ldots, \xi_q)$.

If $\omega(x)$ is a smooth function, then $\omega(x)\delta_n(x)$ is a fundamental sequence, equivalent to $\omega(0)\,\delta_n(x)$. In fact, given $\varepsilon > 0$, there exists an index n_0 such that for $n > n_0$

$$|\omega(x) - \omega(0)| < \varepsilon \quad \text{for} \quad -\alpha_n < \xi_j < \alpha_n \ (j = 1, \ldots, q).$$

Hence

$$\left| \int_{-\infty}^{x} (\omega(t) - \omega(0))\,\delta_n(t)\,dt \right| \leqslant \varepsilon \int_{-\infty}^{\infty} \delta_n(x)\,dx = \varepsilon,$$

which proves that the integral converges uniformly to 0. Hence

$$\omega(x)\,\delta_n(x) - \omega(0)\,\delta_n(x) \sim 0,$$

and consequently

$$\omega(x)\,\delta_n(x) \sim \omega(0)\,\delta_n(x).$$

Since the left- and right-hand sides are fundamental sequences for the products $\omega(x)\,\delta(x)$ and $\omega(0)\,\delta(x)$ respectively, we obtain the formula

$$\omega(x)\,\delta(x) = \omega(0)\,\delta(x).$$

3.1.1. *If $\delta_n(x)$ is a δ-sequence and f is a continuous function in O, then the sequence of smooth functions*

$$f(x) * \delta_n(x)$$

converges to $f(x)$, almost uniformly in O.

In fact, let I be any interval inside O. For every positive number ε there is an index n_0 such that for $n > n_0$

$$|f(x-t) - f(x)| < \varepsilon$$

for x in I and $-\alpha_n < \tau_j < \alpha_n$ $(j = 1, \ldots, q)$, where $t = (\tau_1, \ldots, \tau_q)$. Hence

$$|f(x) * \delta_n(x) - f(x)| \leqslant \int_{-\infty}^{\infty} |f(x-t) - f(x)|\,\delta_n(t)\,dt \leqslant \varepsilon$$

for $n \geqslant n_0$ and x in I.

This proves that $f(x) * \delta_n(x)$ converges to $f(x)$, almost uniformly in O.

The following generalization of 3.1.1 is useful:

3.1.2. *If $\delta_n(x)$ is a δ-sequence and $f_n(x)$ is a sequence of continuous functions, convergent to $f(x)$, almost uniformly in O, then the sequence of smooth functions*

$$f_n(x) * \delta_n(x)$$

converges to $f(x)$, almost uniformly in O.

To prove this, note that

$$f_n(x) * \delta_n(x) = f(x) * \delta_n(x) + \big(f_n(x) - f(x)\big) * \delta_n(x),$$

where the first term on the right hand side converges almost uniformly to $f(x)$, by 3.1.1. It suffices to show that the sequence

$$\varphi_n(x) = \big(f_n(x) - f(x)\big) * \delta_n(x)$$

converges almost uniformly to 0. In fact, given any interval I inside O and any positive number $\varepsilon > 0$, we have, for sufficiently large n,

$$|\varphi_n(x)| \leqslant \varepsilon * \delta_n(x) = \varepsilon \quad \text{in } I.$$

We conclude this section with a simple remark on the product of delta distributions. The product

$$\delta_n(\xi_1) \ldots \delta_n(\xi_q)$$

of one-dimensional δ-sequences is evidently a q-dimensional δ-sequence. Hence, by the definition of the product of distributions with separated variables, we have

$$\delta(x) = \delta(\xi_1) \ldots \delta(\xi_q) \quad \text{for} \quad x = (\xi_1, \ldots, \xi_q).$$

3.2. DISTRIBUTIONS IN SUBSETS

Any distribution in the open set O can be interpreted, if necessary, as a distribution in any open subset O', since the functions of any fundamental sequence representing $f(x)$ can be interpreted by restriction as functions in the subset. Thus every distribution defined in O is also defined in any open subset O'.

Whenever we write

$$f(x) = g(x) \quad \text{in } O',$$

we mean that the open set O' is contained in the intersection of the open sets where the distributions $f(x)$ and $g(x)$ are defined and that $f(x)$ and $g(x)$, when interpreted as distributions in O', are equal.

If we simply write

$$f(x) = g(x),$$

we mean, in the absence of further comment, that the distributions on both sides are equal in the intersection of the open sets where they are defined, and that this intersection is not empty.

3.2.1. *If $f(x) = g(x)$ in every interval inside O, then $f(x) = g(x)$ in O.*

In fact, let $f(x) = [\varphi_n(x)]$ and $g(x) = [\psi_n(x)]$. The equality $f(x) = g(x)$ in every interval inside O implies that the sequences $\varphi_n(x)$, $\psi_n(x)$ satisfy conditions (E_1) and (E_2) in every interval inside an interval inside O, and consequently in any interval inside O. The sequences are thus equivalent in O.

3.3. DISTRIBUTIONS AS A GENERALIZATION OF THE NOTION OF CONTINUOUS FUNCTIONS

Every continuous function may be considered as a distribution, and in this way the theory of distributions includes Classical Analysis.

To obtain the identification of continuous functions with distributions, we need two preparatory lemmas.

3.3.1. *If, in some interval* $a < x < b$, $\varphi_n(x) \rightrightarrows 0$ *and* $\varphi_n^{(k)}(x) \rightrightarrows$, *then*

$$\varphi_n^{(k)}(x) \rightrightarrows 0.$$

This is plainly true for $k = 0$. We now argue by induction. Suppose that the assertion holds for an order k, and that

$$\varphi_n(x) \rightrightarrows 0, \quad \varphi_n^{(k+e_j)} \rightrightarrows f(x) \quad \text{in} \quad a < x < b.$$

Then

$$\varphi_n^{(k)}(x+\eta e_j) - \varphi_n^{(k)}(x) = \int_0^\eta \varphi_n^{(k+e_j)}(x+\zeta e_j)\,d\zeta \rightrightarrows \int_0^\eta f(x+\zeta e_j)\,d\zeta$$

in $a+|\eta|e_j < x < b-|\eta|e_j$. By the induction hypothesis, the last integral vanishes. Since the number η is arbitrary, we have $f(x) = 0$.

3.3.2. *Almost uniformly convergent sequences of smooth functions are equivalent iff they converge to the same continuous function.*

In fact, if the sequences $\varphi_n(x)$ and $\psi_n(x)$ converge almost uniformly to $f(x)$, then they satisfy conditions (E_1) and (E_2) with $k = 0$. Thus $\varphi_n(x) \sim \psi_n(x)$. Conversely, if $\varphi_n(x) \sim \psi_n(x)$, then for every interval I inside O there exist smooth functions $\Phi_n(x)$ and $\Psi_n(x)$ and an order k such that conditions (E_1) and (E_2) are satisfied. Hence

$$\Phi_n(x) - \Psi_n(x) \rightrightarrows 0 \quad \text{on } I.$$

By 3.3.1,

$$\varphi_n(x) - \psi_n(x) \rightrightarrows 0 \quad \text{on } I.$$

Hence the limits of $\varphi_n(x)$ and $\psi_n(x)$ are the same.

We are now in a position to establish the correspondence between continuous functions and certain distributions.

By 3.1.1, for every continuous function $f(x)$ there exists a sequence of smooth functions $\varphi_n(x)$ which converges almost uniformly to $f(x)$. By 1.3.1, this sequence is fundamental. Thus to every continuous function $f(x)$ there corresponds a distribution $[\varphi_n(x)]$. By 3.3.2 the correspondence is one-to-one.

In the sequel we shall always identify the continuous function $f(x)$ with the distribution $[\varphi_n(x)]$.

For instance we can write, by 3.1.1,

$$f(x) = [f(x) * \delta_n(x)] \tag{1}$$

for every continuous function $f(x)$ and any δ-sequence $\delta_n(x)$.

In particular, smooth functions $\varphi(x)$ are distributions, and for them we have the simpler identity

$$\varphi(x) = [\varphi(x)].$$

The zero distribution, i.e., the distribution identified with the function which vanishes everywhere, will be denoted by 0.

Under the identification given here, distributions are seen to be a generalization of the notion of continuous functions. This justifies using for them the notation $f(x)$, $g(x)$, ... as well as for functions.

3.3.3. *The convolution $f(x)*\omega(x)$ of a distribution $f(x)$ with a smooth function $\omega(x)$ is a smooth function.*

In fact, let $f(x) = [\varphi_n(x)]$. By 2.8.1 the sequence $\varphi_n(x)*\omega(x)$ converges almost uniformly to a continuous function $g(x)$. Moreover, for every order m, the sequence $(\varphi_n(x)*\omega(x))^{(m)}$ also converges almost uniformly by 2.8.1 and formula (2) in Section 2.8. By a classical theorem, $g(x)$ has continuous first partial derivatives, namely the limit of $(\varphi_n(x)*\omega(x))^{(e_j)}$ is the jth partial derivative of $g(x)$. By the same argument, all second derivatives, all third derivatives, etc. of $g(x)$ exist. Thus $g(x)$ is a smooth function. On the other hand,

$$f(x)*\omega(x) = [\varphi_n(x)*\omega(x)] = g(x)$$

by the definition of convolution and the identification of continuous function with distributions.

We now have the identity

$$\varphi(x) = \varphi(x) * \delta(x) \tag{2}$$

for every smooth function $\varphi(x)$. In fact, replacing $f(x)$ by $\varphi(x)$ in (1), we get

$$\varphi(x) = [\varphi(x) * \delta_n(x)] = \varphi(x) * [\delta_n(x)] = \varphi(x) * \delta(x).$$

We are also in a position to prove the following generalization of (1):

3.3.4. *If $\delta_n(x)$ is a δ-sequence and $f(x)$ is any distribution, then*

$$f(x) = [f(x) * \delta_n(x)]. \tag{3}$$

In fact, for every interval I inside the set O, where $f(x)$ is defined, there exist an order k and a continuous function $F(x)$ such that $F^{(k)}(x) = f(x)$ in I. By (1),

$$F(x) = [F(x) * \delta_n(x)] \quad \text{in } I.$$

Hence, differentiating k times, we obtain (3) in I. By 3.2.1 formula (3) holds throughout the set O.

Since $0 = \varphi(x)*0$ for every smooth function $\varphi(x)$, it follows from (2) that $\delta(x)$ is not equal to the zero distribution when considered in the whole space. Note, on the other hand, that

$$\delta(x) = 0 \quad \text{for} \quad x \neq 0$$

(i.e. in the open set of all $x \neq 0$) since every δ-sequence $\delta_n(x)$ converges almost uniformly to 0 for $x \neq 0$.

3.4. OPERATIONS ON CONTINUOUS FUNCTIONS

In Sections 2.1–2.8 we have defined several operations on distributions. Now continuous functions are distributions, and so operations defined earlier for distributions are also defined for continuous functions. However, the operations are also defined on continuous functions directly. The question arises as to whether the two definitions are compatible.

Until we prove the compatibility of direct operations and distributional operations, we shall, in this section, use different symbols for them. If A denotes a direct operation, the corresponding distributional operation will be denoted by $\tilde{A}$. This double notation was not necessary before the identification of continuous functions with distributions, since direct operations were performed on continuous functions and distributional operations on distributions, with no possibility of any misunderstanding.

3.4.1. *If $A(\varphi, \psi, \ldots)$ is a regular operation, then*

$$A(\varphi, \psi, \ldots) = \tilde{A}(\varphi, \psi, \ldots) \tag{1}$$

for smooth functions $\varphi, \psi, \ldots$

In fact, because of the identification we can write $\varphi = [\varphi]$, $\psi = [\psi]$, ..., and $A(\varphi, \psi, \ldots) = [A(\varphi, \psi, \ldots)]$. On the other hand, by the definition of distributional operations, we have $\tilde{A}(\varphi, \psi, \ldots) = [A(\varphi, \psi, \ldots)]$. This proves (1).

Given a regular operation A, we shall say that continuous functions $f, g, \ldots$ satisfy the *continuity condition* for A if $A(f, g, \ldots)$ is defined directly for those functions and, moreover, there exist sequences of smooth functions $\varphi_n, \psi_n, \ldots$ almost uniformly convergent to $f, g, \ldots$ respectively, such that $A(\varphi_n, \psi_n, \ldots)$ converges almost uniformly to $A(f, g, \ldots)$.

3.4.2. *If continuous functions $f, g, \ldots$ satisfy the continuity condition for a regular operation A, then*

$$A(f, g, \ldots) = \tilde{A}(f, g, \ldots). \tag{2}$$

In fact, by the identification, we then have $A(f, g, ...) = [A(\varphi_n, \psi_n, ...)]$. On the other hand, by the definition of distributional operations, $\tilde{A}(f, g, ...) = [A(\varphi_n, \psi_n, ...)]$. This proves (2).

The continuity condition is satisfied by all continuous functions for all the operations introduced so far, except derivation. Consequently, these operations coincide with ordinary operations on continuous functions. Moreover, all calculations with continuous functions, except for derivation, may be performed in the usual way.

It is easy to see that every function $f(x)$ which is continuous together with its ordinary (partial) derivative $f^{(e_j)}(x)$ satisfies the continuity condition for derivation. For such functions the ordinary derivative $f^{(e_j)}(x)$ thus coincides with the distributional one. Both notations, $f^{(e_j)}(x)$ and $\frac{\partial}{\partial \xi_j} f(x)$, may be used interchangeably. By induction we have more generally:

3.4.3. *If $f(x)$ is a continuous function and its ordinary partial derivative*

$$\frac{\partial}{\partial \xi_{j_1}} \cdots \frac{\partial}{\partial \xi_{j_m}} f(x) \tag{3}$$

is continuous, and if all the derivatives occurring in the step-by-step differentiation in the arrangement indicated are also continuous, then the derivative (3) *coincides with the distributional derivative of the same order.*

It follows from 2.4.2 that every continuous function has distributional derivatives of all orders. If such a derivative is continuous, and if all the lower order derivatives are also continuous, it coincides with the ordinary derivative. However, it may happen that some distributional derivative of a continuous function $f(x)$ is a continuous function but that the ordinary derivative of the same order does not exist at all, however the symbols $\frac{\partial}{\partial \xi_j}$ are ordered.

For instance, if a continuous function $g(\xi)$ of one real variable is non-differentiable (in the ordinary sense), then for the function

$$f(x) = g(\xi_1) + g(\xi_2)$$

the ordinary derivatives $\frac{\partial^2}{\partial \xi_1 \partial \xi_2} f(x)$, $\frac{\partial^2}{\partial \xi_2 \partial \xi_1} f(x)$ do not exist. The corresponding distributional derivatives are equal, because the distributional derivatives are independent of the ordering of the symbols $\frac{\partial}{\partial \xi_j}$. To find the distributional derivative in question, we first note that

$$\frac{\partial}{\partial \xi_1}\frac{\partial}{\partial \xi_2} g(\xi_1) = 0$$

in the ordinary and consequently also in the distributional sense. Similarly we have

$$\frac{\partial}{\partial \xi_2}\frac{\partial}{\partial \xi_1} g(\xi_2) = 0$$

in the distributional sense. Since the ordering is immaterial, we also have

$$\frac{\partial}{\partial \xi_1}\frac{\partial}{\partial \xi_2} g(\xi_2) = 0$$

in the distributional sense. Hence we obtain, for the distributional derivative, $\frac{\partial^2}{\partial \xi_1 \partial \xi_2} f(x) = 0$. It is interesting to note that neither $\frac{\partial}{\partial \xi_1} f(x)$ nor $\frac{\partial}{\partial \xi_2} f(x)$ are functions. This example shows that there are distributions which are not functions, although some of their derivatives are continuous functions.

3.4.4. *Every distribution in O is, in every interval I inside O, a derivative of some order of a continuous function.*

In fact, let $f(x) = [\varphi_n(x)]$. By (F_1), (F_2) there exist an order k, smooth functions $\Phi_n(x)$, and a continuous function $F(x)$ such that in I

$$\Phi_n^{(k)}(x) = \varphi_n(x) \quad \text{and} \quad \Phi_n(x) \rightrightarrows F(x).$$

Hence $F(x) = [\Phi_n(x)]$ in I and

$$f(x) = [\Phi_n^{(k)}(x)] = [\Phi_n(x)]^{(k)} = F^{(k)}(x) \quad \text{in } I.$$

3.5. LOCALLY INTEGRABLE FUNCTIONS

As we have seen in Section 3.4, distributions are a generalization of continuous functions. We shall now show that they also embrace a larger class of functions, viz. all locally integrable functions. Sections 3.5 and 3.6, which are concerned with such functions, can be omitted by readers not acquainted with the theory of the Lebesgue Integral.

We recall that a function $f(x)$, defined in $O \subset \boldsymbol{R}^q$ is said to be locally integrable in O iff the integral $\int_a^b f(t)\,dt$ exists for every interval $a < x < b$ inside O.

Note first that if $f(x)$ is a continuous function in an interval I, then in I

$$\left(\int_{x_0}^{x} f(t)\,dt\right)' = f(x) \qquad (x_0 \text{ in } I), \tag{1}$$

where the sign $'$ denotes derivation of order $(1, \ldots, 1)$. If we do not assume the continuity of $f(x)$ but only its local integrability, then $\int_{x_0}^{x} f(t)\,dt$ is still a continuous function. In this case equality (1) holds almost everywhere, where the derivative on the left hand side is defined in the usual way as the limit, as $\alpha \to 0$, $\alpha > 0$, of the expression

$$f_\alpha(x) = \frac{1}{\Delta x} \int_{x}^{x+\Delta x} f(t)\,dt, \tag{2}$$

where $\Delta x = (\alpha, \ldots, \alpha)$ and $\dfrac{1}{\Delta x} = \dfrac{1}{\alpha^q}$. The left hand side of (1) can also be interpreted as a distribution, namely the distributional derivative of order $(1, \ldots, 1)$ of the continuous function $\int_{x_0}^{x} f(t)\,dt$. It is easily checked that this distribution is independent of x_0 in I.

This suggests the following identification: A distribution is said to be equal to a function $f(x)$ locally integrable in O iff, for every interval I inside O, the distribution is equal to the distributional derivative

$$\left(\int_{x_0}^{x} f(t)\,dt\right)' \quad (x_0 \text{ in } I).$$

It follows from 3.2.1 that this distribution, if it exists at all, is determined uniquely by the locally integrable function $f(x)$. We shall prove that it always exists, by showing that the distribution

$$[f(x) * \delta_n(x)], \tag{3}$$

has the required property, where $\delta_n(x)$ is any δ-sequence.

In fact, let I be any interval inside O and let

$$F(x) = \int_{x_0}^{x} f(t)\,dt \quad (x_0 \text{ in } I).$$

By 3.1.1, the sequence $F(x) * \delta_n(x)$ converges to $F(x)$, almost uniformly in I. Hence, by the identification of continuous functions with distributions,

$$[F(x) * \delta_n(x)] = F(x) \quad \text{in } I,$$

and consequently $[F'(x) * \delta_n(x)] = F'(x)$, i.e.,

$$[f(x) * \delta_n(x)] = F'(x) \quad \text{in } I.$$

We have thus proved that every locally integrable function $f(x)$ can be identified with the distribution $[f(x) * \delta_n(x)]$.

If $f(x)$ is a continuous function, then $f(x) * \delta_n(x)$ converges almost uniformly to $f(x)$ by 3.1.1; hence the above identification of integrable functions coincides in this case with the identification given in Section 3.3.

The identification of locally integrable functions with distributions leads to the following definition: locally integrable functions $f(x)$ and $g(x)$ are *equal* iff they are equal as distributions, i.e., iff

$$\int_a^b f(t)\,dt = \int_a^b g(t)\,dt$$

for every interval $a < t < b$ inside O, i.e., iff $f(x) = g(x)$ almost everywhere.

3.6. OPERATIONS ON LOCALLY INTEGRABLE FUNCTIONS

As in the case of continuous functions, the question arises as to whether distributional operations on locally integrable functions coincide with operations defined directly.

We shall say that a sequence of smooth functions $\varphi_n(x)$ is L-*convergent* to a locally integrable function $f(x)$ if it converges to $f(x)$ almost everywhere in O, and, moreover, if in interval I inside O

$$\int_a^x \varphi_n(t)\,dt \rightrightarrows \int_a^x f(t)\,dt \qquad (a \text{ in } I). \tag{1}$$

If $\varphi_n(x)$ is L-convergent to $f(x)$, then $\varphi_n(x)$ is fundamental and $[\varphi_n(x)] = f(x)$. In fact, (1) implies that

$$\left[\int_a^x \varphi_n(t)\,dt\right] = \int_a^x f(t)\,dt.$$

Hence, by differentiation of order 1, we get $[\varphi_n(x)] = f(x)$ in every interval I inside O, and consequently in the whole set O.

Given a regular operation A, we say that the locally integrable functions $f, g, \ldots$ satisfy the *integrability condition* for A if the operation $A(f, g, \ldots)$ is defined for those functions and, moreover, there exist sequences of smooth functions $\varphi_n, \psi_n, \ldots$, L-convergent to $f, g, \ldots$ respectively, such that $A(\varphi_n, \psi_n, \ldots)$ is L-convergent to $A(f, g, \ldots)$.

As in Section 3.2, we let $\tilde{A}$ denote the distributional extension of the operation A.

3.6.1. *If the locally integrable functions f, g, ... satisfy the integrability condition for a regular operation A, then*

$$A(f, g, \ldots) = \tilde{A}(f, g, \ldots). \tag{2}$$

In fact, we then have for every interval I inside O

$$\int_a^x A(\varphi_n, \psi_n, \ldots)\,dt \rightrightarrows \int_a^x A(f, g, \ldots)\,dt \qquad (a \text{ in } I).$$

This implies that in I

$$\Big[\int_a^x A(\varphi_n, \psi_n, \ldots)\,dt\Big] = \int_a^x A(f, g, \ldots)\,dt$$

and so by the adopted identification

$$[A(\varphi_n, \psi_n, \ldots)] = A(f, g, \ldots) \quad \text{in } I.$$

On the other hand, by the definition of distributional operations,

$$[A(\varphi_n, \psi_n, \ldots)] = \tilde{A}(f, g, \ldots).$$

Hence equality (2) holds in I. Since I is arbitrary, (2) holds in O.

The integrability condition is always satisfied by all locally integrable functions for all the operations introduced here, except derivation; the proofs owe nothing to the theory of distributions, and we omit the details. Consequently, all calculations with locally integrable functions, except for derivation, may be performed in the usual way.

It can happen that both the ordinary derivative of a locally integrable function and its distributional derivative exist but are different. For instance, the ordinary derivative of the Heaviside function of one real variable

$$H(x) = \begin{cases} 0 & \text{for} \quad x < 0, \\ 1 & \text{for} \quad x > 0 \end{cases}$$

is the zero distribution but the distributional derivative of $H(x)$ is equal to the one-dimensional Dirac delta distribution $\delta(x)$, for if $\delta_n(x)$ is any δ-sequence, then $\int_{-\infty}^{x} \delta_n(t)\,dt$ is L-convergent to $H(x)$ and consequently

$$\delta(x) = [\delta_n(x)] = \Big[\int_{-\infty}^{x} \delta_n(t)\,dt\Big]' = H'(x).$$

In the theory of distributions the ordinary derivative plays a minor role. Therefore, in the absence of further comment, the derivation of functions is always to be understood in the distributional sense.

The only locally integrable functions $f(x)$ of one real variable which satisfy the integrability condition for derivation of order 1 are the absolutely continuous functions, i.e., functions with locally integrable derivative $f'(x)$, such that in every interval I inside O

$$f(x)-f(x_0) = \int_{x_0}^{x} f'(t)\,dt \qquad (x_0 \text{ in } I). \tag{3}$$

We can thus state the following:

3.6.2. *If $f(x)$ is an absolutely continuous function, then its distributional derivative $f'(x)$ coincides with its ordinary derivative.*

Analogous conditions can be given for derivatives of higher orders; we omit the details.

3.7. SEQUENCES OF DISTRIBUTIONS

We say that a sequence of distributions $f_n(x)$ *converges* in O to a distribution $f(x)$ and we write

$$f_n(x) \to f(x) \quad \text{in } O \quad \text{or} \quad \lim_{n\to\infty} f_n(x) = f(x) \quad \text{in } O,$$

iff the distribution $f(x)$ is defined in O and, for every interval I inside O, there exist an order k and continuous functions $F(x)$ and $F_n(x)$ such that in I

$$\begin{aligned} F_n^{(k)}(x) &= f_n(x) \quad \text{for} \quad n > n_0, \\ F^{(k)}(x) &= f(x) \quad \text{and} \quad F_n(x) \rightrightarrows F(x). \end{aligned} \tag{1}$$

According to this definition the limit distribution $f(x)$ is defined in the whole set O, but this is not necessary for the distributions $f_n(x)$ (see Section 1.2).

A useful observation is that the order k which occurs in (1) can be replaced, if necessary, by any order $l \geqslant k$. In fact, if conditions (1) hold, then also

$$\tilde{F}_n^{(l)}(x) = f_n(x) \quad \text{for} \quad n > n_0, \quad \tilde{F}^{(l)}(x) = f(x) \quad \text{and} \quad \tilde{F}_n(x) \rightrightarrows \tilde{F}(x),$$

where

$$\tilde{F}_n(x) = \int_{x_0}^{x} F_n(t)\,dt^{l-k}, \quad \tilde{F}(x) = \int_{x_0}^{x} F(t)\,dt^{l-k} \quad (x_0 \text{ in } I).$$

The limit, if it exists, is unique. To prove this, we need the following auxiliary theorem:

3.7.1. *If a sequence of continuous functions $f_n(x)$ converges almost uniformly to $f(x)$ in O and if $f_n^{(m)}(x) = 0$ for $n = 1, 2, \ldots$, then $f^{(m)}(x) = 0$.*

By 3.1.1, for every interval I inside O, there exist smooth functions $\varphi_{rn}(x)$ such that

$$\varphi_{rn}(x) \rightrightarrows f_n(x) \quad \text{and} \quad \varphi_{rn}^{(m)}(x) = 0 \quad \text{in } I.$$

Let r_n be such that $|\varphi_{r_n,n}(x)-f_n(x)| < 1/n$ in I. Then $\varphi_{r_n,n}(x) \rightrightarrows f(x)$ and therefore $f(x) = [\varphi_{r_n,n}(x)]$ in I. Differentiating m times, we have $f^{(m)}(x) = \varphi_{r_n,n}^{(m)}(x) = 0$ in I. Since I is arbitrary, we have $f^{(m)}(x) = 0$ in the whole set O.

We are now ready to prove the uniqueness of the limit. Let I be any interval inside O. If $f_n(x)$ are distributions such that $f_n(x) \to f(x)$ and $f_n(x) \to g(x)$, then there exist continuous functions $F_n(x)$, $G_n(x)$ and orders k, l such that

$$F_n(x) \rightrightarrows F(x), \quad G_n(x) \rightrightarrows G(x) \quad \text{in } I$$

and

$$F_n^{(k)}(x) = f_n(x), \quad F^{(k)}(x) = f(x),$$

and

$$G_n^{(l)}(x) = f_n(x), \quad G^{(l)}(x) = g(x).$$

We may assume that $k = l$, for otherwise we could replace both orders by a greater order. Since

$$(F_n(x)-G_n(x))^{(k)} = 0 \quad \text{and} \quad F_n(x)-G_n(x) \rightrightarrows F(x)-G(x),$$

we have

$$(F(x)-G(x))^{(k)} = 0,$$

by 3.7.1, which implies $f(x) = g(x)$ in I. Since I is arbitrary, it follows that the limit is unique.

It follows immediately from the definition of the limit that:

3.7.2. *If a sequence of continuous functions converges almost uniformly, then it also converges distributionally, and to the same limit.*

3.7.3. *If $f_n(x) \to f(x)$, then $f_{r_n}(x) \to f(x)$ for every sequence r_n of positive integers such that $r_n \to \infty$.*

3.7.4. *If $f_n(x) \to f(x)$ and $g_n(x) \to f(x)$, then the interlaced sequence*

$$f_1(x), g_1(x), f_2(x), g_2(x), \ldots$$

also converges to $f(x)$.

3.7.5. *If $f_n(x) \to f(x)$, then $\lambda f_n(x) \to \lambda f(x)$ for every number λ. If $f_n(x) \to f(x)$ and $g_n(x) \to g(x)$, then $f_n(x)+g_n(x) \to f(x)+g(x)$.*

3.7.6. *If $f_n(x) \to f(x)$, then $f_n^{(m)}(x) \to f^{(m)}(x)$ for every order m.*

This simple theorem is of great usefulness in calculations with distributions, in contrast to the classical differential calculus, where additional restrictions are necessary.

3.7.7. *If $f_n(x) \to f(x)$ in every interval inside O, then $f_n(x) \to f(x)$ in the whole set O.*

For any interval I inside O there exists an interval I' inside O such that I is inside I'. Since $f_n(x) \to f(x)$ in I', there exist an order k and continuous functions $F_n(x)$, $F(x)$ such that conditions (1) are satisfied; this shows that $f_n(x) \to f(x)$ in O.

We say that a sequence of distributions $f_n(x)$ is *convergent* in O iff for every interval I inside O there exist an order k and continuous functions $F_n(x)$ such that

$$F_n^{(k)}(x) = f_n(x) \quad \text{and} \quad F_n(x) \rightrightarrows \quad \text{in } I.$$

3.7.8. *If a sequence of distributions is convergent in O, then it converges to a distribution in O.*

Suppose that $f_n(x)$ is convergent in O. Let $\delta_n(x)$ be any δ-sequence. We shall prove that the sequence

$$\varphi_n(x) = f_n(x) * \delta_n(x)$$

is fundamental in O and that $f_n(x)$ converges to $[\varphi_n(x)]$.

In fact, let I be an arbitrary interval inside O and let I' be an interval inside O such that I is inside I'. There exist an order k and continuous functions $F_n(x)$, $F(x)$ such that

$$F_n^{(k)}(x) = f_n(x) \quad \text{and} \quad F_n(x) \rightrightarrows F(x) \quad \text{in } I'.$$

By 3.1.2, we have

$$F_n(x) * \delta_n(x) \rightrightarrows F(x) \quad \text{in } I. \tag{2}$$

Since

$$(F_n(x) * \delta_n(x))^{(k)} = \varphi_n(x),$$

the sequence $\varphi_n(x)$ is fundamental in O. It therefore represents a distribution $f(x)$ in O. By (2), we can write

$$F_n(x) * \delta_n(x) \to F(x) \quad \text{and} \quad [F_n(x) * \delta_n(x)] = F(x) \quad \text{in } I.$$

Hence, differentiating k times, we obtain

$$\varphi_n(x) \to F^{(k)}(x) \quad \text{and} \quad [\varphi_n(x)] = F^{(k)}(x) \quad \text{in } I.$$

Consequently

$$\varphi_n(x) \to f(x) \quad \text{in } I.$$

Since $F_n(x) - F_n(x) * \delta_n(x) \rightrightarrows 0$ in I, differentiating k times we obtain

$$f_n(x) - \varphi_n(x) \to 0 \quad \text{in } I.$$

Thus $f_n(x) \to f(x)$ in I. Since I is arbitrary, it follows from 3.7.7 that $f_n(x) \to f(x)$ in O.

3.8. CONVERGENCE AND REGULAR OPERATIONS

For distributions, passage to a limit commutes with all the regular operations introduced so far. In other words, the following equalities hold:

$$\lim_{n\to\infty} \lambda f_n(x) = \lambda \lim_{n\to\infty} f_n(x),$$

$$\lim_{n\to\infty} \big(f_n(x)+g_n(x)\big) = \lim_{n\to\infty} f_n(x)+\lim_{n\to\infty} g_n(x),$$

$$\lim_{n\to\infty} \big(f_n(x)-g_n(x)\big) = \lim_{n\to\infty} f_n(x)-\lim_{n\to\infty} g_n(x),$$

$$\lim_{n\to\infty} f_n^{(m)}(x) = \big(\lim_{n\to\infty} f_n(x)\big)^{(m)},$$

$$\lim_{n\to\infty} \omega(x) f_n(x) = \omega(x)\cdot \lim_{n\to\infty} f_n(x),$$

$$\lim_{n\to\infty} f_n(x) g_n(y) = \lim_{n\to\infty} f_n(x)\cdot \lim_{n\to\infty} g_n(y),$$

$$\lim_{n\to\infty} \big(f_n(x)*\omega(x)\big) = \lim_{n\to\infty} f_n(x)*\omega(x).$$

In the case of substitution, the symbol $\lim\limits_{n\to\infty} f_n(\sigma(x))$ has two interpretations: as the limit of the sequence $f_n(\sigma(x))$ and as the substitution $y = \sigma(x)$ in the distribution $\lim\limits_{n\to\infty} f_n(y)$. The fact that passage to the limit commutes with substitution implies that both interpretations give the same result. Similarly for translation.

The verification of commutativity is trivial for multiplication by a number, addition, subtraction, translation, derivation, multiplication of distributions with separated variables, and convolution with a smooth function vanishing outside an interval. The commutativity of passage to the limit with multiplication by a smooth function and with substitution results from the following two stronger theorems.

3.8.1. *If $\omega_n^{(m)}(x)$ converges to $\omega^{(m)}(x)$ almost uniformly for every order m and if $f_n(x)\to f(x)$, then $\omega_n(x)f_n(x)\to \omega(x)f(x)$.*

For every interval I inside O, there exist continuous functions $F_n(x)$, $F(x)$ and an order k such that $F_n(x) \rightrightarrows F(x)$, $F_n^{(k)}(x) = f_n(x)$, and $F^{(k)}(x) = f(x)$. Thus

$$\omega_n(x)F_n(x) \rightrightarrows \omega(x)F(x) \quad \text{in } I.$$

Since every uniformly convergent sequence is distributionally convergent, we also have

$$\omega_n(x)F_n(x) \to \omega(x)F(x). \tag{1}$$

Similarly we have

$$\omega_n^{(e_j)}(x)F_n(x) \to \omega^{(e_j)}(x)F(x). \tag{2}$$

Differentiating (1), we obtain

$$\omega_n^{(e_j)}(x)F_n(x)+\omega_n(x)F^{(e_j)}(x) \to \omega^{(e_j)}(x)F(x)+\omega(x)F^{(e_j)}(x).$$

Hence, in view of (2),

$$\omega_n(x)F_n^{(e_j)}(x) \to \omega(x)F^{(e_j)}(x).$$

By induction we obtain

$$\omega_n(x)F_n^{(k)}(x) \to \omega(x)F^{(k)}(x),$$

i.e.,

$$\omega_n(x)f_n(x) \to \omega(x)f(x) \quad \text{in } I. \tag{3}$$

Since the interval I is arbitrary, (3) holds in the whole set O.

Another proof of 3.8.1 follows from the formula

$$\omega(x)\varphi^{(k)}(x) = \sum_{0\leqslant m\leqslant k} (-1)^m\binom{k}{m}\big(\omega^{(m)}(x)\varphi(x)\big)^{(k-m)}$$

where $\binom{k}{m} = \binom{\varkappa_1}{\mu_1}\dots\binom{\varkappa_q}{\mu_q}$ and $(-1)^m = (-1)^{\mu_1+\dots+\mu_q}$ for $k = (\varkappa_1, \dots, \varkappa_q)$, $m = (\mu_1, \dots, \mu_q)$. The verification of this formula for smooth functions $\omega(x)$ and $\varphi(x)$ is a matter of routine calculation. If $\omega(x)$ is fixed, both sides of the formula are iterations of regular operations; thus the formula still holds if $\varphi(x)$ is replaced by any distribution or continuous function. In particular, we have

$$\omega_n(x)F_n^{(k)}(x) = \sum_{0\leqslant m\leqslant k} (-1)^m\binom{k}{m}\big(\omega^{(m)}(x)F_n(x)\big)^{(k-m)},$$

and hence (3) follows in every interval I inside O, and consequently in the whole set O.

3.8.2. *If $\sigma_n^{(m)}(x)$ converges to $\sigma^{(m)}(x)$ almost uniformly for every order m, where $\sigma_n(x)$ and $\sigma(x)$ have property* (1) *of Section* 2.6, *and $f_n(y) \to f(y)$, then*

$$f_n\big(\sigma_n(x)\big) \to f\big(\sigma(x)\big).$$

The proof of 3.8.2 will rest on formula (2) of Section 2.6. This formula involves regular operations only, and thus it also holds if $\Phi_n(y)$ is replaced by any distribution $f(y)$. Thus

$$f'\big(\sigma(x)\big) = \frac{\dfrac{\partial}{\partial\xi_1}f\big(\sigma(x)\big)\cdot\dfrac{\partial}{\partial\xi_1}\sigma(x)+\dots+\dfrac{\partial}{\partial\xi_q}f\big(\sigma(x)\big)\cdot\dfrac{\partial}{\partial\xi_q}\sigma(x)}{\left(\dfrac{\partial\sigma(x)}{\partial\xi_1}\right)^2+\dots+\left(\dfrac{\partial\sigma(x)}{\partial\xi_q}\right)^2}. \tag{4}$$

Suppose that $\sigma(x)$ is defined in an open set O and that the values of $\sigma(x)$ belong to an open set O' of real numbers y. The distribution $f(y)$ is taken to be defined in O'. Let I be any interval inside O. The function $\sigma(x)$ maps I onto an interval I' inside O'. The sequence $\sigma_n^{(m)}(x)$ converges to $\sigma^{(m)}(x)$ uniformly in I. There is an interval I'' inside O' such that I' is inside I'' and the values of $\sigma_n(x)$ lie in I'' for sufficiently large n. For that interval I'' there exist functions $F_n(y)$, $F(y)$ and a non-negative integer k such that $F_n(y) \rightrightarrows F(y)$ in I'', $F_n^{(k)}(y) = f_n(y)$, $F^{(k)}(y) = f(y)$. Clearly

$$F_n(\sigma_n(x)) \rightrightarrows F(\sigma(x)) \quad \text{in } I.$$

Since uniformly convergent sequences are distributionally convergent, we also have

$$F_n(\sigma_n(x)) \to F(\sigma(x)) \quad \text{in } I. \tag{5}$$

Applying formula (4) to the distributions $F_n'(\sigma_n(x))$ and $F'(\sigma(x))$, we obtain from (5) and 3.8.1

$$F_n'(\sigma_n(x)) \to F'(\sigma(x)) \quad \text{in } I.$$

By induction we get

$$F_n^{(k)}(\sigma_n(x)) \to F^{(k)}(\sigma(x)) \quad \text{in } I,$$

i.e., $f_n(\sigma_n(x)) \to f(\sigma(x))$ in I. Since I is arbitrary, 3.8.2 follows.

In Part III, Section 12.1, it will by proved that the passage to the limit commutes with any regular operation.

3.9. DISTRIBUTIONALLY CONVERGENT SEQUENCES OF SMOOTH FUNCTIONS

We first prove that:

3.9.1. *A sequence of constant functions converges distributionally iff it converges in the ordinary sense.*

In fact, if constant functions converge in the ordinary sense, then they converge uniformly, and hence distributionally by 3.7.2.

Conversely, suppose that a sequence c_n of constant functions converges distributionally. Then this sequence is bounded, for otherwise there would exist a subsequence c_{r_n} such that $1/c_{r_n}$ converges in the usual sense to 0 and we would have $1 = \dfrac{1}{c_{r_n}} \cdot c_{r_n} \to 0$. Suppose that c_n does not converge in the ordinary sense. Then there exist two subsequences which converge to different limits. These subsequences converge distributionally to different limits, which contradicts 3.7.3.

3.9.2. *A sequence of smooth functions $\varphi_n(x)$ is fundamental in O iff for every interval I inside O there exist continuous functions $F_n(x)$ and an order k such that in I*

$$F_n^{(k)}(x) = \varphi_n(x) \quad \text{and} \quad F_n(x) \rightrightarrows. \tag{1}$$

In fact, if $\varphi_n(x)$ is fundamental, there exist, for every interval I inside O, smooth functions $\Phi_n(x)$ and order k such that in I

$$\Phi_n(x) \rightrightarrows \quad \text{and} \quad \Phi_n^{(k)}(x) = \varphi_n(x). \tag{2}$$

Since smooth functions are continuous, the condition is satisfied.

Suppose, conversely, that (1) holds for every interval I inside O. Let I be fixed arbitrarily inside O and let I' be an interval inside O such that I is inside I'. There exist functions $F_n(x)$ and an order k such that (1) holds in I'. Let

$$\Phi_{nr}(x) = \Big(F_n(x) - \int_{x_0}^{x} \varphi_n(t)\,dt^k\Big) * \delta_r(x) + \int_{x_0}^{x} \varphi_n(t)\,dt^k,$$

where x_0 is in I and $\delta_n(x)$ is a δ-sequence as in Section 3.1. Then $\Phi_{nr}^{(k)}(x) = \varphi_n(x)$ in I for sufficiently large r, say $r > p_n$. Moreover, by 3.1.1 we have $\Phi_{nr}(x) \rightrightarrows F_n(x)$ in I as $r \to \infty$. Let $F(x)$ denote the limit of $F_n(x)$. Since $F_n(x) \rightrightarrows F(x)$, there is a sequence of positive integers $r_n > p_n$ such that

$$\Phi_n(x) = \Phi_{n, r_n}(x) \rightrightarrows F(x) \quad \text{in } I.$$

Clearly $\Phi_n^{(k)}(x) = \varphi_n(x)$ in I; the functions $\Phi_n(x)$ thus have the required properties.

3.9.3. *A sequence of smooth functions converges distributionally to a distribution $f(x)$ iff it is fundamental for $f(x)$.*

In fact, if $\varphi_n(x)$ is a fundamental sequence for $f(x)$, then for every interval I inside O there exist smooth functions $\Phi_n(x)$, a continuous function $F(x)$, and an order k such that

$$\Phi_n(x) \rightrightarrows F(x), \quad \Phi_n^{(k)}(x) = \varphi_n(x)$$

and

$$F^{(k)}(x) = f(x) \quad \text{in } I. \tag{3}$$

The first two conditions follow from the definition of fundamental sequences. The third is obtained by differentiating the equality $F(x) = [\Phi_n(x)]$, which follows from the first condition, k times. Since smooth functions are continuous functions, (3) means that $\varphi_n(x) \to f(x)$ in O.

Conversely, if $\varphi_n(x) \to f(x)$ in O, there exist for every interval I inside O functions $F_n(x)$, $F(x)$ and an order k such that in I

$$F_n(x) \rightrightarrows F(x), \quad F_n^{(k)}(x) = \varphi_n(x) \quad \text{and} \quad F^{(k)}(x) = f(x).$$

The sequence $\varphi_n(x)$ is thus fundamental by 3.9.2. As we have just proved, every fundamental sequence converges to the distribution which it represents. This implies that $f(x) = [\varphi_n(x)]$.

3.10. LOCALLY CONVERGENT SEQUENCES OF DISTRIBUTIONS

It may happen that we know the following property of a sequence of distributions $f_n(x)$: For every point x_0 in O there exists an interval inside O, containing x_0, in which $f_n(x)$ is convergent. The aim of this section is to show that in such a case $f_n(x)$ is convergent in O. We also give some important corollaries.

If the continuous functions $\varphi_1(x)$ and $\varphi_2(x)$ are defined in sets O_1 and O_2, then their product $\varphi_1(x)\varphi_2(x)$ is defined in the intersection of O_1 and O_2. We later adopt the convention that this product is also defined and has the value 0 at all points where at least one of the factors $\varphi_1(x)$ and $\varphi_2(x)$ is defined and has the value 0.

In the next two lemmas, $\omega(x)$ denotes a smooth function defined everywhere and vanishing outside an interval I inside a given open set O, and I' is an interval inside O such that I is inside I'.

3.10.1. LEMMA. *If $\varphi_n(x)$ is a fundamental sequence in O, then $\omega(x)\varphi_n(x)$ is fundamental everywhere.*

In fact, there are smooth functions $\Phi_n(x)$ and an order k such that $\Phi_n^{(k)}(x) = \varphi_n(x)$ and $\Phi_n(x) \rightrightarrows$ in I'. Clearly $\omega(x)\Phi_n(x) \rightrightarrows$ everywhere. Hence the sequence $\omega(x)\Phi_n(x)$ is fundamental. Similarly $\omega^{(e_j)}(x)\Phi_n(x)$ is fundamental. Consequently the sequence

$$\omega(x)\Phi_n^{(e_j)}(x) = \big(\omega(x)\Phi_n(x)\big)^{(e_j)} - \omega^{(e_j)}(x)\Phi_n(x)$$

is fundamental everywhere. By induction, the sequence $\omega(x)\Phi_n^{(k)}(x)$, i.e., $\omega(x)\varphi_n(x)$, is fundamental everywhere.

By 3.10.1, if $f(x) = [\varphi_n(x)]$ in O, then the distribution $\omega(x)f(x) = [\omega(x)\varphi_n(x)]$ is defined everywhere.

3.10.2. LEMMA. *If a sequence of distributions $f_n(x)$ is convergent in O, then the sequence $\omega(x)f_n(x)$ converges everywhere.*

In fact, there are continuous functions $F_n(x)$ and an order k such that $F_n^{(k)}(x) = f_n(x)$ and $F_n(x) \rightrightarrows$ in I'. Clearly $\omega(x)F_n(x) \rightrightarrows$ everywhere. Hence $\omega(x)F_n(x)$ is distributionally convergent everywhere. Similarly $\omega_n^{(e_j)}(x)F_n(x)$ is distributionally convergent everywhere. Consequently the sequence

$$\omega(x)F_n^{(e_j)}(x) = \big(\omega(x)F_n(x)\big)^{(e_j)} - \omega^{(e_j)}(x)F_n(x)$$

is distributionally convergent everywhere. By induction, the sequence $\omega(x)F_n^{(k)}(x)$, i.e., $\omega(x)f_n(x)$, is convergent everywhere.

3.10.3. *If every point x_0 in O is in an interval I_0 such that the sequence of distributions $f_n(x)$ is convergent in I_0, then $f_n(x)$ is convergent in O. In other words: locally convergent sequences of distributions are convergent.*

In fact, let I be any interval inside O. There exists an interval J inside O such that I is inside J. We can divide J into a finite number of subintervals j (in this proof j denotes an interval and not a number) such that each subinterval j is inside an interval I_j in which $f_n(x)$ is convergent.

Let $g_j(x)$ denote the characteristic function of j, i.e., the function given by

$$g_j(x) = \begin{cases} 1 & \text{on } j, \\ 0 & \text{outside } j. \end{cases}$$

Similarly let $g_J(x)$ denote the characteristic function of J. If $\delta_n(x)$ is a δ-sequence (see Section 3.1), we can find an index p such that

$$\varphi_j(x) = g_j(x) * \delta_p(x) = 0 \quad \text{outside } I_j,$$

$$\varphi_J(x) = g_J(x) * \delta_p(x) = 1 \quad \text{on } I.$$

Clearly

$$g_J(x) = \sum_j g_j(x) \quad \text{and} \quad \varphi_J(x) = \sum_j \varphi_j(x).$$

Since $f_n(x)$ is convergent in I_j, the product $\varphi_j(x)f_n(x)$ is convergent everywhere, by 3.10.2. Since the number of intervals j is finite, the sequence

$$\psi_n(x) = \sum_j \varphi_j(x) f_n(x) = \varphi_J(x) f_n(x)$$

is also convergent everywhere. But $\psi_n(x) = f_n(x)$ in I, and so $f_n(x)$ is convergent in I. Since I can be chosen arbitrarily inside O, $f_n(x)$ is convergent in O, by 3.7.7.

By 3.9.3, a sequence of smooth functions converges distributionally iff it is fundamental. The following corollary is therefore immediate:

3.10.4. *If every point x_0 in O is in an interval I_0 such that $\varphi_n(x)$ is fundamental in I_0, then $\varphi_n(x)$ is fundamental in O. In other words: locally fundamental sequences are fundamental.*

We are now in a position to prove the following important theorem:

3.10.5. *Let O be the union of open sets Θ. If in each of the sets Θ there is defined a distribution $f_\Theta(x)$, in such a way that distributions on overlapping sets coincide on the overlap, then there exists a distribution $f(x)$ defined in the whole set O such that $f(x) = f_\Theta(x)$ in every set Θ.*

In fact, let $\delta_n(x)$ be a δ-sequence. For every fixed n, the smooth functions $f_\Theta(x)*\delta_n(x)$ coincide pairwise at all points where they are both defined. These functions can therefore be unified into a single function $\varphi_n(x)$ (depending on n), defined in the union of the open sets in which $f_\Theta(x)*\delta_n(x)$ are defined. The sequence $\varphi_n(x)$ is fundamental in every interval which is inside at least one of the sets Θ. Since the union of all such intervals is O, $\varphi_n(x)$ is fundamental in O by 3.10.4. The distribution $f(x) = [\varphi_n(x)]$ has the required property, for in every set Θ

$$f(x) = [f_\Theta(x)*\delta_n(x)] = f_\Theta(x).$$

4. Extension of the theory

4.1. DISTRIBUTIONS DEPENDING ON A CONTINUOUS PARAMETER

We say that a continuous function $f_\alpha(x)$, depending on a continuous parameter α, *converges uniformly* to $f(x)$ on a set I as $\alpha \to \alpha_0$ and we write

$$f_\alpha(x) \rightrightarrows f(x) \qquad (\alpha \to \alpha_0) \qquad \text{on } I,$$

iff the function $f(x)$ is defined on I and, given any number $\varepsilon > 0$, there is a number $\eta > 0$ such that for every α satisfying $|\alpha-\alpha_0| < \eta$ the function $f_\alpha(x)$ is defined on the whole set I and satisfies there the inequality $|f_\alpha(x)-f(x)| < \varepsilon$.

We say that the function $f_\alpha(x)$ *converges almost uniformly* to $f(x)$ in an open set O as $\alpha\to\alpha_0$ iff $f_\alpha(x) \rightrightarrows f(x)$ $(\alpha \to \alpha_0)$ on every interval inside O.

We say that a distribution $f_\alpha(x)$, depending on a continuous parameter α, *converges to a distribution* $f(x)$ in an open set O as $\alpha \to \alpha_0$ iff $f(x)$ is defined in O, and if for every interval I inside O there exist an order k and continuous functions $F_\alpha(x)$, $F(x)$ such that, for α sufficiently near to α_0,

$$F_\alpha^{(k)}(x) = f_\alpha(x), \qquad F^{(k)}(x) = f(x) \qquad \text{and} \qquad F_\alpha(x) \rightrightarrows F(x) \qquad (\alpha \to \alpha_0) \qquad \text{in } I.$$

We then write

$$f_\alpha(x) \to f(x) \qquad (\alpha \to \alpha_0) \qquad \text{in } O,$$

or

$$f(x) = \lim_{\alpha\to\alpha_0} f_\alpha(x) \qquad \text{in } O.$$

The limit $f(x)$, if it exists, is unique. The proof resembles that for sequences.

In the above definition it does not matter whether α is a real or a complex parameter, or even a variable point of some multidimensional space, as long as the symbol $|\alpha-\alpha_0|$ is taken to be the distance between the points α and α_0. We similarly define the limit when $\alpha_0 = \pm\infty$.

We have at once

4.1.1. *If a continuous function $f(x)$ depending on a parameter converges almost uniformly, then it converges distributionally to the same limit.*

Just as for sequences, we can show that passage to the limit commutes with all the regular operations introduced here. Furthermore the analogues of the theorems of Sections 3.8 and 3.10 are true.

We can now give a definition of the derivative of a distribution which is the same as that usually given for functions. In fact,

4.1.2. *For every distribution $f(x)$,*

$$f^{(e_j)}(x) = \lim_{\alpha\to 0} \frac{f(x+\alpha e_j)-f(x)}{\alpha}.$$

Let I be any interval inside O and let I' be an interval inside O such that I is inside I'. Then there exist an order k and a continuous function $F(x)$ with a continuous derivative $F^{(e_j)}(x)$ such that $F^{(k)}(x) = f(x)$ in I'. Since in I

$$\frac{F(x+\alpha e_j)-F(x)}{\alpha} \rightrightarrows F^{(e_j)}(x) \quad \text{as} \quad \alpha \to 0,$$

we have in I

$$\frac{f(x+\alpha e_j)-f(x)}{\alpha} = \left(\frac{F(x+\alpha e_j)-F(x)}{\alpha}\right)^{(k)} \to \left(F^{(e_j)}(x)\right)^{(k)} = f^{(e_j)}(x).$$

Since the interval I is arbitrary, the convergence holds in the whole set O.

4.2. MULTIDIMENSIONAL SUBSTITUTION

Let $\sigma_1(x), \ldots, \sigma_p(x)$ be smooth functions defined in an open subset O of q-dimensional space such that the transformation

$$\sigma(x) = \left(\sigma_1(x), \ldots, \sigma_p(x)\right)$$

maps O into an open subset O' of p-dimensional space, $p \leqslant q$, and such that at every point x of O at least one of the jacobians

$$J_{j_1,\ldots,j_p}(x) = \frac{\partial(\sigma_1, \ldots, \sigma_p)}{\partial(\xi_{j_1}, \ldots, \xi_{j_p})} = \begin{vmatrix} \frac{\partial\sigma_1}{\partial\xi_{j_1}}, \ldots, \frac{\partial\sigma_1}{\partial\xi_{j_p}} \\ \cdots\cdots\cdots\cdots \\ \frac{\partial\sigma_p}{\partial\xi_{j_1}}, \ldots, \frac{\partial\sigma_p}{\partial\xi_{j_p}} \end{vmatrix} \qquad (j_1 < \ldots < j_p)$$

does not vanish, i.e.,

$$J(x) = \sum_{j_1<\ldots<j_p} J_{j_1,\ldots,j_p}(x)^2 > 0 \quad \text{in } O.$$

We will show that the substitution $\varphi\left(\sigma(x)\right)$, where $\varphi(y)$ is a smooth function defined in O' and $\sigma(x)$ is kept fixed, is a regular operation on $\varphi(y)$. The proof resembles that of the corresponding statement in Section 2.6.

Note first that if, for smooth functions $\Phi_n(y)$, the sequence $\Phi_n(\sigma(x))$ is fundamental in some open set, so is the sequence $\Phi_n^{(e_j)}(\sigma(x))$. In fact, from

$$\Phi_n(\sigma(x))^{(e_i)} = \sum_{j=1}^{p} \Phi_n^{(e_j)}(\sigma(x)) \cdot \frac{\partial \sigma_j(x)}{\partial \xi_i}$$

we find by calculation

$$J_{j_1,\ldots,j_p}(x) \cdot \Phi_n^{(e_j)}(\sigma(x)) = J_{n,j,j_1,\ldots,j_p}(x),$$

where

$$J_{n,j,j_1,\ldots,j_p}(x) = \frac{\partial(\sigma_1, \ldots, \sigma_{j-1}, \Phi_n(\sigma), \sigma_{j+1}, \ldots, \sigma_p)}{\partial(\xi_{j_1}, \ldots, \xi_{j_p})}.$$

Hence

$$\Phi_n^{(e_j)}(\sigma(x)) = \frac{1}{J(x)} \sum_{j_1 < \ldots < j_p} J_{j_1,\ldots,j_p}(x) \cdot J_{n,j,j_1,\ldots,j_p}(x),$$

which establishes the fundamentality of $\Phi_n^{(e_j)}(x)$.

By induction, if a sequence $\Phi_n(\sigma(x))$ is fundamental in some open set, so is the sequence $\Phi_n^{(k)}(\sigma(x))$ for every order k.

Every point x_0 in O is contained in an interval I_0 inside O such that the transformation $\sigma(x)$ maps I_0 into an interval I_0' inside O'. Now let $\Phi_n(y)$ be smooth functions such that, for some order k,

$$\Phi_n^{(k)}(y) = \varphi_n(y) \quad \text{and} \quad \Phi_n(y) \rightrightarrows \quad \text{in } I_0'.$$

Then $\Phi_n(\sigma(x)) \rightrightarrows$ in I_0. Consequently the sequence $\Phi_n^{(k)}(\sigma(x))$, i.e., $\varphi_n(\sigma(x))$, is fundamental in I_0. By 3.10.4, $\varphi_n(\sigma(x))$ is fundamental in O.

Note that both the hypothesis on $\sigma(x)$ and the above proof can be simplified in the case $q = p$. It then suffices to deal with one jacobian only,

$$\frac{\partial(\sigma_1, \ldots, \sigma_q)}{\partial(\xi_1, \ldots, \xi_q)},$$

which should be different from 0 in O.

We have proved that substitution is a regular operation. It can therefore be extended to distributions $f(y) = [\varphi_n(y)]$ defined in O', by setting

$$f(\sigma(x)) = [\varphi_n(\sigma(x))].$$

When $p = 1$, the above definition coincides with that given in Section 2.6. When $p = q$ and $\sigma(x) = x + h$, it coincides with the definition of translation given in Section 2.4. In cases where $f(y)$ is a continuous or locally integrable function, the distributional substitution $f(\sigma(x))$ coincides with the ordinary substitution of functions, provided $p \leqslant q$. If $p > q$, then, in contrast with the case of functions, the substitution $f(\sigma(x))$ is not always feasible.

Theorem 3.8.2 remains true for multidimensional substitutions.

4.3. DISTRIBUTIONS CONSTANT IN SOME VARIABLES

A distribution $f(x)$ in O is said to be *constant in the variables* $\xi_{p+1}, \ldots, \xi_q$ or *independent of* $\xi_{p+1}, \ldots, \xi_q$ $(0 \leqslant p < q)$ iff it can be represented in the form $[\varphi_n(x)]$ where the smooth functions $\varphi_n(x)$ are constant in $\xi_{p+1}, \ldots, \xi_q$.

It follows immediately from the definition that

4.3.1. *If $f(x)$ is constant in $\xi_{p+1}, \ldots, \xi_q$, then $f^{(e_j)}(x) = 0$ for $j = p+1, \ldots, q$.*

The converse statement is not true, even for functions, in the case of an arbitrary open set O.

In fact, let O be the two-dimensional set defined by the inequality $\xi_1 < |\xi_2|$ (Fig. 4.1) and let

$$f(x) = f(\xi_1, \xi_2) = \begin{cases} 0 & \text{for} \quad \xi_1 < 0, \\ \xi_1^2 & \text{for} \quad 0 \leqslant \xi_1 < \xi_2, \\ -\xi_1^2 & \text{for} \quad 0 \leqslant \xi_1 < -\xi_2 \end{cases}$$

(Fig. 4.2). The function $f(x)$ is continuous in O, and

$$f^{(e_2)}(x) = 0 \quad \text{in } O.$$

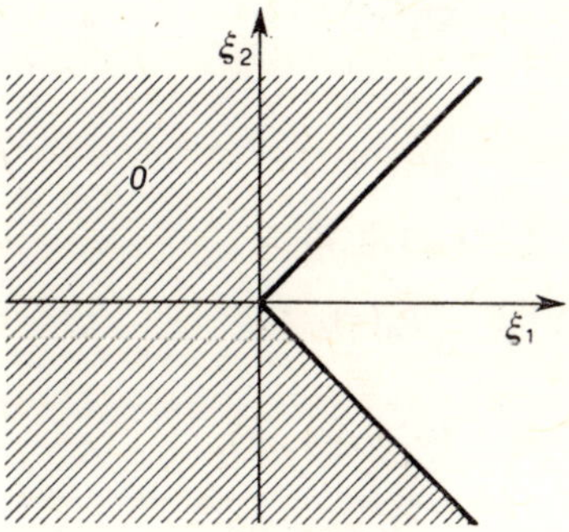

FIG. 4.1

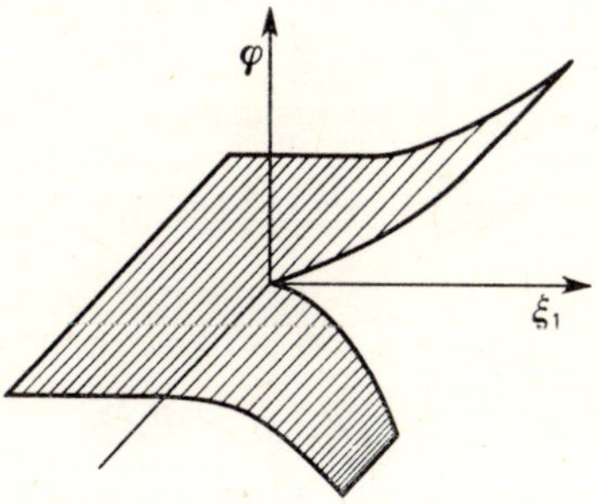

FIG. 4.2

Moreover it is easy to see that $f(x)$ is constant in ξ_2 in every interval I in O, but is not constant in ξ_2 in the whole set O.

The converse statement is true, for functions and for distributions, in the special case of an open interval:

4.3.2. *If $f^{(e_j)}(x) = 0$ for $j = p+1, \ldots, q$ in an interval I, then $f(x)$ is constant in $\xi_{p+1}, \ldots, \xi_q$ in I.*

In fact, for any δ-sequence of functions $\delta_n(x)$ which vanish outside $|x| < \alpha_n$ $(\alpha_n \to 0)$, we have

$$\big(f(x) * \delta_n(x)\big)^{(e_j)} = f^{(e_j)}(x) * \delta_n(x) = 0$$

in intervals I_n such that the distance of points in I_n from points outside I is greater than α_n. The smooth functions $\varphi_n(x) = f(x) * \delta_n(x)$ are thus constant in $\xi_{p+1}, \ldots, \xi_q$ in I_n. Since $f(x) = [\varphi_n(x)]$, the distribution $f(x)$ is constant in $\xi_{p+1}, \ldots$ $\ldots, \xi_q$ in I.

For any order $k = (\varkappa_1, \ldots, \varkappa_q)$, k_p will denote the order

$$k_p = (\varkappa_1, \ldots, \varkappa_p, 0, \ldots, 0) = k - \varkappa_{p+1} e_{p+1} - \ldots - \varkappa_q e_q.$$

The following lemma plays a fundamental role in the investigation of distributions constant in $\xi_{p+1}, \ldots, \xi_q$:

4.3.3. *If $\varphi_n(x)$ are smooth functions constant in $\xi_{p+1}, \ldots, \xi_q$ and if, for some interval I_0, there exist an order k and smooth functions $\Phi_n(x)$ such that*

$$\Phi_n^{(k)}(x) = \varphi_n(x), \qquad \Phi_n(x) \rightrightarrows \qquad \text{in } I_0,$$

then for every interval I inside I_0 there exist smooth functions $\Psi_n(x)$ constant in $\xi_{p+1}, \ldots, \xi_q$ such that

$$\Psi_n^{(k_p)}(x) = \varphi_n(x), \qquad \Psi_n(x) \rightrightarrows \qquad \text{in } I.$$

If $\varkappa_q > 0$, let

$$\bar{\Phi}_n(x) = \frac{1}{\eta} \big(\Phi_n(x + \eta e_q) - \Phi_n(x)\big).$$

We have

$$\bar{\Phi}_n^{(k-e_q)}(x) = \varphi_n(x) \quad \text{and} \quad \bar{\Phi}_n(x) \rightrightarrows .$$

By induction we obtain smooth functions $\tilde{\Phi}_n(x)$ such that

$$\tilde{\Phi}_n^{(k-\varkappa_q e_q)}(x) = \varphi_n(x) \quad \text{and} \quad \tilde{\Phi}_n(x) \rightrightarrows .$$

The functions

$$\tilde{\Psi}_n(x) = \tilde{\Phi}_n(\xi_1, \ldots, \xi_{q-1}, \gamma) \qquad (\gamma \text{ constant})$$

are constant in ξ_q,

$$\tilde{\Psi}_n^{(k-\varkappa_q e_q)}(x) = \varphi_n(x) \quad \text{and} \quad \tilde{\Psi}_n(x) \rightrightarrows . \tag{1}$$

If $\varkappa_q = 0$ we can simply write $\tilde{\Psi}_n(x) = \Phi_n(\xi_1, \ldots, \xi_{q-1}, \gamma)$ and (1) still holds.

Similary, if $p < q-1$, we obtain smooth functions $\bar{\Psi}_n(x)$ constant in ξ_{q-1}, ξ_q, such that

$$\bar{\Psi}_n^{(k-\varkappa_{q-1}e_{q-1}-\varkappa_q e_q)}(x) = \varphi_n(x) \quad \text{and} \quad \bar{\Psi}_n(x) \rightrightarrows .$$

By induction, there exist smooth functions $\Psi_n(x)$ constant in $\xi_{p+1}, \ldots, \xi_q$ and such that

$$\Psi_n^{(k-\varkappa_{p+1}e_{p+1}-\ldots-\varkappa_q e_q)}(x) = \varphi_n(x) \quad \text{and} \quad \Psi_n(x) \rightrightarrows .$$

Provided the number η is sufficiently small in all the inductive steps, the final conditions above will hold in I.

4.3.4. *If a distribution $f(x)$, constant in $\xi_{p+1}, \ldots, \xi_q$ is, in an interval I_1, the derivative of some order k of a continuous function, then in every interval I inside I_1 the distribution $f(x)$ is the derivative of order k_p of a continuous function constant in $\xi_{p+1}, \ldots, \xi_q$.*

Let $\delta_n(x)$ be any δ-sequence. If $f(x) = F^{(k)}(x)$ in I_1, then the smooth functions

$$\varphi_n(x) = f(x) * \delta_n(x), \qquad \Phi_n(x) = F(x) * \delta_n(x)$$

satisfy the hypotheses of 4.3.3, with I_0 an interval inside I_1 such that I is inside I_0. There thus exist smooth functions $\Psi_n(x)$ constant in $\xi_{p+1}, \ldots, \xi_q$ such that

$$\Psi_n^{(k_p)}(x) = \varphi_n(x) \quad \text{and} \quad \Psi_n(x) \rightrightarrows G(x) \quad \text{in } I,$$

where the continuous function $G(x)$ is also constant in $\xi_{p+1}, \ldots, \xi_q$. Since $G(x) = [\psi_n(x)]$, we have

$$G^{(k_p)}(x) = [\varphi_n(x)] = [\Phi_n(x)]^{(k)} = F^{(k)}(x) = f(x) \quad \text{in } I.$$

4.3.5. *A distribution $f(x)$ is constant in $\xi_{p+1}, \ldots, \xi_q$ in an interval I_0 iff, in every interval I inside I_0, $f(x)$ is some derivative of a continuous function constant in $\xi_{p+1}, \ldots, \xi_q$.*

In fact, if $f(x) = F^{(k)}(x)$ in I and $F^{(e_j)}(x) = 0$ for $j = p+1, \ldots, q$, then

$$f^{(e_j)}(x) = \left(F^{(e_j)}(x)\right)^{(k)} = 0 \quad \text{in } I.$$

Since I is arbitrary, we have $f^{(e_j)}(x) = 0$ in the whole interval I_0, for each $j = p+1, \ldots, q$. By 4.3.2, $f(x)$ is constant in $\xi_{p+1}, \ldots, \xi_q$ in I_0.

The remaining part of 4.3.5 follows from 3.4.4 and 4.3.4.

All the remarks of this section naturally remain true if we replace $\xi_{p+1}, \ldots, \xi_q$ by an arbitrary set of variables $\xi_{j_1}, \ldots, \xi_{j_r}$ $(1 \leqslant r \leqslant q)$.

4.4. DIMENSION OF DISTRIBUTIONS

Distributions defined in an open subset of the q-dimensional space are called *q-dimensional distributions* or *distributions of q variables*. If it is necessary to emphasize the number of variables, we write $f(\xi_1, \ldots, \xi_q)$ instead of $f(x)$. We are now going to study relations between p-dimensional distributions and q-dimensional distributions constant in $\xi_{p+1}, \ldots, \xi_q$ $(p < q)$.

Every function $\varphi(\xi_1, \ldots, \xi_p)$ of p variables uniquely determines a corresponding function of q variables $\varphi(\xi_1, \ldots, \xi_q)$ whose value at a point $(\xi_1, \ldots, \xi_q)$ is, for all choices of $\xi_{p+1}, \ldots, \xi_q$, equal to the value of $\varphi(\xi_1, \ldots, \xi_p)$ at the point $(\xi_1, \ldots, \xi_p)$. Hence if a p-dimensional function $\varphi(\xi_1, \ldots, \xi_p)$ is defined in an open subset O' of p-dimensional space, then the corresponding q-dimensional function $\varphi(\xi_1, \ldots, \xi_q)$ is defined in the open set O of all points $(\xi_1, \ldots, \xi_q)$ of q-dimensional space such that $(\xi_1, \ldots, \xi_p)$ is in O'; moreover $\varphi(\xi_1, \ldots, \xi_q)$ is constant in $\xi_{p+1}, \ldots, \xi_q$.

It is easy to see that if $\varphi_n(\xi_1, \ldots, \xi_p)$ is a sequence of p-dimensional smooth functions fundamental in O', then the sequence $\varphi_n(\xi_1, \ldots, \xi_q)$ of the corresponding q-dimensional smooth functions is fundamental in O. The converse follows easily from 4.3.3. We can therefore state

4.4.1. *A sequence of p-dimensional smooth functions $\varphi_n(\xi_1, \ldots, \xi_p)$ is fundamental in O' iff the sequence of corresponding q-dimensional functions $\varphi_n(\xi_1, \ldots, \xi_q)$ is fundamental in O.*

Hence, by the definition of equivalent sequences,

4.4.2. *Two sequences of p-dimensional functions $\varphi_n(\xi_1, \ldots, \xi_p)$ and $\psi_n(\xi_1, \ldots, \xi_p)$ are equivalent in O' iff the corresponding sequences of q-dimensional functions $\varphi_n(\xi_1, \ldots, \xi_q)$ and $\psi_n(\xi_1, \ldots, \xi_q)$ are equivalent in O.*

By 4.4.1 and 4.4.2, every p-dimensional distribution

$$f(\xi_1, \ldots, \xi_p) = [\varphi_n(\xi_1, \ldots, \xi_q)] \quad \text{in } O'$$

determines a corresponding q-dimensional distribution

$$f(\xi_1, \ldots, \xi_q) = [\varphi_n(\xi_1, \ldots, \xi_q)] \quad \text{in } O,$$

constant in $\xi_{p+1}, \ldots, \xi_q$, and this correspondence is one-to-one. Moreover, every distribution in O, constant in $\xi_{p+1}, \ldots, \xi_q$, corresponds to a distribution in O'.

The question arises as to whether operations performed on p-dimensional distributions yield the same result as operations performed on the corresponding q-dimensional distributions. To answer this question, we introduce the notation

$$B(\varphi(\xi_1, \ldots, \xi_p)) = \varphi(\xi_1, \ldots, \xi_q)$$

for the q-dimensional smooth function corresponding to a p-dimensional smooth function $\varphi(\xi_1, \ldots, \xi_p)$. By definition, B is an operation performed on p-dimensional smooth functions, and yields a q-dimensional smooth function. This operation is regular. It thus extends to distributions

$$f(\xi_1, \ldots, \xi_p) = [\varphi_n(\xi_1, \ldots, \xi_p)]$$

on setting

$$B(f(\xi_1, \ldots, \xi_p)) = [B(\varphi_n(\xi_1, \ldots, \xi_p))].$$

By definition, $B(f(\xi_1, \ldots, \xi_p))$ is the q-dimensional distribution $f(\xi_1, \ldots, \xi_q)$ corresponding to the p-dimensional distribution $f(\xi_1, \ldots, \xi_p)$.

Suppose that another regular operation $A(\varphi, \psi, ...)$ is given and that

$$B\big(A(\varphi, \psi, ...)\big) = A\big(B(\varphi), B(\psi), ...\big).$$

This equality expresses in precise terms the fact that the operation A performed on a p-dimensional smooth function yields a function which corresponds to that obtained from the corresponding q-dimensional function. Since both sides of the above equality are iterations of regular operations, the same formula holds for distributions:

$$B\big(A(f, g, ...)\big) = A\big(B(f), B(g), ...\big).$$

Our result can be expressed more intuitively as follows:

Every regular operation performed on p-dimensional distributions yields the corresponding result when performed on the corresponding q-dimensional distributions provided this is also true for smooth functions.

The following theorem shows that the limit of a sequence of p-dimensional distributions exists iff the limit exists for the corresponding q-dimensional distributions and moreover, that the limits correspond to each other.

4.4.3. *A sequence of p-dimensional distributions* $f_n(\xi_1, ..., \xi_p)$ *converges in* O' *to* $f(\xi_1, ..., \xi_p)$ *iff the sequence of the corresponding q-dimensional distributions* $f_n(\xi_1, ..., \xi_q)$ *converges in* O *to the distribution* $f(\xi_1, ..., \xi_q)$ *corresponding to* $f(\xi_1, ..., \xi_p)$.

It is clear that the convergence of the p-dimensional sequence implies the convergence of the q-dimensional sequence to the corresponding limit.

Conversely, suppose that in O

$$f_n(\xi_1, ..., \xi_q) \to f(\xi_1, ..., \xi_q). \tag{1}$$

Let I' be any interval inside O', let I be the interval consisting of all points $(\xi_1, ..., \xi_q)$ such that $(\xi_1, ..., \xi_q)$ is in I' and $|\xi_j| < 1$ for $j = p+1, ..., q$, and let I_0 be any interval inside O such that I is inside I_0. It follows from (1) that all the distributions $f_n(\xi_1, ..., \xi_q)$ are the derivatives of a fixed order $k = (\varkappa_1, ..., \varkappa_q)$ of continuous functions in I_0. By 4.3.4 there exist functions $F_n(\xi_1, ..., \xi_q)$ constant in $\xi_{p+1}, ..., \xi_q$ such that in I

$$F_n^{(k_p)}(\xi_1, ..., \xi_q) = f_n(\xi_1, ..., \xi_q),$$

i.e., in I', for $k' = (\varkappa_1, ..., \varkappa_p)$,

$$F_n^{(k')}(\xi_1, ..., \xi_p) = f_n(\xi_1, ..., \xi_p).$$

Let $\Phi_n(\xi_1, ..., \xi_q)$ be smooth functions such that

$$\Phi_n(\xi_1, ..., \xi_p) - F_n(\xi_1, ..., \xi_p) \rightrightarrows 0, \tag{2}$$

and let

$$\varphi_n(\xi_1, \ldots, \xi_p) = \Phi_n^{(k')}(\xi_1, \ldots, \xi_p).$$

Differentiating (2) k' times, we obtain

$$\varphi_n(\xi_1, \ldots, \xi_p) - f_n(\xi_1, \ldots, \xi_p) \to 0. \tag{3}$$

Hence it follows, by that part of 4.4.3 just proved, that in I

$$\varphi_n(\xi_1, \ldots, \xi_q) - f_n(\xi_1, \ldots, \xi_q) \to 0.$$

Consequently, by (1),

$$\varphi_n(\xi_1, \ldots, \xi_q) \to f(\xi_1, \ldots, \xi_q),$$

and by 3.9.3

$$[\varphi_n(\xi_1, \ldots, \xi_q)] = f(\xi_1, \ldots, \xi_q).$$

Hence, for the corresponding p-dimensional functions and distribution, we have in I'

$$[\varphi_n(\xi_1, \ldots, \xi_p)] = f(\xi_1, \ldots, \xi_p),$$

and by 3.9.3

$$\varphi_n(\xi_1, \ldots, \xi_p) \to f(\xi_1, \ldots, \xi_p).$$

Hence by (3),

$$f_n(\xi_1, \ldots, \xi_p) \to f(\xi_1, \ldots, \xi_p).$$

Since the interval I' is arbitrary, the convergence holds in O'.

If there is no risk of ambiguity, q-dimensional distributions constant in $\xi_{p+1}, \ldots, \xi_q$ can be denoted by symbols $f(\xi_1, \ldots, \xi_p)$ like p-dimensional distributions. A similar convention is widely used for functions.

All the remarks of this section naturally remain true if we replace $\xi_{p+1}, \ldots, \xi_q$ by an arbitrary set of variables $\xi_{j_1}, \ldots, \xi_{j_r}$ $(1 \leqslant r \leqslant q)$.

4.5. DISTRIBUTIONS WITH VANISHING mTH DERIVATIVES

In the course of finding a general form for a distribution satisfying the condition $f^{(m)}(x) = 0$, we shall prove three auxiliary theorems 4.5.1, 4.5.2 and 4.5.3.

4.5.1. *If $f(x)$ is a distribution such that $f^{(\mu e_j)}(x) = 0$ in the interval $a - \varepsilon e_j < x < b + \varepsilon e_j$, then, for $|\eta| < \varepsilon$,*

$$f(x+\eta e_j) = f(x) + \frac{\eta}{1!} f^{(e_j)}(x) + \ldots + \frac{\eta^{\mu-1}}{(\mu-1)!} f^{(\mu e_j - e_j)}(x)$$

in $a < x < b$.

In fact, if $\delta_n(x)$ is a δ-sequence and $\varphi_n(x) = f(x) * \delta_n(x)$, then

$$f(x) = [\varphi_n(x)] \quad \text{and} \quad \varphi_n^{(\mu e_j)}(x) = 0.$$

The above formula therefore follows from the Taylor expansion of $\varphi_n(x+\eta e_j)$, simply by adding the brackets [ ].

4.5.2. *If $f^{(m+e_j)}(x) = 0$ in O, where $m = (\mu_1, \ldots, \mu_q)$ with $\mu_j = 0$, then in every interval I inside O the distribution $f(x)$ can be represented in the form*

$$f(x) = g(x)+h(x),$$

where $g^{(e_j)}(x) = 0$ and $h^{(m)}(x) = 0$ in I. Moreover, if $f(x)$ is a smooth, continuous or integrable function in I, we may assume the same is true of $g(x)$ and $h(x)$.

In fact, the interval I is inside an interval I_0 inside O. If $f^{(m+e_j)}(x) = 0$ in O, then, by 4.3.2, $f^{(m)}(x)$ is constant in ξ_j in I_0. By 4.3.5, there exist an order $k = (\varkappa_1, \ldots, \varkappa_q)$ with $\varkappa_j = 0$, and a continuous function $F(x)$, constant in ξ_j, such that $F^{(k)}(x) = f^{(m)}(x)$ in I. We may further assume that $k \geqslant m$. The distributions $g(x) = F^{(k-m)}(x)$ and $h(x) = f(x)-g(x)$ then have the required properties.

If $f(x)$ is a smooth, continuous or integrable function, we can obtain $g(x)$ directly from $f(x)$, by replacing the variable ξ_j in $f(x)$ by a constant γ. Then $g(x)$ is smooth, continuous or (with suitably chosen γ) integrable, respectively, and so is $h(x) = f(x)-g(x)$. Moreover, $g^{(e_j)}(x) = 0$. It remains to verify that $h^{(m)}(x) = 0$, i.e., that $f^{(m)}(x) = g^{(m)}(x)$. This is clear when $f(x)$ is smooth. If $f(x)$ is continuous, there exists a sequence of smooth functions $\varphi_n(x)$, almost uniformly convergent to $f(x)$, such that $\varphi_n^{(m+e_j)}(x) = 0$. Replacing the variable ξ_j in $\varphi_n(x)$ by γ we get a sequence of smooth functions $\psi_n(x)$ almost uniformly convergent to $g(x)$. Since

$$f(x) = [\varphi_n(x)], \quad g(x) = [\psi_n(x)]$$

and

$$\varphi_n^{(m)}(x) = \psi_n^{(m)}(x),$$

it follows that

$$f^{(m)}(x) = [\varphi_n^{(m)}(x)] = [\psi_n^{(m)}(x)] = g^{(m)}(x).$$

If $f(x)$ is integrable, the proof is the same except that almost uniform convergence must be replaced by L-convergence.

4.5.3. *If $f^{(m+e_j)}(x) = 0$ in O, where $m = (\mu_1, \ldots, \mu_q)$ and $\mu_j > 0$, then in every interval I inside O the distribution $f(x)$ can be represented in the form*

$$f(x) = \xi_j g(x)+h(x),$$

where $g^{(m)}(x) = 0$ and $h^{(m)}(x) = 0$ in I. Moreover, when $f(x)$ is a smooth, continuous or integrable function, we may assume that the same is true of $g(x)$ and $h(x)$.

In fact, let

$$g(x) = \frac{1}{\mu_j \eta}(f(x+\eta e_j)-f(x)), \quad h(x) = f(x)-\xi_j g(x). \tag{1}$$

Provided η is sufficiently small, $g(x)$ and $h(x)$ are defined in I. Moreover, if $f(x)$ is smooth, continuous or integrable, so are $g(x)$ and $h(x)$. Since $f^{(m+e_j)}(x) = 0$, we have $f^{(m)}(x+\eta e_j) = f^{(m)}(x)$ in I, by 4.5.1. This implies that $g^{(m)}(x) = 0$. Differentiating the second equality in (1), we find that

$$h^{(m)}(x) = f^{(m)}(x) - \mu_j g^{(m-e_j)}(x)$$

$$= f^{(m)}(x) + \frac{1}{\eta} f^{(m-e_j)}(x) - \frac{1}{\eta} f^{(m-e_j)}(x+\eta e_j).$$

Since $f^{(m+e_j)}(x) = 0$, the right hand side vanishes, by 4.5.1.

4.5.4. *The equality $f^{(m)}(x) = 0$ holds in O iff in every interval I inside O the distribution $f(x)$ can be represented in the form*

$$f(x) = \sum_{i=0}^{\mu_1-1} \xi_1^i f_{1i}(x) + \ldots + \sum_{i=0}^{\mu_q-1} \xi_q^i f_{qi}(x), \tag{2}$$

where the distributions $f_{ji}(x)$ are constant in ξ_i. Moreover, if $f(x)$ is a smooth or continuous function, we may assume that the same is true of all the coefficients $f_{ji}(x)$. (Note that, if $\mu_j = 0$ for some j in formula (2), then the corresponding sum is taken to be 0).

In fact, it is a matter of easy verification that, if (2) holds in I, then $f^{(m)}(x) = 0$ in I. Since I is arbitrary, we have $f^{(m)}(x) = 0$ in O. Conversely, if $f^{(m)}(x) = 0$ in O, we can prove (2) by induction. We first remark that (2) is trivially satisfied when $m = 0$. Suppose that it is satisfied for some $m \geqslant 0$. It suffices to show that the corresponding formula also holds for $m+e_j$. In fact, if $\mu_j = 0$, this follows from 4.5.2; if $\mu_j > 0$, it follows from 4.5.3.

We finally remark that the representation (2) is not unique. For instance, if $m = (1, 1)$ and $f(x) = \xi_1 + \xi_2$, we can also write $f(x) = (\xi_1 + 1) + (\xi_2 - 1)$.

Part III

Advanced theory of distributions

Introduction to Part III

Chapters 1–5 of Part III are concerned with functions and so they are not concerned with the theory of distributions proper. They have been included in the book because they contain some new results necessary for the understanding of the sequel.

The important concept of the convolution of two distributions is discussed extensively in Chapter 6. Tempered distributions are defined as the tempered derivatives of square integrable functions, in Chapter 7. This approach enables us to simplify the theory of their Hermite expansions and Fourier Transforms, which is the subject of Chapter 8. In turn, Hermite expansions link tempered distributions with the theory of Köthe spaces, presented in Chapters 10 and 11. This enables us to give relatively simple proofs of important theorems on the equivalence of weak and strong convergence of distributions.

The remaining Chapters are mainly devoted to applications. A mathematical justification of a formula containing δ^2 may be of interest to physicists.

In the Appendix the principle of multidimensional induction and recursive definition is carefully explained and illustrated with examples which are useful in the theory of distributions.

It is our pleasant duty to express our gratitude to our colleagues A. Kamiński and K. Skórnik for their assistance in preparing the manuscript of the book and for remarks which led to improvements of the text in several places.

1. Convolution

1.1. CONVOLUTION OF TWO FUNCTIONS

By the *convolution* $f*g$ we mean the integral

$$\int_{R^q} f(x-t)g(t)\,dt. \tag{1}$$

The convolution exists at a point x, whenever the product $f(x-t)g(t)$ is integrable (Lebesgue) with respect to t. In order that the convolution may be defined at as many points as possible we adopt the following convention: if one of the factors $f(x-t)$ or $g(t)$ is 0 for some x and t, then the product $f(x-t)g(t)$ is taken to be 0, even if the second factor is not defined.

Since the integral (1) is taken in the sense of Lebesgue, the existence of the convolution $f*g$ implies the existence of the convolution $|f|*|g|$. Conversely, if the convolution $|f|*|g|$ exists and, moreover, the product $f(x-t)g(t)$ is measurable, then the convolution $f*g$ exists. If we know that both the functions f and g are measurable, then the convolution $f*g$ exists, iff $|f|*|g|$ exists.

By means of a simple substitution in the integral, we get the law of commutativity

$$f*g = g*f, \tag{2}$$

provided at least one of the convolutions exists. Moreover, if the convolution $f*g$ exists, then, given any real number λ, the convolutions $f*(\lambda g)$ and $(\lambda f)*g$ exist and the following equalities hold:

$$(\lambda f)*g = f*(\lambda g) = \lambda(f*g). \tag{3}$$

If the convolutions $f*g$ and $f*h$ exist, then the convolution $f*(g\pm h)$ also exists and we have

$$f*(g\pm h) = f*g\pm f*h. \tag{4}$$

In general, convolution is not associative. In fact, let $q = 1$ and $f(x) = 1$, $g(x) = -xe^{-x^2}$ and $h(x) = \int_{-\infty}^{x} e^{-t^2}dt$. It is easy to see that $f*g = 0$. This implies that $(f*g)*h = 0$ as well. On the other hand, a straightforward calculation gives

$$g*h = \frac{1}{2}\sqrt{\frac{\pi}{2}}\,e^{-x^2/2}$$

and hence $f*(g*h) = \frac{1}{2}\pi$. In this case we thus have $(f*g)*h \neq f*(g*h)$, i.e., convolution is not associative in general. In the next section we state some sufficient conditions for associativity.

1.2. CONVOLUTION OF THREE FUNCTIONS

By the convolution $f*g*h$ of three functions we mean the double integral

$$\iint_{R^{2q}} f(x-t)g(t-u)h(u)\,dt\,du. \tag{1}$$

The convolution exists at a point $x \in \mathbf{R}^q$, whenever the product $f(x-t)g(t-u)h(u)$ is integrable (Lebesgue) over $\mathbf{R}^{2q}$. As before, we understand that if one of the factors $f(x-t)$, $g(t-u)$, or $h(u)$ is 0 for some x, t and u, then the product is always taken to be 0, even if the remaining factors are not defined.

Since the integral (1) is meant in the sense of Lebesgue, the existence of $f*g*h$ implies the existence of $|f|*|g|*|h|$. The converse implication also holds, provided the product $f(x-t)g(t-u)h(u)$ is measurable. If we know that all the functions f, g and h are measurable, then $f*g*h$ exists, iff $|f|*|g|*|h|$ exists.

The convolution of three functions has properties similar to those of the convolution of two functions. In particular, the commutativity law holds:

$$f*g*h = f*h*g = g*f*h = g*h*f = h*f*g = h*g*f, \tag{2}$$

which easily follows by making the appropriate substitutions.

1.2.1. THEOREM. *If the convolution $f*g*h$ exists, then*

$$f*g*h = f*(g*h) = (f*g)*h. \tag{3}$$

More precisely: if $f*g*h$ exists at some point x, then the convolutions $g_1 = f*g$ and $g_2 = g*h$ are defined on sets such that the convolutions $f*g_2$ and g_1*h exist at x. Moreover, the equality $f*g*h = f*g_2 = g_1*h$ holds at x.

PROOF. Suppose $f*g*h$ exists at some point x. Then, by the Fubini theorem, the function

$$G(t) = \int_{R^q} f(x-t)g(t-u)h(u)\,du \tag{4}$$

is defined for almost every t, and is integrable in $\mathbf{R}^q$. Furthermore we have

$$f*g*h = \int_{R^q} G(t)\,dt. \tag{5}$$

If t is a point at which $f(x-t) \neq 0$ and the integral (4) exists, we can write

$$G(t) = f(x-t)\int_{R^q} g(t-u)h(u)\,du \tag{6}$$

and the integral in (6) is defined for almost every t just as in (4). If $f(x-t) = 0$, then the product (6) exists and is equal to 0, according to our convention, no matter whether the integral in (6) exists or not. By (5) and (6) we can write

$$f*g*h = \int_{R^q} f(x-t)\,dt \int_{R^q} g(t-u)h(u)\,du = f*(g*h).$$

From the above and (1) it follows that

$$f*g*h = h*f*g = h*(f*g) = (f*g)*h,$$

which completes the proof.

1.3. ASSOCIATIVITY OF CONVOLUTION

As we saw at the end of Section 1.1, the associativity law

$$f*(g*h) = (f*g)*h \tag{1}$$

does not hold, in general. On the other hand, 1.2.1 tells us that (1) does hold whenever $f*g*h$ exists. For applications it is important to have sufficient conditions for (1) in which the existence of the convolution of three functions is not postulated.

1.3.1. THEOREM. *If the iterated convolution $|f|*(|g|*|h|)$ exists at a point x and the functions f, g, h are measurable, then* (1) *holds at x. More precisely, if the convolution $|g|*|h|$ is defined on a set such that the convolution $|f|*(|g|*|h|)$ exists at x, and f, g, h are measurable, then the convolutions $g_1 = f*g$ and $g_2 = g*h$ are defined on sets such that the convolutions $f*g_2$ and g_1*h exist and are equal at x.*

PROOF. We can write

$$|f|*(|g|*|h|) = \int_{R^q} dt \int_{R^q} |f(x-t)g(t-u)h(u)|\,du.$$

This implies the existence of $f*g*h$ by the Tonelli theorem, and our assertion now follows, by 1.2.1.

1.3.2. THEOREM. *If $|f|*(|g|*|h|)$ exists almost everywhere and the functions f, g, h are measurable, then* (1) *holds almost everywhere. Moreover, if the integral $\int|f|$, $\int|g|$ or $\int|h|$ respectively does not vanish, then the convolution $g*h, f*h$ or $f*g$ respectively must exist almost everywhere.*

PROOF. The first part of the assertion follows immediately from 1.3.1. We can state, furthermore, as in the preceding proof, that the convolution $f*g*h$ exists a.e. (almost everywhere). Assume that $\int|f| > 0$ and that the convolution $g*h$ is not defined almost everywhere, so that if A denotes the set of points at which $g*h$ does not exist, then A is not of measure 0. Let f_1 and f_2 be two functions such that $f_1 = f_2 = g*h$ for $x \notin A$ and $f_1 - f_2 = 1$ for $x \in A$. Since the convolution $|f|*(g*h)$ does not depend on the values of $g*h$ on A, we have

$$|f|*f_1 = |f|*f_2 \quad \text{a.e.}$$

Hence,

$$\int (|f|*(f_1 - f_2)) = 0,$$

where the integral extends over the whole of $\mathbf{R}^q$. On the other hand, we find by the Tonelli theorem that

$$\int (|f|*(f_1 - f_2)) = \int |f| \int (f_1 - f_2) > 0.$$

This contradiction proves the assertion for $\int |f| > 0$. In the cases $\int |g| > 0$ or $\int |h| > 0$, the assertion can be deduced from the preceding result, using commutativity and associativity.

1.4. CONVOLUTION OF A LOCALLY INTEGRABLE FUNCTION WITH A SMOOTH FUNCTION OF BOUNDED CARRIER

Let f be a locally integrable function in an open set $O \subset \mathbf{R}^q$ and φ a smooth function in $\mathbf{R}^q$ such that $\varphi(x) = 0$ for $|x| \geqslant \alpha > 0$.

O_α ($\alpha > 0$) will denote the set of all points $x \in \mathbf{R}^q$ whose distance from the set O is less than α. $O_{-\alpha}$ ($\alpha > 0$) denotes the set of all points $x \in O$ whose distance from the boundary of O is greater than α (if $O = \mathbf{R}^q$ then $O_{-\alpha}$ is taken to be $\mathbf{R}^q$). Provided α is sufficiently small, the set $O_{-\alpha}$ is not empty. Clearly, $O_{-\alpha} \subset O_\alpha$, and the difference $O_\alpha \setminus O_{-\alpha}$ is the set of all $x \in \mathbf{R}^q$ whose distance from the boundary of O is not greater than α. The integral

$$\int f(x-t)\varphi(t)\,dt \tag{1}$$

is defined for $x \in O_{-\alpha}$ and consequently, the convolution $f*\varphi$ exists in $O_{-\alpha}$.

By 1.1(2) and 1.1(3) we have

$$f*\varphi = \varphi*f \quad \text{in } O_{-\alpha}, \tag{2}$$

$$(\lambda f)*\varphi = f*(\lambda\varphi) = \lambda(f*\varphi) \quad \text{in } O_{-\alpha}, \tag{3}$$

where λ is any real number.

If f_1 and f_2 are locally integrable functions in O, then, by 1.1(4), we have

$$(f_1 + f_2)*\varphi = f_1*\varphi + f_2*\varphi \quad \text{in } O_{-\alpha}. \tag{4}$$

If φ_1 and φ_2 are smooth functions in $\mathbf{R}^q$ which vanish for $|x| \geqslant \alpha$, we have

$$f*(\varphi_1 + \varphi_2) = f*\varphi_1 + f*\varphi_2 \quad \text{in } O_{-\alpha}. \tag{5}$$

If φ_1 and φ_2 are smooth functions in R^q such that $\varphi_1(x) = 0$ and $\varphi_2(x) = 0$ for $|x| \geqslant \alpha > 0$, the double integral

$$\iint_{R^{2q}} f(x-t)\varphi_1(t-u)\varphi_2(u)\,dt\,du$$

is defined for $x \in O_{-2\alpha}$ and consequently the convolution $f*\varphi_1*\varphi_2$ exists in $O_{-2\alpha}$. Hence, by 1.2.1, the equalities

$$f*\varphi_1*\varphi_2 = (f*\varphi_1)*\varphi_2 = f*(\varphi_1*\varphi_2) \tag{6}$$

hold in $O_{-2\alpha}$.

For any order k we have

$$(f*\varphi)^{(k)} = f*\varphi^{(k)} \quad \text{in } O_{-\alpha}. \tag{7}$$

All what has been said above, except for formula (7), holds also when the smoothness of φ, φ_1, φ_2 is replaced by the assumption that f, f_1, f_2 are smooth. More exactly, we may assume that f, f_1, f_2 are smooth in O and φ, φ_1, φ_2 are measurable and bounded in R^q, vanishing for $|x| > \alpha > 0$. Then formulae (2)–(6) hold. Instead of (7) we then have for every k

$$(f*\varphi)^{(k)} = f^{(k)}*\varphi \quad \text{in } O_{-\alpha}. \tag{8}$$

Let us still mention that formulae (2)–(8) also hold under much weaker assumptions. However, the present section is, in fact, a preparation to the next chapter on delta sequences and a greater generality is not needed.

2. Delta-sequences and regular sequences

2.1. DELTA-SEQUENCES

We use the following notation:

N = set of all positive integers;

P = set of all non-negative integers;

N^q = set of all positive integer points of R^q (i.e., points all of whose coordinates are positive integers);

P^q = set of all non-negative integer points of R^q;

B^q = set of all integer points in R^q.

By a *delta-sequence* in R^q we mean any sequence of smooth functions δ_n with the following properties:

(i) There is a sequence of positive numbers α_n, converging to 0, such that $\delta_n(x) = 0$ for $|x| \geqslant \alpha_n$, $n \in N$;

(ii) $\int \delta_n = 1$ for $n \in N$;

(iii) For every $k \in P^q$ there is a positive integer M_k such that

$$\alpha_n^k \int |\delta_n^{(k)}| \leqslant M_k \quad \text{for} \quad n \in N.$$

In (i), the symbol $|x|$ denotes the distance of the point $x \in R^q$ from the origin, i.e., $|x| = \sqrt{\xi_1^2 + \ldots + \xi_q^2}$. The symbol α_n^k denotes $\alpha_n^{\varkappa_1 + \cdots + \varkappa_q}$, where $k = (\varkappa_1, \ldots, \varkappa_q)$.

As an example of a delta sequence we can take

$$\delta_n(x) = \alpha_n^{-q} \Omega(\alpha_n^{-1} x), \tag{1}$$

where Ω is any smooth function of bounded carrier, such that $\int \Omega = 1$ and $\alpha_n \neq 0$ is an arbitrary sequence tending to 0.

REMARK. In Part II delta sequences were defined as sequences of non-negative smooth functions with properties (i) and (ii). However, example (1) shows that property (iii) is natural. Moreover, this property is of importance in the more delicate investigations.

It is easy to see that if δ_{1n} and δ_{2n} are delta-sequences, then the interlaced sequence

$$\delta_{11}, \delta_{21}, \delta_{12}, \delta_{22}, \delta_{13}, \ldots \tag{2}$$

is also a delta-sequence.

2.1.1. THEOREM. *The convolution of two delta-sequences is a delta-sequence.*

PROOF. Let δ_{1n} and δ_{2n} be two delta-sequences. We have to prove that $\delta_n = \delta_{1n} * \delta_{2n}$ is another delta-sequence. In fact, δ_n are smooth functions. Moreover, if $\delta_{1n}(x) = 0$ for $|x| \geqslant \alpha_{1n}$ and $\delta_{2n}(x) = 0$ for $|x| \geqslant \alpha_{2n}$, then $\delta_n(x) = 0$ for $|x| \geqslant \alpha_{1n} + \alpha_{2n}$. This implies that δ_n satisfies condition (i). Clearly

$$\int \delta_n = \int dx \int \delta_{1n}(x-t)\,\delta_{2n}(t)\,dt = \int \delta_{2n}(t)\,dt \int \delta_{1n}(x-t)\,dx = 1 \cdot 1,$$

so that condition (ii) is satisfied. Finally, we have

$$\int |\delta_n^{(k)}| = \int dx \int |\delta_{1n}^{(k)}(x-t)|\,|\delta_{2n}(t)|\,dt = \int |\delta_{2n}(t)|\,dt \int |\delta_{1n}^{(k)}(x-t)|\,dx.$$

Hence

$$\alpha_{1n}^k \int |\delta_n^{(k)}| \leqslant M_{1k} M_{20}.$$

Similarly we obtain

$$\alpha_{2n}^k \int |\delta_n^{(k)}| \leqslant M_{10} M_{2k}.$$

Since $(\alpha_{1n} + \alpha_{2n})^k \leqslant 2^k(\alpha_{1n}^k + \alpha_{2n}^k)$, the two last inequalities imply

$$(\alpha_{1n} + \alpha_{2n})^k \int |\delta_n^{(k)}| \leqslant M_k = 2^k(M_{1k} M_{20} + M_{10} M_{2k}),$$

which proves condition (iii).

2.2. REGULAR SEQUENCES

If δ_n is a delta-sequence in $\boldsymbol{R}^q$ and f is a locally integrable function in an open set O, then the sequence $f_n = f * \delta_n$ consists of smooth functions in open subsets $O_{-\alpha_n} \subset O$ such that $O_{-\alpha_n} \to O$, i.e., for each $x \in O$, we have $x \in O_{-\alpha_n}$ for sufficiently large n. The sequence f_n will be called a *regular sequence* for the locally integrable function f.

2.2.1. THEOREM. *If f is a continuous function in O, then its regular sequence $f * \delta_n$ converges to f, almost uniformly in O.*

This theorem was proved in Part II as 3.1.1. It implies

2.2.2. THEOREM. *If f is a smooth function in O, then, for every $k \in \boldsymbol{P}^q$, the sequence of derivatives $(f * \delta_n)^{(k)}$ converges to $f^{(k)}$, almost uniformly.*

To prove this, it suffices to remark that $f_n^{(k)} = f^{(k)} * \delta_n$.

2.2.3. THEOREM. *If f is a locally integrable function in O, then $f*\delta_n$ converges locally in mean to f, i.e., given any interval I inside O, we have*

$$\int_I |f*\delta_n - f| \to 0.$$

PROOF. Let I be any given interval inside O and let α be a positive number such that I_α (see Section 1.4) is inside O. There is an index n_0 such that $\alpha_n < \alpha$ and $I_{\alpha_n} \subset O_{-\alpha_n}$ for $n > n_0$. Clearly $I_{\alpha_n} \subset I_\alpha$. Now a well known theorem of Lebesgue tells us that, if f is integrable in I_α and I is inside I_α, then

$$\int_I |f(x-t)-f(x)|\,dx \to 0, \quad \text{as} \quad t \to 0.$$

Thus, if $|t| \leqslant \alpha$ and $n > n_0$, we have

$$\int_I |f(x-t)-f(x)|dx \leqslant \varepsilon_n \quad \text{where} \quad \varepsilon_n \to 0.$$

For $x \in I$ and $n > n_0$ we obtain

$$|f*\delta_n - f| = \left|\int (f(x-t)-f(x))\,\delta_n(t)\,dt\right| \leqslant \int |f(x-t)-f(x)|\,|\delta_n(t)|\,dt$$

and hence

$$\int_I |f*\delta_n - f| \leqslant \int |\delta_n(t)|\,dt \int_I |f(x-t)-f(x)|dx \leqslant M_0 \varepsilon_n,.$$

for $\delta_n(t) = 0$ if $|t| \geqslant \alpha_n$, which implies our assertion.

It is known that if f and g are integrable (Lebesgue) functions in $\mathbf{R}^q$, then the convolution $f*g$ is also an integrable function in $\mathbf{R}^q$. Hence in particular, if f is integrable in $\mathbf{R}^q$, all the functions $f*\delta_n$ are also integrable in $\mathbf{R}^q$.

2.2.4. THEOREM. *If f is a function, integrable in $\mathbf{R}^q$, then $f*\delta_n$ converges in mean to f, i.e.,*

$$\int |f*\delta_n - f| \to 0.$$

PROOF. We first have

$$\int |f*\delta_n - f| \leqslant \int \left|\int (f(x-t)-f(x))\,\delta_n(t)\,dt\right| dx.$$

By the Lebesgue theorem, we have

$$\int |f(x-t)-f(x)|\,dx \to 0, \quad \text{as} \quad t \to 0.$$

Hence

$$\int |f(x-t)-f(x)|dx < \varepsilon_n, \quad \text{for} \quad |t| \leqslant \alpha_n,$$

where $\alpha_n \to 0$. Since $\delta_n(t) = 0$ for $|t| \geqslant \alpha_n$, we obtain

$$\int |f*\delta_n - f| \leqslant \int |\delta_n(t)|dt\,\varepsilon_n = M_0\,\varepsilon_n \to 0.$$

2.3. CONVOLUTION OF A CONVERGENT SEQUENCE WITH A DELTA-SEQUENCE

We now generalize the preceding theorem, replacing f by a convergent sequence f_n. As before, δ_n continues to denote a delta-sequence.

2.3.1. THEOREM. *If a sequence of continuous functions f_n converges to f, almost uniformly in O, then the sequence of convolutions $f_n * \delta_n$ also converges to f, almost uniformly in O.*

The proof of this theorem is given in Part II (3.1.2).

2.3.2. THEOREM. *If f_n and f are smooth functions such that, for every $k \in \boldsymbol{P}^q$, the sequence $f_n^{(k)}$ converges to $f^{(k)}$, almost uniformly in O, then $(f_n * \delta_n)^{(k)}$ converges to $f^{(k)}$, almost uniformly in O.*

PROOF. It suffices to note that $(f_n * \delta_n)^{(k)} = f_n^{(k)} * \delta_n$ and to apply 2.3.1.

2.3.3. THEOREM. *If f_n is a sequence of locally integrable functions which converges, in O, to f locally in mean, then the sequence $f_n * \delta_n$ converges locally in mean to f in O.*

PROOF. Clearly,

$$f_n * \delta_n = f * \delta_n + (f_n - f) * \delta_n .$$

By 2.2.3, we have to show that $(f_n - f) * \delta_n$ converges locally in mean to 0.

Let I be any given interval inside O and let α be a positive real number such that I_α is inside O. There is an index n_0 such that $\alpha_n < \alpha$ and $I_{\alpha_n} \subset O_{-\alpha_n}$ for $n > n_0$. Clearly $I_{\alpha_n} \subset I_\alpha$ for $n > n_0$. Since $\delta_n(x-t) = 0$ for $x \in I$ and $t \notin I_{\alpha_n}$, we have

$$\int_I |(f_n - f) * \delta_n| \, dx \leqslant \int_I dx \int_{I_{\alpha n}} |f_n(t) - f(t)| |\delta_n(x-t)| \, dt$$

$$= \int_{I_{\alpha n}} |f_n(t) - f(t)| \, dt \int_I |\delta_n(x-t)| \, dx$$

$$\leqslant \int_{I_\alpha} |f_n(t) - f(t)| \, dt M_0 \to 0, \quad \text{as } n \to \infty .$$

2.3.4. THEOREM. *If a sequence of functions f_n, integrable in $\boldsymbol{R}^q$ converges in mean to f, then the sequence $f_n * \delta_n$ also converges in mean to f.*

PROOF. Abbreviating $\int_{R^q}$ to $\int$, we have

$$\int |(f_n - f) * \delta_n| \leqslant \int dx \int |f_n(t) - f(t)| |\delta_n(x-t)| dt$$

$$= \int |f_n(t) - f(t)| dt \int |\delta_n(x-t)| dx$$

$$\leqslant \int |f_n - f| M_0 \to 0, \quad \text{as } n \to \infty.$$

Thus $(f_n - f) * \delta_n$ converges in mean to 0. This, together with 2.2.4, implies our assertion.

It is worth remarking that, in the theorems of this section and of Section 2.2, hypothesis (iii) was used in a weaker form only, viz., with $k = 0$. This means that a larger class could be admitted as delta sequences for the purposes of these sections.

3. Existence theorems for convolutions

3.1. CONVOLUTIVE DUAL SETS

All functions in this section are assumed measurable.

Given any set U of functions in $\boldsymbol{R}^q$, we denote by U^* the set of all measurable functions g such that the convolution $f*g$ exists a.e. for each $f \in U$. U^* is called the *convolutive dual set* of U. Note that, if $f*g$ exists, then $|f|*|g|$ exists, because the integral $\int\limits_{R} f(x-t)g(t)\,dt$ defining the convolution is taken in the sense of Lebesgue. More generally, if $|g| \leqslant |g_1|$, the existence of the convolution $f*g_1$ implies the existence of both the convolutions $f*g$ and $|f|*|g|$, so that if $|g| \leqslant |g_1|$ and $g_1 \in U^*$, then $g \in U^*$.

A set V of functions g will be called *standard* iff the conditions $|g| \leqslant |g_1|$, $g_1 \in V$ imply $g \in V$. We thus have

3.1.1. THEOREM. *A convolutive dual set is standard.*

EXAMPLE 1. Let U be the set of all constant functions. We show that in this case U^* is the set of all integrable functions. In fact, if $f \in U$ and g is integrable, we have

$$\int\limits_{R^q} f(x-t)g(t)\,dt = f(0)\int\limits_{R^q} g(t)\,dt$$

and consequently, the integral

$$\int\limits_{R^q} f(x-t)g(t)\,dt \tag{1}$$

exists for every x. Thus $g \in U^*$. Conversely, if $g \in U^*$, then the integral (1) exists for every $f \in U$, and in particular for $f = 1$. Hence the integral $\int\limits_{R^q} g(t)\,dt$ exists, i.e., g is integrable.

EXAMPLE 2. Let U be the set of all bounded functions. Then U^* is the set of all integrable functions. In fact, if $f \in U$ and g is integrable, then the integral

$$\int\limits_{R^q} f(x-t)g(t)\,dt \tag{2}$$

exists for every x. Thus $g \in U^*$. The converse is proved as in the preceding example.

It is worth noting that in Examples 1 and 2 we obtain the same dual set, although the initial set U is much larger in Example 2.

3.1.2. THEOREM. *If U has the property that $f \in U$, $g \in U^*$ imply $|f| * |g| \in U$, then U^* is a commutative semi-group under convolution.*

PROOF. Let $f \in U$ and $g, h \in U^*$. By hypothesis, we have $|f| * |g| \in U$. Hence also $(|f| * |g|) * |h| \in U$. By 1.3.1, it follows that the convolution $f * (g * h)$ exists. Since f is arbitrary within U, we have $g * h \in U^*$, by the definition of U^*.

If $f, g, h \in U^*$, then also $|f|, |g|, |h| \in U^*$ and $(|f| * |g|) * |h| \in U^*$. By 1.3.1 we therefore have $(f * g) * h = f * (g * h)$. The commutativity was stated earlier in 1.1(2) and the proof is complete.

REMARK. From Example 1 and 3.1.2 it follows that the set of all integrable functions is a comutative semi-group under convolution, i.e., the convolution of two integrable functions is another integrable function.

Let U_0 be a class of functions in $\boldsymbol{R}^q$ such that for any $u \in U_0$ there is a positive function $v \in U_0$ such that

$$|u(x+y)| \leqslant v(x)v(y).$$

The class U_0 may consist of a single function, e.g., e^x. Another example of a class U_0 is the set of all polynomials.

In the sequel, $\bar{v}$ denotes the function given by $\bar{v}(x) = v(-x)$.

3.1.3. THEOREM. *Let U be a standard set and U^* its convolutive dual set (which is also standard). Let V be the set of functions in $\boldsymbol{R}^q$ defined by*

$$f \in V, \quad \textit{iff } \frac{f}{u} \in U \textit{ for some positive } u \in U_0.$$

Then V is standard and V^* is the set of functions in $\boldsymbol{R}^q$ such that

$$g \in V^*, \quad \textit{iff } \bar{v}g \in U^* \textit{ for every positive } v \in U_0.$$

Moreover, if U has the property that $f \in U$, $g \in U^*$ imply $f * g \in U$, then V has a similar property, i.e., $f \in V$, $g \in V^*$ imply $f * g \in V$.

PROOF. Let g be a function such that $\bar{v}g \in U^*$ for any positive $v \in U_0$. We prove that $g \in V^*$. In fact, for any $f \in V$, we have $\dfrac{f}{u} \in U$ for some positive $u \in U_0$. There is a positive $v \in U_0$ such that $|u(x-t)| \leqslant v(x)v(-t)$. Hence

$$|f*g| \leqslant \int_{R^q} |f(x-t)g(t)|\,dt = \int_{R^q} \left|\frac{f(x-t)}{u(x-t)}\right| |u(x-t)g(t)|\,dt$$

$$\leqslant v(x) \int_{R^q} \left|\frac{f(x-t)}{u(x-t)}\right| |v(-t)g(t)|\,dt.$$

Since $\frac{f}{u} \in U$ and $\bar{v}g \in U^*$, the last integral exists a.e., which implies the existence of the convolution $f*g$. Thus $g \in V^*$.

Assume now that $g \in V^*$. This means that $f*g$ exists a.e. for each $f \in V$.

Let $u \in U_0$. We show that $\bar{u}g \in U^*$, i.e., that the convolution $f*\bar{u}g$ exists a.e. for any $f \in U$. In fact, there is a function $v \in U_0$ such that $|u(t-x)| \leqslant v(t)v(-x)$. Hence

$$|(f*\bar{u}g)(x)| \leqslant \int_{R^q} |u(t-x)g(x-t)f(t)|\,dt$$

$$\leqslant v(-x) \int_{R^q} |g(x-t)v(t)f(t)|\,dt = v(-x)(|g|*|vf|)(x).$$

But $\frac{vf}{v} \in U$, and so $vf \in V$ and $vf*g$ exists a.e. This implies that $f*\bar{u}g$ exists a.e.; thus $\bar{u}g \in U^*$, which completes the proof of the first part of our theorem.

Now, let $g \in V^*$ and $f \in V$. There are $u, v \in U_0$ such that $\frac{f}{u} \in U$, $|u(x-t)| \leqslant v(x)v(-t)$ and v is a positive function, and we have

$$\frac{|f*g|}{v} \leqslant \frac{1}{v(x)} \int_{R^q} \left|\frac{f(x-t)}{u(x-t)}\right| |u(x-t)g(t)|\,dt$$

$$\leqslant \int_{R^q} \left|\frac{f(x-t)}{u(x-t)}\right| |\bar{v}(t)g(t)|\,dt = \left|\frac{f}{u}\right| * |\bar{v}g|.$$

Moreover, we have $\bar{v}g \in U^*$, by the first part of 3.1.3. Since $\frac{f}{u} \in U$, $\bar{v}g \in U^*$ and the sets U, U^* are standard, we have $\left|\frac{f}{u}\right| \in U$ and $|\bar{v}g| \in U^*$. We can thus write $\left|\frac{f}{u}\right| * |\bar{v}g| \in U$, by hypothesis. Since $\left|\frac{f*g}{v}\right| \leqslant \left|\frac{f}{u}\right| * |\bar{v}g| \in U$ and U is standard, we have $\frac{f*g}{v} \in U$. This means that $f*g \in V$. Since $f*g \in V$ for any $f \in V$ and $g \in V^*$, V^* is a semi-group, by 3.1.2, and the proof of our theorem is complete.

EXAMPLE 3. If U_0 consists of all polynomials and U consists of all bounded functions, then V is the set of all so-called *slowly increasing* functions, i.e., $f \in V$, iff there is a polynomial p such that $|f| \leqslant p$. The convolutive semi-group V^* consists of all so-called *rapidly decreasing* functions, i.e., $g \in V^*$, iff the product gp is bounded for every polynomial p. Since $f * g \in U$ for every $f \in U$ and $g \in U^*$ (U^* is the set of all integrable functions), the convolution of a slowly increasing function with a rapidly decreasing function is a slowly increasing function, by the second part of 3.1.3. Moreover, the convolution of two rapidly decreasing functions is again a rapidly decreasing function.

3.2. CONVOLUTION OF FUNCTIONS WITH COMPATIBLE CARRIERS

Given two functions f and g in $\boldsymbol{R}^q$, it may happen that, for every $x \in \boldsymbol{R}^q$, the set S^x of all points t such that $f(x-t)g(t) \neq 0$ is bounded. If the functions f and g are continuous, the convolution $f * g$ exists trivially, for the integral

$$\int_{\boldsymbol{R}^q} f(x-t)g(t)\,dt$$

in this case extends over the bounded set S^x. The set S^x depends on the points x and, clearly, on the carriers X and Y of f and g, i.e., on the sets of points at which f and g are different from 0, respectively. The set S^x is completely determined by the point x and the sets X, Y, whatever the functions f and g are. If we denote by $X(t)$ and $Y(t)$ the characteristic functions of X and Y, then S^x is the set consisting of the points t such that $X(x-t)Y(t) = 1$.

We say that the sets X and Y are *compatible*, iff for every bounded interval $I \subset \boldsymbol{R}^q$ there is another bounded interval $J \subset \boldsymbol{R}^q$ such that $x \in I$ implies $S^x \subset J$.

EXAMPLE 1. If the set X is arbitrary and the set Y is bounded, then X, Y are compatible.

EXAMPLE 2. If all the points of X and Y are positive, i.e., have positive coordinates, then X, Y are compatible.

3.2.1. THEOREM. *Let f and g be locally integrable functions in $\boldsymbol{R}^q$ whose carriers are contained in the compatible sets X and Y, respectively. Then the convolution $f * g$ exists almost everywhere and represents a locally integrable function in $\boldsymbol{R}^q$ whose carrier is contained in $X+Y$ (i.e., in the set of all points $x+y$ where $x \in X$ and $y \in Y$).*

PROOF. Let I be any given bounded interval and J another bounded interval such that $x \in I$ implies $S^x \subset J$. There is a bounded interval K such that $J \subset K$ and $I-J \subset K$, where $I-J$ is the set of all points $x-t \in \boldsymbol{R}^q$ with $x \in I$ and $t \in J$.

Let $f_1(x) = f(x)K(x)$ and $g_1(x) = g(x)K(x)$, where $K(x)$ is the characteristic function of K. Since the functions f_1 and g_1 vanish outside K, they are integrable in $\boldsymbol{R}^q$. The convolution $f_1 * g_1$ therefore exists almost everywhere in $\boldsymbol{R}^q$ and represents an integrable function in $\boldsymbol{R}^q$ (cf. Remark of 3.1). If $x \in I$, then $f(x-t)g(t) = 0$ for $t \notin J$, and $K(x-t)K(t) = 1$ for $t \in J$. We thus have

$$\int_{\boldsymbol{R}^q} f(x-t)g(t)\,dt = \int_{\boldsymbol{R}^q} f(x-t)K(x-t)g(t)K(t)\,dt = \int_{\boldsymbol{R}^q} f_1(x-t)g_1(t)\,dt. \tag{1}$$

Thus, the integral on the left hand side exists a.e. in I and represents an integrable function there. Since the interval I was chosen arbitrarily, $f*g$ exists almost everywhere in $\boldsymbol{R}^q$ and represents a locally integrable function in $\boldsymbol{R}^q$.

It remains to verify that the integral (1) vanishes if $x \notin X+Y$. In fact, given any $t \in \boldsymbol{R}^q$ we then have either $x-t \notin X$ or $t \notin Y$, which implies $f(x-t)g(t) = 0$, and so, the integral is equal to 0 in this case.

The set of points where $f \neq 0$ is called the *carrier* of f. If this set is bounded, we say that the function f is of bounded carrier.

3.2.2. COROLLARY. *If f and g are locally integrable functions in $\boldsymbol{R}^q$ and one of them is of bounded carrier, then the convolution $f*g$ exists a.e. and is a locally integrable function.*

PROOF. The assertion follows from the fact that the carrier which is bounded is compatible with every other set.

3.3. PROPERTIES OF COMPATIBLE SETS

In this section we prove a few theorems on compatible sets and formulate the property of compatibility for three sets.

3.3.1. THEOREM. *The following condition is necessary and sufficient for sets X and Y to be compatible*:

(*) *If $x_n \in X$, $y_n \in Y$ and $|x_n|+|y_n| \to \infty$, then $|x_n+y_n| \to \infty$.*

PROOF. First suppose that condition (*) is not satisfied, i.e., that there exist sequences x_n and y_n such that $x_n \in X$, $y_n \in Y$, $|x_n|+|y_n| \to \infty$ and $|x_n+y_n| < M < \infty$. Put $z_n = x_n+y_n$. Then $X(z_n-y_n)Y(y_n) = 1$ and $|z_n| < M$. The points $x = z_n$ lie in a bounded interval I, but the set of points $t = y_n$ is not bounded. This shows that X and Y are not compatible. Condition (*) is thus necessary.

Now assume that the sets X and Y are not compatible. Then there is a bounded interval I and sequences z_n, t_n such that $z_n - t_n \in X$, $t_n \in Y$, $z_n \in I$ and $t_n \to \infty$. Let $x_n = z_n - t_n$ and $y_n = t_n$. Then $x_n \in X$, $y_n \in Y$, $|x_n| + |y_n| \to \infty$ and $|x_n + y_n| = |z_n| < M < \infty$, which shows that condition $(*)$ is not satisfied. This proves that condition $(*)$ is sufficient.

Using condition $(*)$ we can easily prove the following properties of compatible sets:

3.3.2. PROPERTY. *If X, Y are compatible sets and $V \subset X$, $W \subset Y$, then V, W are compatible. I.e., subsets of compatible sets are compatible.*

3.3.3. PROPERTY. *If X, Y are compatible sets and α, β are positive numbers, then the neighbourhoods X_α, Y_β are compatible* (see Section 1.4).

Property 3.3.2 is evident. To prove Property 3.3.3, suppose that $x_n \in X_\alpha$ and $y_n \in Y_\beta$. Then there are points $\bar{x}_n$ in X and $\bar{y}_n$ in Y such that

$$|\bar{x}_n - x_n| < \alpha \quad \text{and} \quad |\bar{y}_n - y_n| < \beta .$$

If $|x_n| + |y_n| \to \infty$, then

$$|\bar{x}_n| + |\bar{y}_n| > |x_n| + |y_n| - \alpha - \beta \to \infty$$

and, since X and Y satisfy $(*)$, $|\bar{x}_n + \bar{y}_n| \to \infty$. Hence

$$|x_n + y_n| > |\bar{x}_n + \bar{y}_n| - \alpha - \beta \to \infty ,$$

which proves that X_α and Y_β satisfy $(*)$ and so are compatible by 3.3.1.

Condition $(*)$ has the advantage that its symmetry with respect to the sets X, Y is self-evident. Moreover, it has a simple geometrical interpretation. In fact, since $x + y = 2\dfrac{x+y}{2}$, condition $(*)$ may be reformulated as follows: if point x varies in X and the point y varies in Y so that at least one of them tends to infinity, then their arithmetic mean $\dfrac{x+y}{2}$ must also tend to infinity.

Another advantage is that condition $(*)$ can easily be formulated for three (or more) sets X, Y, Z:

$(**)$ *If $x_n \in X$, $y_n \in Y$, $z_n \in Z$ and $|x_n| + |y_n| + |z_n| \to \infty$, then $|x_n + y_n + z_n| \to \infty$.*

As before we see that condition $(**)$ is symmetric with respect to all three sets X, Y, Z.

If the sets X, Y, Z satisfy $(**)$, we say that they are compatible.

3.3.4. THEOREM. *The sets X, Y, Z are compatible, iff X, Y are compatible and $X+Y$, Z are compatible.*

PROOF OF SUFFICIENCY. The compatibility of $X+Y$ and Z means, by 3.3.1, that

$$x_n \in X, y_n \in Y, z_n \in Z \text{ and } |x_n+y_n|+|z_n| \to \infty \text{ imply } |x_n+y_n+z_n| \to \infty. \quad (1)$$

Let p_n be the increasing sequence consisting of all positive integers such that $|x_{p_n}|+|y_{p_n}| \geqslant |z_{p_n}|$, and let q_n be the increasing sequence consisting of all the remaining positive integers, so that we have $|x_{q_n}|+|y_{q_n}| < |z_{q_n}|$. If both the sequences p_n, q_n are infinite and $|x_n|+|y_n|+|z_n| \to \infty$ holds, then

$$|x_{p_n}|+|y_{p_n}| \to \infty \quad \text{and} \quad |z_{q_n}| \to \infty.$$

Since the sets X and Y are compatible, it follows that $|x_{p_n}+y_{p_n}| \to \infty$. Hence $|x_n+y_n|+|z_n| \to \infty$. Since the sets $X+Y$ and Z are compatible, we obtain by (1), $|x_n+y_n+z_n| \to \infty$, which proves the sufficiency of the condition. If one of the sequences p_n or q_n is finite, the proof is trivial.

PROOF OF NECESSITY. If $|x_n|+|y_n| \to \infty$, then for any fixed $z_0 \in Z$ we have $|x_n|+|y_n|+|z_0| \to \infty$ and, by (**),

$$|x_n+y_n| \geqslant |x_n+y_n+z_0|-|z_0| \to \infty.$$

Hence X and Y are compatible by 3.3.1. If $|x_n+y_n|+|z_n| \to \infty$, then $|x_n|+|y_n|+$ $+|z_n| \to \infty$ and, by (**), $|x_n+y_n+z_n| \to \infty$. Thus (**) implies (1), which means that the sets $X+Y$ and Z are compatible.

3.4. ASSOCIATIVITY OF CONVOLUTION OF FUNCTIONS WITH RESTRICTED CARRIERS

The concept of compatibility of three sets enables us to formulate the following associativity theorem.

3.4.1. THEOREM. *Let f, g and h be locally integrable functions in $\boldsymbol{R}^q$ whose carriers are X, Y and Z respectively. If X, Y, Z are compatible, then all the convolutions involved in the equality*

$$(f*g)*h = f*(g*h) \quad (1)$$

exist a.e. and the equality holds.

PROOF. The moduli $|f|$, $|g|$ and $|h|$ are also locally integrable and have the same carriers X, Y and Z. The convolution $|f|*|g|$ therefore exists and represents a locally integrable function, by 3.2.1. Since the carrier of $|f|*|g|$ lies in $X+Y$, the convolution $(|f|*|g|)*|h|$ also exists, by 3.3.4. The assertion now follows, by 1.3.2.

REMARK. Theorem 3.4.1 would be false, if we replaced its hypothesis concerning the sets $X+Y$ and Z by the hypothesis that the carrier of $f*g$ be compatible with Z. In fact, let f be a function of bounded carrier such that $\int f = 0$, $\int |f| > 0$, and let $g = 1$, $h = 1$. Then $(f*g)*h = (f*1)*1 = 0*1 = 0$. On the other hand $f*(g*h)$ does not make sense, since $g*h = \infty$.

From condition $(**)$ and 3.4.1, we immediately obtain the following

3.4.2. COROLLARY. *If f, g and h are locally integrable functions in $\boldsymbol{R}^q$ and at least two of them have bounded carriers, then all the convolutions involved in* (1) *exist a.e. and the equality holds a.e.*

However, the following much stronger corollary follows directly from 1.3.2.

3.4.3. COROLLARY. *If f, g and h are locally integrable functions in $\boldsymbol{R}^q$ such that one of them is of bounded carrier and the convolution of the moduli of the other two exists a.e. and is a locally integrable function, then all the convolutions involved in* (1) *exist a.e. and the equality holds a.e.*

PROOF. Assume that f is of bounded carrier and that the convolution $|g|*|h|$ is locally integrable. Then the convolution $|f|*(|g|*|h|)$ exists a.e., by 3.2.1. This implies, by 1.3.2, that the two outer convolutions in equality (1) exist a.e. and equality holds. Furthermore the inner convolutions in (1) exist a.e.; in fact, $g*h$ exists by hypothesis, and $f*g$ exists by 3.2.1, since f is of bounded carrier.

If we assume that g is of bounded carrier and that $|f|*|h|$ exists a.e. and is locally integrable, we can apply the preceding argument to the equality $g*(f*h) = (g*f)*h$. (1) then follows by the commutativity of convolution. Finally, if we assume that h is of bounded carrier and that $|f|*|g|$ exists and is locally integrable, we can apply the preceding argument to the equality $h*(f*g) = (h*f)*g$. Equality (1) again follows by the commutativity of convolution.

3.4.4. COROLLARY. *If f, g, h, k are locally integrable functions in $\boldsymbol{R}^q$ such that the convolution $|f|*|g|$ exists and represents a locally integrable function and h, k are of bounded carriers, then*

$$(f*h)*(g*k) = (f*g)*(h*k)$$

and all the convolutions appearing in the equality exist.

PROOF. We have

$$\begin{aligned}
(f*g)*(k*h) &= [(f*g)*k]*h && \text{(by Corollary 3.4.2)} \\
&= [f*(g*k)]*h && \text{(by Corollary 3.4.3)} \\
&= [(g*k)*f]*h && \\
&= (g*k)*(f*h) && \text{(by Corollary 3.4.3)} \\
&= (f*h)*(g*k).
\end{aligned}$$

3.4.5. THEOREM. *If f and g are locally integrable functions in $\boldsymbol{R}^q$ such that the convolution of their moduli exists a.e. and represents a locally integrable function in $\boldsymbol{R}^q$, then for every delta-sequence δ_n the sequence $(f*\delta_n)*(g*\delta_n)$ converges locally in mean to $f*g$.*

PROOF. By 3.4.4, we have

$$(f*\delta_n)*(g*\delta_n) = (f*g)*(\delta_n*\delta_n).$$

Since, by 2.1.1, $\delta_n*\delta_n$ is a delta-sequence, the sequence $(f*g)*(\delta_n*\delta_n)$ converges locally in mean to $f*g$, by 2.2.3. This proves our assertion.

3.5. A PARTICULAR CASE

Let ε^* and ε_* be real numbers such that $-\varepsilon_* < \varepsilon^* \leqslant \varepsilon_* < 1$ and let w be a point of $\boldsymbol{R}^q$ such that $|w| = 1$. Let X be the set of points $x \in \boldsymbol{R}^q$ such that $wx \geqslant \varepsilon^*|x|$ and Y the set of points $y \in \boldsymbol{R}^q$ such that $wy \geqslant \varepsilon_*|y|$, where wx and wy are scalar products. Then X and Y satisfy condition $(*)$ of Section 3.3. In fact, let $x_n \in X$, $y_n \in Y$ and $m_n = \min(|x_n|, |y_n|)$. If $|x_n|+|y_n| \to \infty$ and the sequence m_n is bounded, then the assertion $|x_n+y_n| \to \infty$ follows trivially. If m_n is not bounded, then we apply the general inequality

$$|x_n+y_n| \geqslant m_n|u_n|, \quad \text{where} \quad u_n = \frac{x_n}{|x_n|}+\frac{y_n}{|y_n|}.$$

(If $x = 0$, we assign $\dfrac{x}{|x|}$ the value 0.) Since $wu_n \geqslant \varepsilon_*+\varepsilon^* > 0$, each term of the sequence $|u_n|$ is greater than a fixed positive number. Since m_n is not bounded, it follows that $|x_n+y_n|$ is not bounded either. This shows that X and Y satisfy $(*)$.

It is easy to give a geometrical description of the sets X and Y. Namely, we have $wx = |x|\cos(w,x)$, and so X is the set of all points x such that the angle $\theta = (w,x)$ between the vectors w and x, emanating from the origin satisfies the inequality $\cos\theta \geqslant \varepsilon^*$, i.e., is not greater than $\arccos \varepsilon^*$. Hence X is a cone whose apex is at the origin and whose axis is the line joining the origin to the point w. The cone is convex, if $\varepsilon^* \geqslant 0$ and concave, if $\varepsilon^* < 0$. The set Y is a similar cone, but it is always convex and is contained in X.

Note that, if Z is the set of all points $x+y$ with $x \in X$ and $y \in Y$, then $Z = X$. In fact,

$$w(x+y) \geqslant \varepsilon^*|x|+\varepsilon_*|y| \geqslant \varepsilon^*|x+y|,$$

and so $Z \subset X$. On the other hand, given any point $x \in X$, we can write $x = x+y \in Z$ where $y = 0 \in Y$. Thus $X \subset Z$.

Let U^* actually denote the set of all locally integrable functions whose carriers are contained in the cone X. Similarly, U_* will denote the set of all locally integrable functions whose carriers are contained in the cone Y. If $f \in U^*_*$ and $g \in U_*$, then by Property 3.3.2 the carriers of f and g satisfy ($*$). Thus, by 3.2.1 the convolution $f*g$ exists and represents a locally integrable function. Moreover, the carrier of $f*g$ is contained in the cone X, which implies $f*g \in U^*$.

3.6. CONVOLUTION OF TWO SMOOTH FUNCTIONS

If φ and ψ are smooth functions in $\boldsymbol{R}^q$, then the convolution $\varphi*\psi$ is not necessarily a smooth function, even if it exists at every point $x \in \boldsymbol{R}^q$.

EXAMPLE. Let $\mathscr{M}$ be a smooth function in $\boldsymbol{R}^1$ such that $\mathscr{M}(x) = 0$ if $|x| \leqslant \frac{1}{4}$ or $|x| \geqslant \frac{1}{2}$, and let

$$\varphi(x) = \psi(x) = \sum_{n=-\infty}^{\infty} \mathscr{M}[2^{|n|}(x-n)].$$

Then

$$\varphi'(x) = \psi'(x) = \sum_{n=-\infty}^{\infty} 2^{|n|}\mathscr{M}'[2^{|n|}(x-n)].$$

If $x \neq 0$, the infinite integrals

$$\int_{-\infty}^{\infty} \varphi(x-t)\psi(t)\,dt \quad \text{and} \quad \int_{-\infty}^{\infty} \varphi(x-t)\psi'(t)\,dt$$

reduce to integrals over a bounded interval. The convolutions $\chi = \varphi*\psi$ and $\chi' = \varphi*\psi'$ thus exist for every $x \neq 0$. On the other hand, we have

$$\chi(0) = \sum_{n=-\infty}^{\infty} 2^{-|n|}a \quad \text{and} \quad \chi'(0) = \sum_{n=-\infty}^{\infty} b,$$

where

$$a = \int_{-\infty}^{\infty} \mathscr{M}(-t)\mathscr{M}(t)\,dt \quad \text{and} \quad b = \int_{-\infty}^{\infty} \mathscr{M}(-t)\mathscr{M}'(t)\,dt.$$

This implies that the convolution $\varphi*\psi$ is also defined at $x = 0$, and is thus defined everywhere in $\boldsymbol{R}'$. Whether or not $\chi' = \varphi*\psi'$ is defined at $x = 0$, depends on b, and we can choose $\mathscr{M}$ in such a way that $b \neq 0$. Then $\chi'(0) = \pm\infty$ and $\chi'(x)$ does not tend to any finite limit as $x \to 0$. Thus, although the convolution $\varphi*\psi$ of the smooth functions φ and ψ exists at $x = 0$, it is not a smooth function there.

We shall say that the convolution $\varphi*\psi$ of the smooth functions φ, ψ in $\boldsymbol{R}^q$ exists smoothly, iff for any $k, l \in \boldsymbol{P}^q$, the convolutions $\varphi^{(k)}*\psi^{(l)}$ exist in $\boldsymbol{R}^q$

and are continuous, and the convolutions $|\varphi^{(k)}|*|\psi^{(l)}|$ are locally integrable. Then we have

$$(\varphi*\psi)^{(k)} = \varphi^{(k)}*\psi = \varphi*\psi^{(k)} \tag{1}$$

for every $k \in \boldsymbol{P}^q$. In fact, the assumption of continuity implies that the function

$$\Phi_j(x) = \int_0^x du^{e_j} \int_{\boldsymbol{R}^q} \varphi^{(e_j)}(u-t)\psi(t)\,dt$$

is continuous. The integrability of $|\varphi^{(e_j)}|*|\psi|$ allows us to interchange the order of integration, which leads to $\Phi_j = \Psi - \Psi_j$ a.e. in $\boldsymbol{R}^q$, where $\Psi = \varphi*\psi$ and Ψ_j is the function, constant in ξ_j (jth coordinate of x), which equals to Ψ for $\xi_j = 0$. Evidently, both the functions Ψ, Ψ_j are continuous, which implies that the formula $\Phi_j = \Psi - \Psi_j$ holds everywhere in $\boldsymbol{R}^q$. Taking the e_jth derivative on both sides of this formula, we obtain $\varphi^{(e_j)}*\psi = (\varphi*\psi)^{(e_j)}$, which proves the first equality in (1), when $k = e_j$. Moreover, we have proved that $(\varphi*\psi)^{(e_j)}$ is continuous. For arbitrary k, the equality in (1) and the continuity of $(\varphi*\psi)^{(k)}$ follows by q-dimensional induction. The second equality follows from the first one by the commutativity.

We thus have proved that *if the convolution of two smooth functions exists smoothly, then it is itself a smooth function.*

If λ is a real number and the convolution $\varphi*\psi$ exists smoothly, then the convolutions $(\lambda\varphi)*\psi$ and $\varphi*(\lambda\psi)$ exist smoothly and the equalities $(\lambda\varphi)*\psi = \varphi*(\lambda\psi) = \lambda(\varphi*\psi)$ hold everywhere. If φ, ψ and χ are smooth functions and the convolutions $\varphi*\psi$ and $\varphi*\chi$ exist smoothly, then the convolution $\varphi*(\psi+\chi)$ also exists smoothly and is everywhere equal to the sum $\varphi*\psi+\varphi*\chi$.

The concept of smooth convolution will be used in the definition of the convolution of two distributions (Chapter 6). It will ensure that the important property (1) is preserved for convolution of distributions.

3.6.1. Theorem. *If f, g are locally integrable functions such that the convolution $|f|*|g|$ is a locally integrable function and h, k are smooth functions whose carriers are bounded, then the convolution $(f*h)*(g*k)$ exists smoothly.*

Proof. By 1.4(7) and 3.4.4 we have, for any $r, s \in \boldsymbol{P}^q$,

$$\Phi = (f*h)^{(r)}*(g*k)^{(s)} = (f*|h^{(r)})*(g*k^{(s)}) = (f*g)*(h^{(r)}*k^{(s)})$$

which implies that Φ is continuous. Moreover,

$$\Psi = |(f*h)^{(r)}|*|(g*k)^{(s)}| \leqslant (|f|*|h|^{(r)}|)*(|g|*|k^{(s)}|) = (|f|*|g|)*(|h^{(r)}|*|k^{(s)}|),$$

which implies that Ψ is locally integrable (for it is bounded by a continuous function).

4. Square integrable functions

4.1. FUNDAMENTAL DEFINITIONS AND THEOREMS

In this section, we recall a few well known facts, the proofs of which can be found in the standard text books. Let O be any open subset of $\boldsymbol{R}^q$. Then the following inequalities are well known

$$\Big(\int_O |f+g|^2\Big)^{1/2} \leqslant \Big(\int_O |f|^2\Big)^{1/2} + \Big(\int_O |g|^2\Big)^{1/2}, \tag{1}$$

$$\int_O |fg| \leqslant \Big(\int_O |f|^2\Big)^{1/2} \Big(\int_O |g|^2\Big)^{1/2}. \tag{2}$$

These inequalities hold for any measurable functions f, g, provided we allow the integral to take infinite values. Inequality (1) is known as the *Minkowski inequality*, and (2) the *Schwarz inequality*. However, inequality (2) was found earlier by Buniakowski and still earlier by Cauchy, and is therefore also called the Buniakowski–Schwarz inequality or the Cauchy–Buniakowski–Schwarz inequality. In the sequel, it will be called the *geometric inequality*. Similarly, (1) may be called the *arithmetic inequality*.

If we introduce the symbol

$$||f|| = \Big(\int_O |f|^2\Big)^{1/2},$$

inequalities (1) and (2) may be expressed as

$$||(f+g)|| \leqslant ||f|| + ||g|| \quad \text{(arithmetic inequality)}, \tag{3}$$

$$\left\|\sqrt{|fg|}\right\| \leqslant \sqrt{||f|| \cdot ||g||} \quad \text{(geometric inequality)}. \tag{4}$$

Moreover, we trivially have

$$||\lambda f|| = |\lambda|\, ||f||$$

for any real number λ. From above equality and (3) it follows that if λ, μ are numbers, then

$$||\lambda f + \mu g|| \leqslant |\lambda|\, ||f|| + |\mu|\, ||g||. \tag{5}$$

If $||f|| < \infty$, then the function f is said to be *square integrable* over O. The set of all functions f which are square integrable over O will be denoted by $L^2(O)$.

From (5) it follows that if $f \in L^2(O)$ and $g \in L^2(O)$, then also $\lambda f + \mu g \in L^2(O)$. In other words, $L^2(O)$ is a linear space. From (4) it follows that, if $f, g \in L^2(O)$, then the product fg is an integrable function.

We say that a sequence of functions $f_n \in L^2(O)$ *converges in square mean* to a function $f \in L^2(O)$ and we write $f_n \overset{2}{\to} f$, iff $\|f_n - f\| \to 0$ as $n \to \infty$. If $f_n \overset{2}{\to} f$ and $g_n \overset{2}{\to} g$, then

$$\int_O |f_n g_n - fg| \to 0 \quad \text{and} \quad \int_O f_n g_n \to \int_O fg.$$

We say that a sequence of functions $f_n \in L^2(O)$ is a *Cauchy sequence*, iff, for any given $\varepsilon > 0$, there is an index n_0 such that $m, n > n_0$ implies $\|f_m - f_n\| < \varepsilon$. The space $L^2(O)$ is complete, i.e., every Cauchy sequence converges in square mean to some function $f \in L^2(O)$.

If $f_A \in L^2(O)$ for every index A belonging to the interval $0 < A < \infty$, and if $|f_A - f\| \to 0$ as $A \to \infty$, we say that f_A converges in square mean to f as $A \to \infty$.

4.2. REGULAR SEQUENCES

In the sequel, we shall mainly be concerned with functions which are square integrable in $\boldsymbol{R}^q$.

4.2.1. THEOREM. *If $f, g \in L^2(\boldsymbol{R}^q)$, then the convolution $f * g$ exists everywhere and is a bounded continuous function.*

PROOF. The existence everywhere is obvious. Moreover

$$|(f*g)(x+h) - (f*g)(x)|^2 = \left|\int [f(x+h-t) - f(x-t)]g(t)\,dt\right|^2$$
$$\leqslant \int |f(x+h-t) - f(x-t)|^2\,dt \int |g(t)|^2\,dt \to 0, \text{ as } h \to 0,$$

and so $f * g$ is continuous. Finally, the boundedness follows from the estimate:

$$|(f*g)(x)|^2 = \left|\int f(x-t)g(t)\,dt\right|^2 \leqslant \int |f(x-t)|^2\,dt \int |g(t)|^2\,dt = \int |f|^2 \cdot \int |g|^2.$$

4.2.2. THEOREM. *If f is a square integrable function in $\boldsymbol{R}^q$ and φ is a smooth function of bounded carrier, then the convolution $f * \varphi$ is a smooth function, square integrable in $\boldsymbol{R}^q$.*

PROOF. Since f is square integrable in $\boldsymbol{R}^q$, it is locally integrable in $\boldsymbol{R}^q$. Hence $f * \varphi$ is a smooth function, by equality (7) of 1.4. Moreover, if A is the carrier of φ, we have, by the geometric inequality,

$$|f*\varphi|^2 \leqslant \Big(\int_A |f(x-t)\varphi(t)|\,dt\Big)^2 \leqslant \int_A |f(x-t)|^2\,dt\cdot M,$$

where $M = \int|\varphi|^2$. Hence

$$\int_{R^q} |f*\varphi|^2 \leqslant M \int_{R^q} dx \int_A |f(x-t)|^2\,dt = M\int_A dt \int_{R^q} |f(x-t)|^2\,dx = M\int_A dt\,||f||^2 < \infty,$$

which shows that $f*\varphi$ is square integrable.

From 4.2.2 it follows that if f is square integrable in $\boldsymbol{R}^q$, then all the terms of the regular sequence $f*\delta_n$ (see Chapter 2) are also square integrable in $\boldsymbol{R}^q$.

4.2.3. THEOREM. *If f is a square integrable function in $\boldsymbol{R}^q$, then*

$$f*\delta_n \overset{2}{\to} f.$$

In other words, if $f\in L^2(\boldsymbol{R}^q)$, then

$$||f*\delta_n - f|| \to 0 \quad \textit{as} \quad n\to\infty.$$

PROOF. We have

$$|f*\delta_n - f| \leqslant \int_{R^q} |f(x-t)-f(t)|\,\sqrt{|\delta_n(t)|}\cdot\sqrt{|\delta_n(t)|}\,dt,$$

and by the geometric inequality

$$|f*\delta_n - f|^2 \leqslant \int |f(x-t)-f(x)|^2|\delta_n(t)|\,dt\cdot M_0. \tag{1}$$

Since $\int|f(x-t)-f(x)|^2\,dx \to 0$, as $t\to 0$, we have

$$\int|f(x-t)-f(x)|^2\,dx < \varepsilon_n \quad \text{for} \quad |t|\leqslant \alpha_n,$$

where $\varepsilon_n \to 0$, as $n\to\infty$.

Inequality (1) therefore implies

$$||f*\delta_n - f||^2 \leqslant \int|\delta_n(t)|\,dt \int|f(x-t)-f(x)|^2\,dx\cdot M_0$$

$$\leqslant \int|\delta_n(t)|\,dt\cdot\varepsilon_n M_0 = M_0^2\,\varepsilon_n \to 0,$$

which implies the assertion.

4.2.4. THEOREM. *If a sequence of functions f_n, square integrable in $\boldsymbol{R}^q$, converges in square mean to f, then the sequence $f_n*\delta_n$ also converges in square mean to f. In other words, if $f_n\in L^2(\boldsymbol{R}^q)$ and $||f_n - f||\to 0$, then $||f_n*\delta_n - f||\to 0$.*

PROOF. We have

$$|(f_n - f) * \delta_n| \leqslant \int |f_n(t) - f(t)| \sqrt{|\delta_n(x-t)|}\, \sqrt{|\delta_n(x-t)|}\, dt,$$

and by the geometric inequality

$$|(f_n - f) * \delta_n|^2 \leqslant \int |f_n(t) - f(t)|^2 |\delta_n(x-t)|\, dt \cdot M_0.$$

Hence $\|(f_n - f) * \delta_n\|^2 \leqslant \|f_n - f\|^2 M_0^2$, which implies, by the Minkowski inequality,

$$\|f_n * \delta_n - f\| \leqslant \|(f_n - f) * \delta_n\| + \|f * \delta_n - f\| \to 0,$$

in view of 4.2.3.

4.2.5. THEOREM. *If $f_n,\ g_n \in L^2(\boldsymbol{R}^q)$, $f_n \overset{2}{\to} f$ and $g_n \overset{2}{\to} g$, then $f_n * \delta_n \to f * g$ uniformly.*

PROOF. In fact, by the geometric inequality, we have

$$\begin{aligned}
|f_n * g_n - f * g| &\leqslant |(f_n - f) * g_n| + |f * (g_n - g)| \\
&\leqslant \int |f_n(x-t) - f(x-t)| \cdot |g_n(t)|\, dt + \int |f(x-t)| \cdot |g_n(t) - g(t)|\, dt \\
&\leqslant \left(\int |f_n(x-t) - f(x-t)|^2 dt \int |g_n(t)|^2 dt\right)^{1/2} \\
&\qquad + \left(\int |f(x-t)|^2 dt \int |g_n(t) - g(t)|^2 dt\right)^{1/2} \\
&= \|f_n - f\| \cdot \|g_n\| + \|f\| \cdot \|g_n - g\| \to 0,
\end{aligned}$$

which implies our assertion.

4.3. THE FOURIER TRANSFORM OF SQUARE INTEGRABLE FUNCTIONS

The theory presented here is due to Plancherel and really belongs in Classical Analysis. The proofs are omitted here, but can be found in any of the more advanced text books on integration.

All functions considered in this section are assumed to belong to $L^2(\boldsymbol{R}^q)$, and are therefore defined in the whole of $\boldsymbol{R}^q$.

Let

$$F_A(x) = \left(2\sqrt{\pi}\right)^{-q} \int_{-A}^{A} e^{ixt/2} f(t)\, dt, \tag{1}$$

where xt is the scalar product $\xi_1\tau_1 + \ldots + \xi_q\tau_q$ of $x = (\xi_1, \ldots, \xi_q)$ and $t = (\tau_1, \ldots, \tau_q)$, and the integral extends over the q-dimensional interval $-A < t < A$, i.e., over the set of all points t such that $-A < \tau_j < A$, $j = 1, \ldots, q$.

We assume that A is a positive number so that the notation $\int_{-A}^{A}$ and $-A < t < A$ is purely symbolic. In the same spirit we assign a more general meaning to the inequality

$$a < b. \tag{2}$$

Namely, if a and b are real numbers, the inequality is understood in the ordinary sense. If $a, b \in \boldsymbol{R}^q$, then we understand by (2) that inequalities similar to (2) hold for the coordinates of a and b. I.e., that if $a = (\alpha_1, \ldots, \alpha_q)$ and $b = (\beta_1, \ldots, \beta_q)$, then $\alpha_i < \beta_i$ for $i = 1, \ldots, q$. Finally, if a is a real number and $b \in \boldsymbol{R}^q$, then inequality (2) is to be read as $a < \beta_i$ for $i = 1, \ldots, q$. In other words, a is less than any coordinate of b. Similarly, if $a \in \boldsymbol{R}^q$ and b is a real number, then inequality (2) means that every coordinate of a is less than b.

A similar convention will be adopted for the weak inequality $a \leqslant b$. Hence each of the intervals

$$a < x < b, \quad a < x \leqslant b, \quad a \leqslant x < b, \quad a \leqslant x \leqslant b$$

has a clear meaning if $x \in \boldsymbol{R}^q$, whether a, b are real numbers or vectors of $\boldsymbol{R}^q$.

Since the integral in (1) is taken in the sense of Lebesgue, it is immaterial over which one of the intervals $-A<x<A$, $-A<x\leqslant A$, $-A\leqslant x<A$ or $-A \leqslant x \leqslant A$ it extends.

It is well known that, if $f \in L^2(\boldsymbol{R}^q)$, F_A converges in square mean as $A \to \infty$ to a function $F \in L^2(\boldsymbol{R}^q)$.

Moreover

$$\|F\| = \|f\|. \tag{3}$$

We write $F = \mathcal{F}(f)$ and call F the *Fourier transform* of f.

REMARK. Instead of starting with (1), it would be simpler to start with a similar equality, but with e^{ixt} instead of $e^{ixt/2}$. This would be the usual way of introducing the Fourier transform. Some authors, e.g., L. Schwartz, take $e^{2\pi ixt}$ as the exponential function in (1). Each way has its advantages and disadvantages. Our definition has been chosen with a view to obtaining, in the sequel, the simplest possible connection with the theory of Hermite expansions. The reader should be aware that formulae appearing in the sequel may well differ in their coefficients from similar formulae in other treatments of the subject.

If $f \in L^1(\boldsymbol{R}^q)$, i.e., if f is integrable over $\boldsymbol{R}^q$, the *Fourier transform* of f is given by the ordinary Lebesgue integral

$$F(x) = \left(2\sqrt{\pi}\right)^{-q} \int_{\boldsymbol{R}^q} e^{ixt/2} f(t)\,dt.$$

If f belongs both to $L^2(\boldsymbol{R}^q)$ and to $L^1(\boldsymbol{R}^q)$, the two definitions are equivalent.

In particular, the Fourier transform of $e^{-\alpha x^2}$ $(\alpha>0)$ is $2^{-q}\alpha^{-q/2}e^{-x^2/16\alpha}$ and, so taking $\alpha = \frac{1}{4}$ and $q = 1$, we have

$$\mathscr{F}(e^{-x^2/4}) = e^{-x^2/4}. \tag{4}$$

The formulae below follow immediately from the definition, provided all the functions involved belong to $L^2(\boldsymbol{R}^q)$

$$\xi_j F(x) = 2i\mathscr{F}\left(f^{(e_j)}(x)\right), \tag{5}$$

$$F^{(e_j)}(x) = \frac{i}{2}\mathscr{F}\left(\xi_j f(x)\right), \tag{6}$$

$$e^{-iax/2}F(x) = \mathscr{F}\left(f(x+a)\right), \tag{7}$$

$$F(x+a) = \mathscr{F}\left(e^{iax/2}f(x)\right), \tag{8}$$

where $a = (\alpha_1, \ldots, \alpha_q)$ and $ax = \alpha_1\xi_1 + \ldots + \alpha_q\xi_q$ is the inner product of a and x. Moreover, if f, g and the product fg all belong to $L^2(\boldsymbol{R}^q)$, then

$$\mathscr{F}(fg) = (2\sqrt{\pi})^{-q}\mathscr{F}(f)*\mathscr{F}(g). \tag{9}$$

This last equality also holds without assuming that $fg \in L^2(\boldsymbol{R}^q)$, but the Fourier transform on the left hand side has then to be taken as the ordinary Lebesgue integral

$$(2\sqrt{\pi})^{-q}\int_{\boldsymbol{R}^q} e^{ixt/2}f(t)g(t)\,dt$$

and not as the limit in mean of $\int_{-A}^{A}$.

It follows from (3) that if a sequence of square integrable functions f_n converges in square mean to f, then the sequence $\mathscr{F}(f_n)$ converges in square mean to $\mathscr{F}(f)$.

If $F = \mathscr{F}(f)$, then the function $G(x) = F(-x)$ is called the *inverse Fourier transform* of f,

$$G = \mathscr{G}(f).$$

This name is justified on account of the properties

$$\mathscr{G}[\mathscr{F}(f)] = f \quad \text{and} \quad \mathscr{F}[\mathscr{G}(f)] = f.$$

4.4. TWO APPROXIMATION THEOREMS

Our aim is now to show that every function $f \in L^2(\boldsymbol{R}^q)$ can be approximated in square mean by functions $e^{-x^2/2}P(x)$, where the $P(x)$ are polynomials. We first prove two auxiliary theorems.

4.4.1. Theorem. *If $f \in L^2(\mathbf{R}^q)$, then for each $\varepsilon > 0$ there exists a number $\alpha > 0$ and a linear combination g of functions of the form*

$$e^{-\alpha x^2} e^{ibx} \qquad (b \in \mathbf{R}^q), \tag{1}$$

such that $\|f-g\| < \varepsilon$. Further, number α can be chosen arbitrarily small.

Proof. Let $F = \mathscr{F}(f)$. Then $f = \mathscr{G}(F)$. I.e., if

$$f_A = (2\sqrt{\pi})^{-q} \int_{-A}^{A} e^{-xt/2} F(t)\, dt, \tag{2}$$

then $f_A \xrightarrow{2} f$. Hence, given $\varepsilon > 0$, there is an A such that

$$\|f - f_A\| < \varepsilon/3. \tag{3}$$

There is also a number $\alpha > 0$, arbitrarily small, such that

$$\|f_A - E^\alpha f_A\| < \varepsilon/3 \tag{4}$$

where $E^\alpha(x) = e^{-\alpha x^2}$ and $\alpha x^2 = \alpha(\xi_1^2 + \ldots + \xi_q^2)$.

In the sequel we denote by $\mathbf{B}$ the set of all integers and by $\mathbf{B}^q$ the set of all points of $\mathbf{R}^q$ whose coordinates are integers.

For each natural number μ, we can partition the interval $-A < x \leqslant A$ into $(2\mu)^q$ subintervals $I_{\mu m}$ defined by

$$\frac{mA}{\mu} < x \leqslant \frac{(m+1)A}{\mu} \qquad (m \in \mathbf{B}^q)$$

and write

$$f_A = (2\sqrt{\pi})^{-q} \sum_m \int_{I_{\mu m}} F(t) \exp\left(-\frac{i}{2} xt\right) dt \qquad (-\mu \leqslant m < \mu).$$

If we replace the value of the exponential function by its value at the left hand end point of the interval $I_{\mu m}$, we obtain the following sum

$$S_\mu = (2\sqrt{\pi})^{-q} \sum_m \exp\left(-\frac{i}{2}\frac{A}{\mu} mx\right) \int_{I_{\mu m}} F(t)\, dt.$$

We now show that there is an index μ such that

$$\|E^\alpha f_A - E^\alpha S_\mu\| < \varepsilon/3. \tag{5}$$

In fact

$$E^\alpha f_A - E^\alpha S_\mu = (2\sqrt{\pi})^{-q} e^{-\alpha x^2/2} \sum_m \int_{I_{\mu m}} F(t) \left[\exp\left(-\frac{\alpha}{2}x^2 - \frac{i}{2}tx\right) - \exp\left(-\frac{\alpha}{2}x^2 - \frac{i}{2}\frac{A}{\mu}mx\right)\right] dt.$$

We can easily check that all the derivatives

$$\frac{\partial}{\partial \tau_j} \exp\left(-\frac{\alpha}{2}x^2 - \frac{i}{2}tx\right)$$

are bounded for all $x, t \in \boldsymbol{R}^q$. This implies that, for every fixed $M > 0$, there is an index μ such that

$$\left|\exp\left(-\frac{\alpha}{2}x^2 - \frac{i}{2}tx\right) - \exp\left(-\frac{\alpha}{2}x^2 - \frac{i}{2}\frac{A}{\mu}mx\right)\right| < \frac{\varepsilon}{M}$$

for $t \in I_{\mu m}$, $x \in \boldsymbol{R}^q$. Hence

$$|E^\alpha f_A - E^\alpha S_\mu| \leqslant (2\sqrt{\pi})^{-q} e^{-\alpha x^2/2} \sum_m \int_{I_{\mu m}} |F(t)| \frac{\varepsilon}{M} dt$$

and

$$\|E^\alpha f_A - E^\alpha S_\mu\| \leqslant (2\sqrt{\pi})^{-q} \left(\int_{R^q} e^{-\alpha x^2} dx\right)^{1/2} \int_{-A}^{A} |F(t)|\, dt \cdot \frac{\varepsilon}{M}.$$

Taking M sufficiently large, we see that the right hand side can be made less than $\varepsilon/3$ for suitably chosen μ, which proves (5).

From (3), (4) and (5) it follows that

$$\|f - E^\alpha S_\mu\| < \varepsilon.$$

4.4.2. THEOREM. *For any real number α such that $0 < \alpha < \frac{1}{4}$, and any $b \in \boldsymbol{R}^q$, the sequence*

$$e^{-x^2/4} P_\nu(x)$$

where

$$P_\nu(x) = \sum_{\varkappa=0}^{\nu} \frac{1}{\varkappa!} (\eta x^2 + ibx)^\varkappa, \quad \eta = \frac{1}{4} - \alpha,$$

converges in square mean to

$$e^{-\alpha x^2 + ibx}.$$

PROOF. We have

$$e^{-\alpha x^2 + ibx} - e^{-x^2/4} P_\nu(x) = e^{-x^2/4} r_\nu(x)$$

where

$$r_\nu(x) = e^{\eta x^2+ibx} - P_\nu(x).$$

It is clear that $r_\nu(x)$ tends, almost uniformly in $\boldsymbol{R}^q$, to 0. Since

$$|\eta x^2+ibx| \leqslant \eta x^2+b^2/2\eta,$$

we have

$$|r_\nu(x)| \leqslant \sum_{\varkappa=\nu+1}^{\infty} \frac{1}{\varkappa!} |\eta x^2+ibx|^\varkappa \leqslant e^{\eta x^2+b^2/\eta}.$$

Hence

$$e^{-x^2/2}|r_\nu(x)|^2 \leqslant e^{-2\alpha x^2+2b^2/\eta}$$

and, it follows by a standard argument, that

$$||e^{-x^2/4}r_\nu(x)||^2 = \int_{\boldsymbol{R}^q} e^{-x^2/2}|r_\nu(x)|^2\,dx \to 0, \quad \text{as} \quad \nu \to \infty.$$

4.5. THE MAIN APPROXIMATION THEOREM

By a polynomial $P(x)$ we mean, generally, any linear combination of powers

$$x^n = \xi_1^{\nu_1} \ldots \xi_q^{\nu_q} \quad \text{where} \quad x = (\xi_1, \ldots, \xi_q) \in \boldsymbol{R}^q \text{ and } n = (\nu_1, \ldots, \nu_q) \in \boldsymbol{P}^q$$

(if $\nu_j = 0$ for some j, we always take $\xi_j^0 = 1$, even when $\xi_j = 0$).

Since

$$(\tfrac{1}{4}-\alpha)x^2+ibx = (\tfrac{1}{4}-\alpha)(\xi_1^2+ \ldots +\xi_q^2)+i(\beta_1\xi_1+ \ldots +\beta_q\xi_q),$$

it is plain that the functions $P_\nu(x)$ in 4.4.2 are polynomials.

Note that if $q = 1$, the symbol x^2 has the same meaning, whether it is thought of as an inner product or as a power of x. There is therefore no scope for any misunderstanding.

4.5.1. THEOREM. *For every $f \in L^2(\boldsymbol{R}^q)$ and every number $\varepsilon > 0$ there is a polynomial P such that*

$$||f(x)-e^{-x^2/4}P(x)|| < \varepsilon. \tag{1}$$

Moreover, if f is real valued, P can be chosen so that its coefficients are real.

PROOF. The existence of a polynomial with complex coefficients follows directly from 4.4.1 and 4.4.2. We can write this polynomial as $P(x)+iQ(x)$, where the coefficients of the polynomials P and Q are real, so that

$$||f(x)-e^{-x^2/4}P(x)-ie^{-x^2/4}Q(x)|| < \varepsilon.$$

If f is real, this implies (1), since $||g|| \leqslant ||g+ih||$ for arbitrary real valued functions g and h.

REMARK. Theorem 4.5.1 is usually proved only for the one-dimensional case $q = 1$ and is called a *density theorem*, because it says that the functions $e^{-x^2/4}P(x)$ are dense in $L^2(\mathbf{R}^q)$. The classical proof is quite different. However, in the classical formulation, the coefficient $\frac{1}{2}$ appears in the exponent rather than $\frac{1}{4}$. We prefer $\frac{1}{4}$, because this makes for simpler formulae in the sequel. However, we can always pass from $\frac{1}{4}$ to $\frac{1}{2}$ or vice versa, by means of a simple substitution. We remark that all the above work is also valid for functions f taking values in an arbitrary Banach space over the complex field.

4.6. HERMITE POLYNOMIALS OF A REAL VARIABLE

By *Hermite polynomials* of a real variable x we mean the functions

$$H_n(x) = (-1)^n e^{x^2/2}(e^{-x^2/2})^{(n)}, \tag{1}$$

where $n = 0, 1, 2, \ldots$ Carrying out the differentiation, we find

$$H_0(x) = 1, \quad H_1(x) = x, \quad H_2(x) = x^2-1, \quad H_3(x) = x^3-3x,$$
$$H_4(x) = x^4-6x^2+3, \quad \ldots$$

Let

$$k_n(x) = (-1)^n e^{x^2/4}(e^{-x^2/2})^{(n)}. \tag{2}$$

If we introduce the notation $E^\alpha(x) = e^{\alpha x^2/4}$, (1) and (2) may be rewritten as follows

$$H_n = (-1)^n E^2(E^{-2})^{(n)} \quad \text{and} \quad k_n = E^{-1}H_n. \tag{3}$$

It is easy to see that

$$(E^{-2})' = -xE^{-2}$$

and, by induction,

$$(E^{-2})^{(n+1)} = -x(E^{-2})^{(n)} - n(E^{-2})^{(n-1)}.$$

Multiplying this all through by $(-1)^{n+1}E^2$, we see that

$$H_{n+1} = xH_n - nH_{n-1}, \tag{4}$$

by (3). Multiplying (4) by E^{-1} we have

$$k_{n+1} = xk_n - nk_{n-1}.$$

Since $H_0 = 1$ and $H_1 = x$, it follows by induction from (4) that H_n is a polynomial of degree n exactly, and that the coefficient of x^n is 1. Since H_n is a polynomial and $k_n = E^{-1}H_n$, we have

$$\lim_{|x|\to\infty} k_n^{(p)}(x) = 0 \quad \text{for} \quad n, p = 0, 1, \ldots$$

Differentiating (1), we obtain

$$H_n' = -H_{n+1} + xH_n,$$

and substituting (4),

$$H_n' = nH_{n-1} \quad \text{for} \quad n = 1, 2, \ldots \tag{5}$$

Clearly, we have by (3) and (2)

$$k_m k_n = (-1)^n H_m (E^{-2})^{(n)} = (-1)^m H_n (E^{-2})^{(m)}.$$

Integrating by parts, we get

$$\int k_m k_n = \int H_m^{(n)} E^{-2} = \int H_n^{(m)} E^{-2}, \tag{6}$$

where the range of integrations is from $-\infty$ to $+\infty$. Since H_m and H_n are polynomials of degrees m and n respectively, and their leading coefficients are both equal to 1, we have by (6) and (5)

$$\int k_m k_n = 0 \quad \text{for} \quad m \neq n \quad \text{and} \quad \int k_n^2 = n! \int E^{-2} = n! \sqrt{2\pi}\,. \tag{7}$$

The first of the above equalities (7) means that the k_n are *orthogonal.*

If we set

$$h_n = (\sqrt{2\pi}\, n!)^{-1/2} k_n, \tag{8}$$

i.e.,

$$h_n(x) = (-1)^n (\sqrt{2\pi}\, n!)^{-1/2} e^{x^2/4} (e^{-x^2/2})^{(n)} \tag{9}$$

or

$$h_n = (-1)^n (\sqrt{2\pi}\, n!)^{-1/2} E(E^{-2})^{(n)}, \tag{10}$$

then (7) implies that

$$\int h_m h_n = 0 \quad \text{for} \quad m \neq n \quad \text{and} \quad \int h_n^2 = 1.$$

In other words, the functions $h_0, h_1, \ldots$ are *orthonormal*.

4.7. HERMITE POLYNOMIALS OF SEVERAL VARIABLES

We now return to the q-dimensional case. If

$$x = (\xi_1, \ldots, \xi_q) \in \boldsymbol{R}^q \quad \text{and} \quad n = (\nu_1, \ldots, \nu_q) \in \boldsymbol{P}^q,$$

we define:

$$H_n(x) = H_{\nu_1}(\xi_1) \ldots H_{\nu_q}(\xi_q).$$

Hence, by 4.6 (5), we have $H_n^{(e_j)} = \nu_j H_{n-e_j}$ and, by induction,

$$\frac{1}{n!} H_n^{(k)} = \frac{1}{(n-k)!} H_{n-k} \text{ for } 0 \leqslant k \leqslant n \quad \text{and} \quad H_n^{(k)} = 0 \text{ for } k \nleqslant n, \tag{1}$$

where $n! = \nu_1! \ldots \nu_q!$ (see Appendix). The coefficients in H_n of the powers x^n are equal to 1, and the exponents in the remaining powers are less than n. Furthermore x^n is the only term of H_n whose degree is equal to the degree of H_n. We recall that, generally, the degree of a polynomial

$$P(x) = \sum_m \alpha_m x^m \quad (m = (\mu_1, \ldots, \mu_q))$$

is taken to be the greatest of the numbers $\mu_1 + \ldots + \mu_q$, for all m appearing in the sum. Thus the degree of H_n is $\bar{n}$. The only polynomials of degree 0 are the constant functions.

We now show that every polynomial $P(x)$ can be expressed as a linear combination of Hermite polynomials. In fact, this assertion is trivially true for polynomials P of degree 0. Assume that it is true for all polynomials of degree less than or equal to a non-negative integer $\varkappa$. Now, if P is of degree $\varkappa+1$, and ε_m denotes the coefficient of x^m in $P(x)$, then the difference

$$P(x) - \sum_m \varepsilon_m H_m(x),$$

where the sum extends over all m such that $\mu_1 + \ldots + \mu_q = \varkappa+1$, is a polynomial of degree $\leqslant \varkappa$ and can be represented, by the induction hypothesis, as a linear combination of Hermite polynomials. It follows that $P(x)$ can be represented as a linear combination of Hermite polynomials, and the general assertion now follows by induction.

4.8. SERIES OF HERMITE FUNCTIONS

Let

$$h_n(x) = h_{\nu_1}(\xi_1) \ldots h_{\nu_q}(\xi_q).$$

Then

$$\int_{R^q} h_m h_n = 0 \quad \text{for} \quad m \neq n \quad \text{and} \quad \int_{R^q} h_n^2 = 1,$$

i.e., the functions h_n are orthonormal in $\boldsymbol{R}^q$. Moreover,

$$h_n(x) = (2\pi)^{-q/4} \frac{1}{\sqrt{n!}} e^{-x^2/4} H_n(x), \quad n \in \boldsymbol{P}^q. \tag{1}$$

We now consider series of the form

$$\sum_{n \in P^q} c_n h_n, \quad \text{where the } c_n \text{ are complex numbers.} \tag{2}$$

We say that such a series converges unconditionally in square mean to a function $f \in L^2(\boldsymbol{R}^q)$ and write $f \overset{2}{=} \sum_{n \in \boldsymbol{P}^q} c_n h_n$, iff, given any number $\varepsilon > 0$, there is a finite set $N_0 \subset \boldsymbol{P}^q$ such that

$$\Big\| f - \sum_{n \in N} c_n h_n \Big\| < \varepsilon$$

for every finite set N such that $N_0 \subset N \subset \boldsymbol{P}^q$.

4.8.1. THEOREM. *The series* (2) *converges unconditionally in square mean* (*to some square integrable function*), *iff* $\sum_{n \in \boldsymbol{P}^q} |c_n|^2 < \infty$.

PROOF. Suppose first that all the numbers c_n are real. Let N_ν $(\nu = 1, 2, \ldots)$ be the set of all $n = (\nu_1, \ldots, \nu_q) \in \boldsymbol{P}^q$ such that $\nu_j \leqslant \nu$ $(j = 1, \ldots, q)$ and let $f_\nu = \sum_{n \in N_\nu} c_n h_n$, so that the functions f_ν are real valued.

By the orthonormality of the h_n, we have

$$\int_{\boldsymbol{R}^q} (f_\mu - f_\nu)^2 = \int_{\boldsymbol{R}^q} \Big(\sum_{n \in M_{\mu\nu}} c_n h_n \Big)^2 = \sum_{n \in M_{\mu\nu}} c_n^2, \tag{3}$$

where $M_{\mu\nu} = N_\mu \diagdown N_\nu$ with $\mu > \nu$, i.e., $M_{\mu\nu}$ is the set of points n which belong to N_μ but not to N_ν. If $\sum_{n \in \boldsymbol{P}^q} c_n^2 < \infty$, f_ν is a Cauchy sequence by (3). Since the space $L^2(\boldsymbol{R}^q)$ is complete, there is a function $f \in L^2(\boldsymbol{R}^q)$ such that f_ν converges in square mean to f. Hence there is an index ν_0 such that $\|f - f_\nu\| < \varepsilon/2$ for $\nu > \nu_0$. We can choose ν_0 in such a way that, moreover, $\sum_{n \in \boldsymbol{P}^q \diagdown N_{\nu_0}} c_n^2 < \varepsilon/2$. If N is a finite subset of $\boldsymbol{P}^q$ such that $N_{\nu_0} \subset N$, then

$$\Big\| f - \sum_{n \in N} c_n h_n \Big\| \leqslant \frac{\varepsilon}{2} + \Big\| \sum_{n \in N} c_n h_n - f_{\nu_0} \Big\| \leqslant \frac{\varepsilon}{2} + \sum_{n \in \boldsymbol{P}^q \diagdown N_{\nu_0}} c_n^2 < \varepsilon.$$

This means that (2) converges unconditionally in square mean to f.

Assume now, conversely, that the series (2) converges unconditionally in square mean to f. Then the sequence f_ν satisfies the Cauchy condition. Hence, by (3), the series $\sum_{n \in \boldsymbol{P}^q} c_n$ satisfies the Cauchy condition and is therefore convergent. This completes the proof for real c_n.

If the c_n are complex, we write $c_n = a_n + ib_n$, where a_n and b_n are real. Since $\sum_{n \in \boldsymbol{P}^q} |c_n|^2 < \infty$ iff $\sum_{n \in \boldsymbol{P}^q} a_n^2 < \infty$ and $\sum_{n \in \boldsymbol{P}^q} b_n^2 < \infty$, and the series $\sum_{n \in \boldsymbol{P}^q} c_n h_n$ is unconditionally convergent in square mean iff this is true of the series $\sum_{n\in P^q} a_n h_n$ and $\sum_{n\in P^q} b_n h_n$, the theorem for complex c_n follows from what we have just proved.

4.8.2. THEOREM. *The series* (2) *converges unconditionally in square mean to* f, *iff* $f \in L^2(\boldsymbol{R}^q)$ *and* $c_n = \int fh_n$.

PROOF. We retain the notation of the proof of the preceding theorem. To begin with, we assume, that the numbers c_n are real and that the series $\sum_{n\in P^q} c_n h_n$ converges unconditionally in square mean to f, so that $f \in L^2(\boldsymbol{R}^q)$. By the geometric inequality, we have

$$\int |(f-f_\nu)h_n| \leqslant \left(\int |f-f_\nu|^2\right)^{1/2} \cdot 1.$$

Since $\int |f-f_\nu|^2 \to 0$, we have $\int f_\nu h_n \to \int fh_n$, as $\nu \to \infty$. From the orthonormality of the h_m it follows that $\int f_\nu h_n = c_n$ for sufficiently large ν. Hence $c_n = \int fh_n$.

Assume now, conversely, that $f \in L^2(\boldsymbol{R}^q)$, $c_n = \int fh_n$ and f is real valued, so that the c_n are also real and the functions f_ν are real valued. Thus

$$0 \leqslant \int (f-f_\nu)^2 = \int f^2 - 2\int ff_\nu + \int f_\nu^2, \tag{4}$$

where all the integrals extend over all $\boldsymbol{R}^q$. Remembering that the functions h_n are orthonormal, it is easy to see that

$$\int f_\nu^2 = \sum_{n\in N_\nu} c_n^2 \quad \text{and} \quad \int ff_\nu = \sum_{n\in N} c_n^2.$$

We therefore obtain from (4),

$$0 \leqslant \int f^2 - \sum_{n\in N\nu} c_n^2,$$

and letting $\nu \to \infty$, we now obtain

$$\sum_{n\in P^q} c_n^2 \leqslant \int f^2,$$

which is the well known *Bessel inequality*. Hence it follows from 4.8.1 that series (2) converges unconditionally in square mean to some function $g \in L^2(\boldsymbol{R}^q)$. We show that $g = f$ almost everywhere. To this end, note that by what we have proved, the Hermite coefficients c_n of f are the same as the Hermite coefficients of g. Hence the Hermite coefficients of the difference $f-g$ are all 0. It remains to show that, if all the Hermite coefficients of a function h are 0, then $h = 0$ almost everywhere.

By 4.5.1, there is a sequence of polynomials P_ν such that the sequence $p_\nu(x) = e^{-x^2/4}P_\nu(x)$ converges in square mean to h. But every polynomial p_ν is a linear combination of the functions h_n. Since $\int hh_n = 0$, this implies that $\int hp_\nu = 0$ for $\nu = 1, 2, \ldots$ Moreover $\int hp_\nu \to \int h^2$, and so $\int h^2 = 0$. Hence $h = 0$ a.e.

This argument shows that $g=f$ a.e. This completes the proof of 4.8.2 for real valued functions f.

If f is complex valued and $f \in L^2(\boldsymbol{R}^q)$, we can write $f = g+ik$, where g, k are real valued and $g, k \in L^2(\boldsymbol{R}^q)$. In this way, the proof reduces easily to that for the real case.

4.8.3. THEOREM. *If* $f \in L^2(\boldsymbol{R}^q)$ *and* $c_n = \int fh_n$, *then*

$$\int |f|^2 = \sum_{n \in \boldsymbol{P}^q} |c_n|^2 \quad \textit{(Parseval identity).}$$

PROOF. First suppose that the function f is real valued. In this case the numbers c_n are real. From the orthonormality of the h_n it follows that for every finite subset S of $\boldsymbol{P}^q$, we have

$$\int \Big(f - \sum_{n \in S} c_n h_n\Big)^2 = \int f^2 - \sum_{n \in S} c_n^2 .$$

Since, by 4.8.2, the series $\sum_{n \in \boldsymbol{P}^q} c_n h_n$ converges unconditionally in square mean to f, our assertion for real f follows from the above equality.

If f is complex valued and $f \in L^2(\boldsymbol{R}^q)$, we write $f = g+ik$, where g, k are real valued and $g, k \in L^2(\boldsymbol{R}^q)$. In this way, the proof reduces easily to that for the real case.

4.9. THE FOURIER TRANSFORM OF AN HERMITE EXPANSION

Let $E(x)$ denote, as before, the function $e^{x^2/4}$, where $x^2 = \xi_1^2 + \ldots + \xi_q^2$ for $x = (\xi_1, \ldots, \xi_q)$, and let f be a smooth function in $\boldsymbol{R}^q$. We adopt the following definition:

$$d^n f = E(E^{-1} f)^{(n)} \quad \text{for} \quad n \in \boldsymbol{P}^q. \tag{1}$$

Then we obtain from 4.6(2),

$$d^n E^{-1} = (-1)^n k_n . \tag{2}$$

By (1), we find

$$d^{e_j} f = f^{(e_j)} - \tfrac{1}{2} \xi_j f$$

and in view of 4.3(5), 4.3(6) simple calculations yield, $\mathscr{F}(d^{e_j}f) = id^{e_j}\mathscr{F}(f)$, provided $d^{e_j}f, f^{(e_j)} \in L^2(\boldsymbol{R}^q)$. Hence, by induction, we get

$$\mathscr{F}(d^k f) = i^k d^k \mathscr{F}(f), \tag{3}$$

whenever $d^m f, f^{(m)} \in L^2(\boldsymbol{R}^q)$ for $0 \leqslant m \leqslant k$. Moreover, we have

$$\begin{aligned} \mathscr{F}(k_n) &= (-1)^n \mathscr{F}(d^n E^{-1}) && \text{(by (2))} \\ &= (-i)^n d^n \mathscr{F}(E^{-1}) && \text{(by (3))} \\ &= (-i)^n d^n E^{-1} && \text{(by 4.3(4))} \\ &= i^n k_n && \text{(by (2)).} \end{aligned}$$

Hence, by 4.6(8)

$$\mathscr{F}(h_n) = i^n h_n. \tag{4}$$

4.9.1. THEOREM. *We have*

$$f \overset{2}{=} \sum_{n \in P^q} c_n h_n, \quad \textit{iff} \quad \mathscr{F}(f) \overset{2}{=} \sum_{n \in P^q} i^n c_n h_n.$$

PROOF. Let N be an arbitrary finite subset of $\boldsymbol{P}^q$. Then, by 4.3(3) and (4), we have

$$\left\| \mathscr{F}(f) - \sum_{n \in N} i^n c_n h_n \right\| = \left\| f - \sum_{n \in N} c_n h_n \right\|,$$

and the theorem follows.

5. Inner product

5.1. INNER PRODUCT OF TWO FUNCTIONS

In the inner product

$$(f, g) = \int_{R^q} fg = \int_{R^q} f(x)g(x)\,dx$$

of two functions f, g the integral is always to be taken in the sense of Lebesgue. Moreover, as in the case of convolution, the following convention is adopted: the (ordinary) product fg takes the value 0 at a point, whenever one of its factors is 0 at that point, no matter whether the other factor is finite, infinite or even undetermined. This convention implies that, e.g., the inner product (f, g) exists, when f is defined in an open set O, g is defined in $\boldsymbol{R}^q$, the carrier of g is inside O and the product fg is locally integrable in O.

Since the integral is taken in the sense of Lebesgue, the existence of the inner product (f, g) implies the existence of the inner product $(|f|, |g|)$. Conversely, if the inner product $(|f|, |g|)$ exists and, moreover, the product fg is measurable, then the inner product (f, g) exists.

Note that some particular inner products have already been used in Chapter 4. The following equalities clearly hold

$$\begin{aligned} (f, g) &= (g, f), \\ (\lambda f, g) = (f, \lambda g) &= \lambda(f, g), \\ (f+g, h) &= (f, h)+(g, h), \\ (f, g+h) &= (f, g)+(f, h), \\ (fg, h) &= (f, gh). \end{aligned} \tag{1}$$

Using the notation $\bar{f}(x) = f(-x)$, we can write

$$(f, g) = \int \bar{f}(0-t)g(t)\,dt.$$

Hence the inner product (f, g) exists whenever the convolution $\bar{f}*g$ exists at 0, and we have the equality

$$(f, g) = (\bar{f}*g)(0). \tag{2}$$

This fact enables us to reduce theorems concerning the inner product to theorems on convolution.

Instead of (2), we can equally write

$$(f, g) = (f * \bar{g})(0) \tag{3}$$

which follows from the commutativity both of the inner product (the first equality of (1)) and of convolution.

5.1.1. THEOREM. *We have*

$$(\bar{f}, g * h) = (f * g, \bar{h}), \tag{4}$$

whenever the convolution of f, g, h is associative at the origin.

PROOF. By (2), the left hand side of (4) is equal to $[f*(g*h)](0)$. By (3), the right hand side of (4) is equal to $[(f*g)*h](0)$. Hence the equality of the two sides follows by the associativity.

5.1.2. COROLLARY. *Equality* (4) *holds whenever* $(f*g*h)(0)$ *exists.*

5.2. INNER PRODUCT OF THREE FUNCTIONS

By the inner product (f, g, h) of three functions f, g, h we mean the value of the convolution $f*g*h$ at the origin,

$$(f, g, h) = (f*g*h)(0). \tag{1}$$

5.2.1. THEOREM. *The inner product (f, g, h) exists, iff the integral*

$$\int_{R^{2q}} f(t)g(u)h(-t-u)\,dtdu \tag{2}$$

exists in the sense of Lebesgue.

PROOF. In fact, by (1) and 1.2(1), we have

$$(f, g, h) = \int_{R^{2q}} f(-t)g(t-u)h(u)\,dudt = \int_{R^{2q}} f(t)g(u)h(-t-u)\,dtdu$$

which implies our assertion.

Since the integral (2) is taken in the sense of Lebesgue the existence of the inner product (f, g, h) implies the existence of the inner product $(|f|, |g|, |h|)$. Conversely, if the inner product $(|f|, |g|, |h|)$ exists and, moreover, the product $f(x)g(y)h(-x-y)$ is a measurable function in $\boldsymbol{R}^{2q}$, then the inner product (f, g, h) exists.

5.2.2. THEOREM. *If the inner product* (f, g, h) *exists, then*

$$(f, g, h) = (f*g, \bar{h}) = (\bar{f}, g*h) = (f*h, \bar{g}) \tag{3}$$

holds.

PROOF. In fact, if (f, g, h) exists, then the equalities

$$f*g*h = (f*g)*h = f*(g*h) = (f*h)*g$$

hold at the origin by Theorem 1.2.1 and 1.2 (2). Hence, by the definition of the inner product of two functions, the equalities (3) hold.

The properties of (f, g, h) follow from properties of the convolution $f*g*h$. Their detailed study is omitted here.

6. Convolution of distributions

6.1. DISTRIBUTIONS OF FINITE ORDER

In our study of the convolution of distributions, knowledge of a few properties of distributions of finite order will be useful. A distribution f in $\boldsymbol{R}^q$ is said to be of *finite order*, iff there is a continuous function F in $\boldsymbol{R}^q$ such that $f = F^{(k)}$ in $\boldsymbol{R}^q$ for some $k \in \boldsymbol{P}^q$.

6.1.1. THEOREM. *If G is a locally integrable function in $\boldsymbol{R}^q$ and $k \in \boldsymbol{P}^q$, then the distribution $f = G^{(k)}$ is of finite order.*

PROOF. The function $F(x) = \int_0^x G(t)dt$ is continuous and we have $f = F^{(k+1)}$, where $k+1$ is the vector each of whose coordinates is greater by 1 than the corresponding coordinate of k.

6.1.2. THEOREM. *The set of all distributions of finite order in $\boldsymbol{R}^q$ is a linear space.*

PROOF. It is clear that if f is of finite order, so is λf for every number λ. It therefore suffices to show that if f and g are of finite order, so is their sum $f+g$. Let $f = F^{(k)}$, $g = G^{(l)}$, where F and G are continuous functions. Let $p \in \boldsymbol{P}^q$, with $p \geqslant k$ and $p \geqslant l$, and let

$$\bar{F}(x) = \int_0^x F(t)\,dt^{p-k}, \qquad \bar{G}(x) = \int_0^x G(t)\,dt^{p-l}.$$

Then $f+g = (\bar{F}(x)+\bar{G}(x))^{(p)}$.

6.1.3. THEOREM. *If a distribution f vanishes outside a bounded interval I, it is of finite order.*

PROOF. There is a continuous function F in $\boldsymbol{R}^q$ such that $F^{(k)} = f$ in the neighbourhood $I_{2\alpha}$ $(\alpha > 0)$ of I, for some $k \in \boldsymbol{P}^q$. Let φ be a smooth function such that $\varphi(x) = 1$ on I_α and $\varphi(x) = 0$ outside $I_{2\alpha}$. Then

$$f = F^{(k)}\varphi = \sum_m (-1)^m \binom{k}{m} (F\varphi^{(m)})^{(k-m)}, \tag{1}$$

and the assertion now follows by 6.1.2.

By the *carrier* of a distribution we mean the smallest closed subset of $\boldsymbol{R}^q$ outside which the distribution vanishes.

6.1.4. THEOREM. *Let X be the carrier of a distribution f of order $k \in \boldsymbol{P}^q$, i.e., $f = F^{(k)}$ where F is a continuous function. Then for any $\alpha > 0$, f can be represented as a sum of derivatives of orders $\leqslant k$ of continuous functions whose carriers are contained in the neighbourhood X_α of X.*

PROOF. Let F be a continuous function in $\boldsymbol{R}^q$ such that $f = F^{(k)}$ and let φ be a smooth function which takes the value 1 on $X_{\alpha/2}$ and vanishes outside X_α. The assertion now follows from the equality (1).

6.2. CONVOLUTION OF A DISTRIBUTION WITH A SMOOTH FUNCTION OF BOUNDED CARRIER

Let φ be a smooth function which vanishes for $|x| > \alpha > 0$. If f_n is a fundamental sequence in an open set $O \subset \boldsymbol{R}^q$, then the convolutions $f_n * \varphi$ represent a fundamental sequence in another open set O' (see Part II, Section 2.8). It can be verified that $O' \subset O_{-\alpha}$, where $O_{-\alpha}$ is the set of all points $x \in O$ whose distance from the boundary of O is greater than α or is the whole set O if the boundary of O is empty (see Section 1.4). Given any distribution f in O, the convolution $f * \varphi$ is defined, as a regular operation, in O'. I.e., if f_n is a fundamental sequence for f, then $f * \varphi$ is given in O' by the fundamental sequence $f_n * \varphi$. The convolution $f * \varphi$ is uniquely defined in O' and represents a smooth function there (see Part II, Section 2.8).

Since convolution with a fixed φ is a regular operation and all the other operations in 1.4, formulae (2)–(8), are also regular, it follows that all those formulae are meaningful and also hold, for any distributions f, f_1, f_2.

6.2.1. THEOREM. *If f is any distribution in O and δ_n is a delta sequence, then $f * \delta_n$ converges distributionally to f in O.*

This theorem is given in Part II, as Theorem 3.9.3. Also, it follows trivially from the following, more general theorem.

6.2.2. THEOREM. *If a sequence of distributions f_n converges to f in O and δ_n is a delta sequence, then $f_n * \delta_n$ also converges to f in O (i.e., $f_n * \delta_n$ is a fundamental sequence for f).*

PROOF. Let I be an interval inside O. There are continuous functions F_n, F and $k \in \boldsymbol{P}^q$ such that the sequence F_n converges uniformly to F in I and $F_n^{(k)} = f_n$, $F^{(k)} = f$ in I. Hence, by Theorem 2.3.1 the sequence $F_n * \delta_n$ converges almost uniformly to F in I. By 1.4 (8),

$$(F_n * \delta_n)^{(k)} = f_n * \delta_n \quad \text{in} \quad I_{-\alpha_n}.$$

Hence the sequence $f_n * \delta_n$ converges distributionally to f in I. Since I is arbitrary, we have

$$f_n * \delta_n \to f \quad \text{in } O$$

and the theorem is proved.

In the next section, we shall also need the following theorem.

6.2.3. THEOREM. *Let f_n be any sequence of distributions which converges distributionally to f in $\boldsymbol{R}^q$. Let φ_n and φ be smooth functions in $\boldsymbol{R}^q$ such that $\varphi_n(x) = 0$ for $|x| > \alpha > 0$ and such that for every $m \in \boldsymbol{P}^q$, $\varphi_n^{(m)}$ converges uniformly to $\varphi^{(m)}$ as $n \to \infty$. Then, for every $m \in \boldsymbol{P}^q$, the sequence $(f_n * \varphi_n)^{(m)}$ converges almost uniformly to $(f * \varphi)^{(m)}$ in $\boldsymbol{R}^q$.*

PROOF. Let I be any given interval and let I' be an interval such that $I \subset I'_{-\alpha}$. There exist an order $k \in \boldsymbol{P}^q$ and continuous functions F_n, F such that the sequence F_n converges uniformly to F in I' and $F_n^{(k)} = f_n$, $F^{(k)} = f$ in I'. Now, for any given $m \in \boldsymbol{P}^q$, we have

$$(f_n * \varphi_n)^{(m)} = F_n * \varphi_n^{(m+k)} \quad \text{and} \quad (f * \varphi)^{(m)} = F * \varphi^{(m+k)} \quad \text{in } I.$$

Moreover, we have in I

$$\begin{aligned} |F_n * \varphi_n^{(m+k)} - F * \varphi^{(m+k)}| &\leqslant |F_n - F| * |\varphi_n^{(m+k)}| + |F| * |\varphi_n^{(m+k)} - \varphi^{(m+k)}| \\ &\leqslant \varepsilon_n \int |\varphi_n^{(m+k)}| + \eta_n \int_I |F|, \end{aligned}$$

where $\varepsilon_n \to 0$ and $\eta_n \to 0$ as $n \to \infty$. Thus the sequence $(f_n * \varphi_n)^{(m)}$ converges uniformly to $(f * \varphi)^{(m)}$ in I, which implies our assertion.

The sequence $f_n = f * \delta_n$, where f is a distribution in O and δ_n is a delta sequence, is called a *regular sequence* for f. Since its terms f_n are smooth functions, it is a fundamental sequence for f, by 6.2.1. The class of regular sequences is therefore a special subclass of fundamental sequences.

6.3. CONVOLUTION OF TWO DISTRIBUTIONS

Let f and g be distributions in $\boldsymbol{R}^q$ and let $f_n = f * \delta_n$ and $g_n = g * \delta_n$ be their regular sequences (with the same delta sequence). We say that the convolution of f and g exists, iff, for every delta sequence δ_n, the corresponding convolutions $f_n * g_n$ exist smoothly and represent a fundamental sequence. The distribution determined by that fundamental sequence is, by definition, the convolution of f and g.

The consistency of the definition follows from the following reasoning. Assume that we have two different delta sequences δ_{1n} and δ_{2n}. If the convolution $f*g$ exists, then both the sequences

$$(f*\delta_{1n})*(g*\delta_{1n}) \quad \text{and} \quad (f*\delta_{2n})*(g*\delta_{2n}) \tag{1}$$

are fundamental. We have to show that they represent the same distribution. Now, if δ_n is the nth element of the interlaced sequence 2.1(2), then the sequence δ_n is also a delta sequence. This implies that the sequence $(f*\delta_n)*(g*\delta_n)$ is fundamental. But the sequences (1) are subsequences of this last sequence, and they represent the same distribution.

Convolution is not a regular operation. If it were, the sequence f_n*g_n would be fundamental for any fundamental sequences f_n and g_n corresponding to distributions f and g. However, in our present definition, we restrict ourselves to those particular fundamental sequences which are of the form $f_n = f*\delta_n$ and $g_n = g*\delta_n$. This is one way in which convolution fails to be a regular operation. Another is that the convolution $f*g$ is not defined for all pairs of distributions f, g.

In particular, if f and g are locally integrable functions in $\boldsymbol{R}^q$ such that the convolution of their moduli exists a.e. and represents a locally integrable function in $\boldsymbol{R}^q$, then, by 3.6.1, the convolutions f_n*g_n exist smoothly. Moreover, by 3.4.5, the sequence f_n*g_n converges locally in mean, and thus distributionally, to $f*g$. Hence $f*g$ is compatible with the general definition of convolution of distributions. The use of the symbol $f*g$ for the convolution of distributions therefore does not lead to any confusion, provided we adopt in the sequel the convention that, in the particular case of locally integrable functions f and g, the symbol $f*g$ can equally be interpreted as the integral $\int_{\boldsymbol{R}^q} f(x-t)g(t)\,dt$ only if the integral $\int_{\boldsymbol{R}^q} |f(x-t)g(t)|\,dt$ exist a.e. and represents a locally integrable function.

We have to show further that our present definition is compatible with the definition in Section 6.2, if the distribution g reduces to a smooth function of bounded carrier. It is not difficult to check that the convolutions f_n*g_n exist then smoothly, it thus suffices to verify that $f_n*g_n - f_n*g \to 0$, i.e., that $f_n*(g_n-g) \to 0$. In fact, by 2.2.2, we have $(g_n-g)^{(k)} \to 0$ almost uniformly for every $k \in \boldsymbol{P}^q$. Since $f_n \to f$, this implies that $f_n*(g_n-g) \to f*0 = 0$, by 6.2.3.

6.3.1. THEOREM. *If the convolution $f*g$ of the distributions f and g exists, then the convolutions $(\lambda f)*g$ and $f*(\lambda g)$ also exist for every real number λ and we have*

$$(\lambda f)*g = g*(\lambda f) = \lambda(f*g).$$

6.3.2. THEOREM. *If the convolutions* $f*g$ *and* $f*h$ *exist, then the convolution* $f*(g+h)$ *also exists, and we have*

$$f*(g+h) = f*g+f*h.$$

6.3.3. THEOREM. *If the convolution* $f*g$ *exists, then* $g*f$ *also exists and we have*

$$f*g = g*f.$$

6.3.4. THEOREM. *If the convolution* $f*g$ *exists, the convolutions* $f^{(k)}*g$ *and* $f*g^{(k)}$ *also exist for every* $k \in \boldsymbol{P}^q$ *and we have*

$$(f*g)^{(k)} = f^{(k)}*g = f*g^{(k)}.$$

PROOFS OF THEOREMS 6.3.1–6.3.4. Let δ_n be a delta sequence and let $f_n = f*\delta_n$, $g_n = g*\delta_n$, $h_n = h*\delta_n$. If the convolution $f*g$ exists, then the convolutions f_n*g_n exist smoothly and we have $(\lambda f_n)*g_n = f_n*(\lambda g_n) = \lambda(f_n*g_n)$, $f_n*g_n = g_n*f_n$, $(f_n*g_n)^{(k)} = f_n^{(k)}*g_n = f_n*g_n^{(k)}$ and all the convolutions involved exist smoothly. This proves 6.3.1, 6.3.3 and 6.3.4. If, moreover $f*h$ exist, then $f_n*(g_n+h_n) = f_n*g_n+f_n*h_n$ and all the convolutions exist smoothly. This proves 6.3.2.

6.3.5. THEOREM. *If the convolution* $f*g$ *exists, then* $f^{(k)}*g^{(l)}$ *also exists for any* $k,l \in \boldsymbol{P}^q$ *and we have*

$$f^{(k)}*g^{(l)} = (f*g_n)^{(k+l)}.$$

PROOF. By 6.3.4, we have

$$(f*g)^{(k+l)} = [(f*g)^{(k)}]^{(l)} = (f^{(k)}*g)^{(l)} = f^{(k)}*g^{(l)}.$$

This last theorem is particularly important, as it enables us to construct various classes of distributions within which convolution is always feasible.

EXAMPLE 1. A distribution f in $\boldsymbol{R}^q$ is said to be *tempered*, if it is the derivative of some order $k \in \boldsymbol{P}^q$ of a slowly increasing function F, that is, F is a measurable function bounded by a polynomial. (Another, but equivalent, definition of tempered distributions will be given in Chapter 7.)

As we saw in Section 3.1 (Example 3) the convolution $F*G$ of F with any rapidly decreasing function G exists and is a slowly increasing function. We now prove that $F*G$ also exists in the distributional sense and represents the same slowly increasing function. In fact, if δ_n is any delta sequence, then the convolution

$$(F*\delta_n)*(G*\delta_n)$$

exists smoothly, by 3.6.1. Moreover, by 3.4.5, the above sequence tends locally in mean and thus distributionally to $F*G$. This proves our assertion.

Now, by 6.3.5, the convolution $F^{(k)}*G^{(l)}$ exists for any $k, l \in \boldsymbol{P}^q$.

Since $f = F^{(k)}$, this implies, by 6.3.5, that the convolution $f*g$ exists for every distribution $g = G^{(l)}$.

Hence if f is a tempered distribution, the convolution $f*g$ exists for every distribution g which is the derivative of some order k of a rapidly decreasing function.

If we have any finite number of distributions $g_1, \ldots, g_r$ which are the derivatives of some orders of rapidly decreasing functions, then their sum $g = g_1 + \ldots + g_r$ is called a *rapidly decreasing distribution.* Since the convolutions $f*g_1, \ldots, f*g_r$ exist, so does the convolution $f*g$. Thus, the *convolution $f*g$ exists, if one of the factors is a tempered distribution and the other one is a rapidly decreasing distribution.*

Example 2. A measurable function F in $\boldsymbol{R}^q$ is said to belong to the class L^p, iff $|F|^p$ is integrable over $\boldsymbol{R}^q$. If $F \in L^p$ with $p \geqslant 1$, then F is locally integrable. There is a theorem, due to Young, which states that if $F \in L^p$ and $G \in L^s$, where

$$p, s \geqslant 1 \quad \text{and} \quad r = 1 \Bigg| \left(\frac{1}{p} + \frac{1}{s} - 1\right) > 0, \tag{2}$$

then $F*G \in L^r$.

Applying 6.3.5, we see that if f is the derivative of order k of a function belonging to L^p and g is the derivative of order l of a function belonging to L^s and if (2) holds, then $f*g$ is the derivative of order $k+l$ of a function belonging to L^r.

If $f = f_1 + \ldots + f_r$, where the f_i are the derivatives of some orders of functions belonging to L^p, we write, following L. Schwartz, $f \in \mathscr{D}'_{L^p}$. Since L^q is a linear space, so is $\mathscr{D}'_{L^p}$. From the foregoing result it follows that if $f \in \mathscr{D}'_{L^p}$ and $g \in \mathscr{D}'_{L^s}$ and (2) holds, then $f*g \in \mathscr{D}'_{L^r}$.

6.4. CONVOLUTION OF DISTRIBUTIONS WITH COMPATIBLE CARRIERS

The concept of the *carrier* or *support* of a distribution is very familiar. L. Schwartz defines it as the smallest closed set outside which the distribution vanishes. This definition will also be used in this book; for a more sharper definition see Section 12.2.

We prove the following theorems:

6.4.1. Theorem. *Let f and g be any distributions in $\boldsymbol{R}^q$ whose carriers are contained in the compatible closed sets X and Y, respectively. Then the convolution $f*g$ exists in $\boldsymbol{R}^q$ and its carrier is contained in $X+Y$.*

6.4.2. THEOREM. *Let X and Y be two compatible sets. Then, for any fixed bounded interval I there is a smooth function φ of bounded carrier such that, if the carriers of given distributions f and g are contained in X and Y respectively, then* $f*g = f\varphi * g\varphi$ *in I.*

The proofs of these theorems will be based on

6.4.3. LEMMA. *Let X and Y be two bounded sets, and suppose that the carriers of the distributions f_n and g_n are contained in X and Y, respectively. If $f_n \to f$ and $g_n \to g$, then the convolutions $f_n * g_n$, $f*g$ exist and $f_n * g_n \to f*g$.*

PROOF OF LEMMA 6.4.3. Case 1. Assume first that, in addition, f_n and g_n are continuous functions and that the convergence is uniform. Then there exist sequences of numbers ε_n and η_n, converging to 0, such that $|f_n - f| < \varepsilon_n$ and $|g_n - g| < \eta_n$. There exists further a number M such that $|g_n| < M$ and $|f| < M$. Since the carriers of $|g_n|$ and of $|g_n - g|$ are contained in Y, we have

$$|f_n * g_n - f*g| \leqslant |f_n - f| * |g_n| + |f| * |g_n - g| \leqslant \int_Y \varepsilon_n M + \int_Y M\eta_n = (\varepsilon_n + \eta_n)\int_Y M,$$

which implies that $f_n * g_n \rightrightarrows f*g$. The existence of the convolutions follows by 3.2.2.

Case 2. Assume now that $f_n = F_n^{(m)}$, $g_n = G_n^{(k)}$, where F_n, G_n are continuous functions whose carriers are contained in X and Y respectively, and $F_n \rightrightarrows F$, $G_n \rightrightarrows G$. Then $F^{(m)} = f$ and $G^{(k)} = g$. Since

$$f_n * g_n = (F_n * G_n)^{(m+k)},$$

the result just proved in Case 1 implies $f_n * g_n \to (F*G)^{(m+k)} = f*g$.

Case 3. In the general case, we take a smooth function φ such that $\varphi = 1$ in $X \cup Y$ and $\varphi = 0$ outside an interval I such that $X \cup Y$ is inside I. Then $f_n = f_n\varphi$, $f = f\varphi$, $g_n = g_n\varphi$ and $g = g\varphi$. Let I_0 be an interval such that I is inside I_0. There exist continuous functions F_n, F in I_0 such that $F_n^{(k)} = f_n$ and $F^{(k)} = f$ for some order k and $F_n \rightrightarrows F$ in I_0. We have

$$f_n = F_n^{(k)}\varphi = \sum_{0 \leqslant m \leqslant k} F_{mn}^{(m)} \quad \text{where} \quad F_{mn} = (-1)_m \binom{k}{m} F_n \varphi^{(k-m)}.$$

Now the functions F_{mn} are continuous and we have $F_{mn} \to F_m$ (as $n \to \infty$), where $F_m = (-1)^m \binom{k}{m} F\varphi^{(k-m)}$. The sequence f_n can thus be expressed as a finite sum of sequences of distributions of the form considered in Case 2, and the same is true of the sequence g_n. Since the sums are finite, the general case thus reduces to Case 2.

PROOF OF THEOREMS 6.4.1 AND 6.4.2. Given any $\alpha > 0$, there exists a delta sequence δ_n such that $\delta_n(x) = 0$ for $|x| > \alpha$. The carriers of the smooth functions $f_n = f * \delta_n$ and $g_n = g * \delta_n$ are contained in X_α and Y_α, respectively. By 3.3.3, the sets X_α and Y_α are compatible. Thus, given any I, there is a bounded interval J such that, for each $x \in I$, the product $X_\alpha(x-t)Y_\alpha(t)$ vanishes outside J. Consequently so do the products $f_n^{(r)}(x-t)g_n^{(s)}(t)$ for $r,s \in \boldsymbol{P}^q$. This implies the existence of the convolutions $f_n^{(r)} * g_n^{(s)}$ and $|f_n^{(r)}| * |g_n^{(s)}|$ in I and, consequently, in $\boldsymbol{R}^q$ (for I is arbitrary). Let I' be a bounded interval such that each of the sets J and $I-J$ (i.e., the set of the differences $x-t$ with $x \in I$, $t \in J$) are inside I'. For φ we take a smooth function which vanishes outside I' and admits the value 1 on the sets J and $I-J$. The functions $\bar{f}_n = f_n\varphi$ and $\bar{g}_n = g_n\varphi$ are then smooth and of bounded carrier so that their convolutions $\bar{f}_n * \bar{g}_n$ exist smoothly. I.e., the $\bar{f}_n^{(r)} * \bar{g}_n^{(s)}$ are continuous in $\boldsymbol{R}^q$ and the $|\bar{f}_n^{(r)}| * |\bar{g}_n^{(s)}|$ are locally integrable in $\boldsymbol{R}^q$. Evidently, for any $x \in I$ and $t \in \boldsymbol{R}^q$ we have $f_n^{(r)}(x-t)\, g_n^{(s)}(t) = \bar{f}_n^{(r)}(x-t)\bar{g}_n^{(s)}(t)$. This implies that $f_n^{(r)} * g_n^{(s)} = \bar{f}_n^{(r)} * \bar{g}_n^{(s)}$ in I and $|f_n^{(r)}| * |g_n^{(s)}| = |\bar{f}_n^{(r)}| * |\bar{g}_n^{(s)}|$ in I. The convolutions $f_n^{(r)} * g_n^{(s)}$ are thus continuous in I and $|f_n^{(r)}| * |g_n^{(s)}|$ integrable in I. Consequently, they are so in $\boldsymbol{R}^q$ (for I is arbitrary), which means that $f_n * g_n$ exist smoothly.

Since $\bar{f}_n \to f\varphi$ and $\bar{g}_n \to g\varphi$ and the carriers of $\bar{f}_n$, $\bar{g}_n$ are included in the bounded set I, we have $\bar{f}_n * \bar{g}_n \to f\varphi * g\varphi$, by 6.4.3. This implies that $f_n * g_n \to f\varphi * g\varphi$ in I and, consequently, in $\boldsymbol{R}^q$ (for I is arbitrary). Since the $f_n * g_n$ exist smoothly, their limit in $\boldsymbol{R}^q$ is, by definition, $f * g$. Thus the formula $f * g = f\varphi * g\varphi$ holds in I and 6.4.2 is proved.

To complete the proof of 6.4.1, it remains to show that, if the sets X and Y are closed, the carrier of $f * g$ is included in $X+Y$. This easy task is left to the reader.

6.4.4. Corollary. *If f, g are distributions in $\boldsymbol{R}^q$ and at least one of them is of bounded carrier, then the convolution $f * g$ exists in $\boldsymbol{R}^q$.*

Proof. The carriers are then compatible, and so the assertion follows from 6.4.1.

6.4.5. Theorem. *Let X and Y be two compatible sets. We assume that the carriers of the distributions f_n and g_n are contained in X and Y, respectively. If $f_n \to f$ and $g_n \to g$, then $f_n * g_n \to f * g$.*

Proof. The existence of the convolutions $f_n * g_n$ and $f * g$ follows from 3.3.2 and 6.4.1. Let I be any given bounded interval. By 6.4.2, there is a smooth function φ of bounded carrier K such that $f_n * g_n = f_n\varphi * g_n\varphi$ in I. Similarly, we have $f * g = f\varphi * g\varphi$ in I, because the carriers of f, g are contained in X, Y as well as the carriers of f_n, g_n. Since $f_n\varphi \to f\varphi$ any $g_n\varphi \to g\varphi$, it follows, by 6.4.3, that $f_n\varphi * g_n\varphi \to f\varphi * g\varphi$. This implies that $f_n * g_n \to f * g$ in I. Since I is arbitrary, the convergence holds in the whole space $\boldsymbol{R}^q$.

6.4.6. THEOREM. *Let f, g and h be distributions in $\boldsymbol{R}^q$ with carriers X, Y and Z respectively. If X, Y, Z are compatible, then all the convolutions involved in the equality*

$$(f*g)*h = f*(g*h) \tag{1}$$

exist and the equality holds.

PROOF. The existence of all the convolutions in the equality (1) follows from 3.3.4 and 6.4.1. It remains to show that the equality (1) holds. Let $f_n = f*\delta_n$, $g_n = g*\delta_n$ and $h_n = h*\delta_n$. If the carriers X, Y, Z of f, g, h are compatible, the carriers of f_n, g_n, h_n are contained in compatible sets X_α, Y_α, Z_α and therefore, are themselves compatible. Hence, by 3.4.1 all the convolutions in the equality

$$(f_n*g_n)*h_n = f_n*(g_n*h_n) \tag{2}$$

exist and the equality holds. Hence equality (1) follows, by 6.4.5.

6.4.7. COROLLARY. *If two of the distributions f, g, h are of bounded carrier, all the convolutions involved in* (1) *exist and the equality holds.*

PROOF. The corollary follows immediately as a particular case of 6.4.6.

6.4.8. THEOREM. *If f, g, h are distributions in $\boldsymbol{R}^q$, and one of them is of bounded carrier, while the convolution of the remaining two exists, then all the convolutions involved in the equality* (1) *exist and the equality holds.*

PROOF. Assume that $f*g$ exists and that h is of bounded carrier. The existence of the convolutions $(f*g)*h$ and $g*h$ follows from 6.4.4. We also prove that $f*(g*h)$ exists and that (1) holds. Let $f_n = f*\delta_n$, $g_n = g*\delta_n$. Then the convolutions f_n*g_n exist smoothly and tend to $f*g$, as $n \to \infty$, by the hypothesis that $f*g$ exists. Now, by 6.4.5 we have $(f_n*g_n)*h \to (f*g)*h$. Assume that h is a continuous function. Letting $u_n = (g*h)*\delta_n$, we then have

$$f_n^{(r)}*u_n^{(s)} = f_n^{(r)}*(g_n*h)^{(s)} = f_n^{(r)}*(g_n^{(s)}*h) = (f_n^{(r)}*g_n^{(s)})*h\,; \tag{3}$$

here, the first equality follows from the fact that, by 6.4.7, $(g*h)*\delta_n = (g*\delta_n)*h$; the second follows by 1.4 (8), and the last equality follows, by 3.4.3, from the fact that $|f_n^{(r)}|*|g_n^{(s)}|$ is locally integrable (which in turn is due to the smooth existence of the f_n*g_n). Similarly we can prove that

$$|f_n^{(r)}|*|u_n^{(s)}| \leqslant |f_n^{(r)}|*(|g_n^{(s)}|*|h|) = (|f_n^{(r)}|*|g_n^{(s)}|)*|h|\,. \tag{4}$$

It is easily seen from (3) and (4) that the $f_n^{(r)}*u_n^{(s)}$ are continuous and the $|f_n^{(r)}|*|u_n^{(s)}|$ are locally integrable so that the convolutions f_n*u_n exist smoothly. But (3) also implies that the sequence f_n*u_n converges to $(f*g)*h$. This implies the existence of the convolutions $f*(g*h)$ and equality (1). If h is not a continuous function, the assertion follows from what we have proved, by 6.1.4.

If we assume the boundedness of the carrier of f or g (instead of h), the assertion still follows, by the commutativity of convolution.

6.4.9. Theorem. *If f, g, h, k are distributions such that the convolution $f*g$ exists and each of h, k is of bounded carrier, then*

$$(f*h)*(g*k) = (f*g)*(h*k)$$

and all the convolutions appearing in the equality exist.

The proof is word for word the same as that of 3.4.4. The only difference is that it is in fact based on 6.4.6 and 6.4.7 instead of on 3.4.2 and 3.4.3.

7. Tempered distributions

7.1. TEMPERED DERIVATIVES

By the *kth tempered derivative of distribution* f defined in $\boldsymbol{R}^q$, we mean the distribution

$$D^k f = E^{-1}(Ef)^{(k)} \qquad (k \in \boldsymbol{P}^q),$$

where

$$E(x) = e^{x^2/4} = e^{(\xi_1^2+\ldots+\xi_q^2)/4}$$

with $x = (\xi_1, \ldots, \xi_q)$. We may also make use of the complementary derivative

$$d^k f = E(E^{-1}f)^{(k)}.$$

It is easy to check that

$$D^0 f = f, \qquad d^0 f = f,$$

$$D^k D^m f = D^{k+m} f, \quad d^k d^m f = d^{k+m} f.$$

Both derivatives are plainly linear operations, i.e.,

$$D^k(f+g) = D^k f + D^k g, \quad d^k(f+g) = d^k f + d^k g,$$

$$D^k(cf) = cD^k f, \qquad d^k(cf) = cd^k f$$

for any number c. Moreover,

$$D^k(fg) = \sum_m \binom{k}{m} D^m f \cdot g^{(k-m)}, \quad d^k(fg) = \sum_m \binom{k}{m} d^m f \cdot g^{(k-m)}, \tag{1}$$

where $0 \leqslant m \leqslant k$: these formulae are easily obtained from formula 16.6 (1) of the Appendix, by applying it to $((Ef)g)^{(k)}$ and to $((E^{-1}f)g)^{(k)}$.

Since $E^{(e_j)} = \frac{1}{2}\xi_j E$, we obtain

$$D^{e_j} f = f^{(e_j)} + \tfrac{1}{2}\xi_j f, \quad d^{e_j} f = f^{(e_j)} - \tfrac{1}{2}\xi_j f, \tag{2}$$

and hence

$$D^{e_j}(\xi_j f) = \xi_j D^{e_j} f + f, \quad d^{e_j}(\xi_j f) = \xi_j d^{e_j} f + f.$$

The last two equalities are special cases of (1).

We now give a few equalities connecting the derivatives D and d. In view of (2), it is easy to check that

$$D^{e_j} f - d^{e_j} f = \xi_j f, \quad D^{e_j} f + d^{e_j} f = 2f^{(e_j)}, \quad D^{e_i} d^{e_j} f = d^{e_j} D^{e_i} f \quad \text{for} \quad i \neq j. \tag{3}$$

We shall also need the formulae

$$D^k d^{e_j} f = d^{e_j} D^k f - \varkappa_j D^{k-e_j} f, \quad d^k D^{e_j} f = D^{e_j} d^k f + \varkappa_j d^{k-e_j} f, \tag{4}$$

where $k = (\varkappa_1, \ldots, \varkappa_q) \in \boldsymbol{P}^q$ (if $\varkappa_j = 0$ for some j, the symbol D^{k-e_j} is meaningless but, according to the general convention (see Section 1.1), we nevertheless set $\varkappa_j D^{k-e_j} f = 0$, because one of the factors is 0). To prove (4), we apply, in turn, formulae (3), (1) and (3) again: thus

$$D^k d^{e_j} f = D^k(D^{e_j} f - \xi_j f) = D^{k+e_j} f - \xi_j D^k f - \varkappa_j D^{k-e_j} f = d^{e_j} D^k f - \varkappa_j D^{k-e_j} f.$$

The proof of the second of the equalities (4) is similar.

7.2. TEMPERED INTEGRALS

Our aim is to introduce *tempered distributions*. They can be defined as tempered derivatives of square integrable functions (see Section 7.3). It should be noted that, in this definition, the tempered derivative D cannot be replaced by the complementary derivative d (see Section 8.3), and so in this context the derivative D turns out to be more important than d, whose role is of a more auxiliary nature. We therefore study the properties of D in somewhat greater detail. In particular, we introduce the inverse operation S.

If F is a continuous function in $\boldsymbol{R}^q$, we write

$$S^k F(x) = E^{-1}(x) \int_0^x E(t) F(t)\, dt^k. \tag{1}$$

Clearly

$$D^k S^k F = F.$$

S^k is therefore, in some sense, an inverse operation with respect to D^k. It cannot be defined for arbitrary distributions, since the integral in (1) is not meaningful for all distributions.

If $k = (1, \ldots, 1)$, we simply write S instead of S^k.

In particular, we can state that if F is a locally integrable function, $S^k F$ is also locally integrable (or, better still, continuous, whenever $k \geqslant 1$).

S^k is clearly a linear operation, i.e.,

$$S^k(F+G) = S^k F + S^k G \quad \text{and} \quad S^k(cF) = cS^k F$$

for every number c. Moreover

$$|S^k F| \leqslant \big|S^k |F|\big|.$$

We introduce the auxiliary functions

$$B(\varrho, \xi) = E^{-1}(\xi)(1+\xi)^{1-\varrho} \int_0^\xi E(\tau)(1+\tau)^\varrho d\tau \quad (\xi \geqslant 0)$$

and

$$\beta(\varrho) = \sup_{0 \leqslant \xi < \infty} B(\varrho, \xi).$$

Using l'Hospital's rule, it is easy to show that $B(\varrho, \xi)$ tends to 2, as $\xi \to \infty$, independently of the value of the real number ϱ. This implies that the function $\beta(\varrho)$ takes a finite positive value for every real ϱ.

For $x = (\xi_1, \ldots, \xi_q) \in \boldsymbol{R}^q$, we write

$$\hat{x} = (1+|\xi_1|, \ldots, 1+|\xi_q|).$$

$\hat{x}$ is thus a point (or vector) in $\boldsymbol{R}^q$ none of whose coordinates is less than one. If $r = (\varrho_1, \ldots, \varrho_q) \in \boldsymbol{R}^q$, we mean, by the power $\hat{x}^r$, the positive number

$$\hat{x}^r = (1+|\xi_1|)^{\varrho_1} \ldots (1+|\xi_q|)^{\varrho_q}.$$

We now prove that

$$|S^{e_j}\hat{x}^r| \leqslant \beta(\varrho_j)\hat{x}^{r-e_j}. \tag{2}$$

In fact, we have

$$S^{e_j}\hat{x}^r = E^{-1}(x)\int_0^x E(t)\hat{t}^r dt^{e_j}.$$

It is easy to verify that

$$E(t)\hat{t}^r = E(t-\tau_j e_j)\hat{t}^{r-\varrho_j e_j}E(\tau_j e_j)(1+|\tau_j|)^{\varrho_j},$$

where the first factor on the right hand side is constant with respect to τ_j. This implies that, for $\xi_j \geqslant 0$,

$$S^{e_j}\hat{x}^r = E^{-1}(x)E(x-\xi_j e_j)\hat{x}^{r-\varrho_j e_j}\int_0^{\xi_j} E(\tau)(1+\tau)^{\varrho_j}d\tau = \hat{x}^{r-e_j}B(\varrho_j, \xi_j).$$

Hence inequality (2) follows for $\xi_j \geqslant 0$. However, both sides of (2) are even functions of ξ_j so that (2) holds for every $x \in \boldsymbol{R}^q$.

7.2.1. THEOREM. *For any given $r \in \boldsymbol{P}^q$, $m \in \boldsymbol{P}^q$ and $s \in \boldsymbol{P}^q$, there exists a number $\beta = \beta(r, m, s)$ with the following property: if G is a continuous function such that $|G(x)| < \hat{x}^r$ for all $x \in \boldsymbol{R}^q$, there is a continuous function F such that $D^{m+s}F = G^{(m)}$ and $|F(x)| < \beta\hat{x}^{r-s}$.*

PROOF. We trivially have $\beta(r, 0, 0) = 1$. We now use $2q$-dimensional induction with respect to the vector $(m, s) = (\mu_1, \ldots, \mu_q, \sigma_1, \ldots, \sigma_q)$. It is thus enough to show that if the assertion is true for all $m \leqslant k \in \boldsymbol{P}^q$ and all $s \leqslant l \in \boldsymbol{P}^q$ then it is also true for (i) $m = k+e_j$, $s = l$, and for (ii) $m = k$, $s = l+e_j$.

Case (i). From the inequality $|G(x)| < \hat{x}^r$, it follows that $|\xi_j G(x)| < \hat{x}^{r+e_j}$. Hence by the induction hypothesis, there are continuous functions F_1, F_2 and, if $\varkappa_j \geqslant 1$, F_3, such that

$$D^{k+l}F_1 = G^{(k)}, \qquad D^{k+l}F_2 = (\xi_j G)^{(k)}, \qquad D^{k-e_j+l}F_3 = G^{(k-e_j)}$$

and

$$|F_1(x)| < \beta_1 \hat{x}^{r-l}, \qquad |F_2(x)| < \beta_2 \hat{x}^{r+e_j-l}, \qquad |F_3(x)| < \beta_3 \hat{x}^{r-l}$$

with $\beta_1 = \beta(r, k, l)$, $\beta_2 = \beta(r+e_j, k, l)$, $\beta_3 = \beta(r, k-e_j, l)$.

We have

$$D^{k+e_j+l}F_1 = G^{(k+e_j)} + \tfrac{1}{2}\xi_j G^{(k)} = G^{(k+e_j)} + \tfrac{1}{2}(\xi_j G)^{(k)} - \tfrac{1}{2}\varkappa_j G^{(k-e_j)}$$

and hence

$$G^{(k+e_j)} = D^{k+e_j+l}F_1 - \tfrac{1}{2}D^{k+l}F_2 + \tfrac{1}{2}\varkappa_j D^{k-e_j+l}F_3, \tag{3}$$

where the products by $\varkappa_j$ are neglected if $\varkappa_j = 0$.

Let $F_4 = S^{e_j}F_2$. Then, by (2),

$$|F_4(x)| < \beta_2 |S^{e_j}\hat{x}^{r+e_j-l}| \leqslant \beta_4 \hat{x}^{r-l}$$

with $\beta_4 = \beta_2 \cdot \beta(\varrho_j + 1 - \lambda_j)$, where λ_j is the jth coordinate of l. If $\varkappa_j \geqslant 1$, we put $F_5 = S^{2e_j}F_3$ and obtain, applying inequality (2) twice,

$$|F_5(x)| < \beta_3 |S^{2e_j}\hat{x}^{r-l}| \leqslant \beta_5 \hat{x}^{r-l-2e_j} \leqslant \beta_5 \hat{x}^{r-l}$$

with $\beta_5 = \beta_3 \cdot \beta(\varrho_j - \lambda_j) \cdot \beta(\varrho_j - \lambda_j - 1)$.

Finally, we put $F = F_1 - \tfrac{1}{2}F_4 + \tfrac{1}{2}\varkappa_j F_5$. Then, by (3),

$$D^{k+e_j+l}F = G^{(k+e_j)}$$

and, moreover,

$$|F(x)| < (\beta_1 + \beta_4 + \varkappa_j \beta_5)\hat{x}^{r-l}.$$

Case (ii). As before, there is a continuous function F_1 such that $D^{k+l}F_1 = G^{(k)}$ and $|F_1(x)| < \beta_1 \hat{x}^{r-l}$. We put $F = S^{e_j}F_1$. Then

$$D^{k+e_j+l}F = G^{(k)}$$

and

$$|F(x)| < \beta_1 |S^{e_j}\hat{x}^{r-l}| \leqslant \beta_1 \cdot \beta(\varrho_j - \lambda_j)\hat{x}^{r-l-e_j}.$$

This completes the proof of 7.2.1.

7.2.2. THEOREM. *If f is measurable and bounded, Sf is square integrable.*

PROOF. There is a number M such that $|f| \leqslant M$. We have

$$|Sf| \leqslant |S|f|| \leqslant |SM| = M|S1| \leqslant M\hat{x}^{-1}.$$

Since $\hat{x}^{-1}$ is square integrable, so is Sf.

7.2.3. THEOREM. *If a sequence of bounded measurable functions f_n converges uniformly to f, then the sequence Sf_n converges in square mean to Sf.*

PROOF. There is a sequence $\varepsilon_n \to 0$ such that $|f_n - f| \leqslant \varepsilon_n$ in $\boldsymbol{R}^q$. We thus have

$$||Sf_n - Sf|| = ||S(f_n - f)|| \leqslant ||S|f_n - f||| \leqslant ||S\varepsilon_n|| \leqslant \varepsilon_n ||S1|| = \varepsilon_n ||\hat{x}^{-1}|| \to 0.$$

7.3. TEMPERED DISTRIBUTIONS

We say that a distribution f, defined in $\boldsymbol{R}^q$, is *tempered*, iff there is a function F, square integrable in $\boldsymbol{R}^q$, such that $D^k F = f$ for some $k \in \boldsymbol{P}^q$. In other words, a distribution is tempered iff it is a tempered derivative of some order of a square integrable function in $\boldsymbol{R}^q$.

7.3.1. THEOREM. *A distribution f is tempered, iff there exist $m \in \boldsymbol{P}^q$, $r \in \boldsymbol{P}^q$ and a continuous function G such that*

$$G^{(m)} = f \quad \textit{and} \quad \hat{x}^{-r} G \textit{ is bounded in } \boldsymbol{R}^q. \tag{1}$$

REMARK. It is easy to see that a function G is slowly increasing (see Example 3 in Section 3.1) iff there is an $r \in \boldsymbol{P}^q$ such that $\hat{x}^{-r} G$ is bounded (see Example 1 in Section 6.3). Thus tempered distributions are derivatives (of some order) of slowly increasing functions. Evidently, polynomials are very particular tempered distributions.

PROOF OF THEOREM 7.3.1. Assume that f is tempered, i.e., that $f = D^k F$ where $F \in L^2$. ($F \in L^2$ means that F is square integrable over $\boldsymbol{R}^q$.) Let $G(x) = \int_0^x F(t)\,dt$. Then, for a positive constant M, we have

$$|G(x)| \leqslant \left| \int_0^x dt \cdot \int_0^x |F(t)|^2 dt \right|^{1/2} \leqslant \sqrt{|\xi_1 \ldots \xi_q|} \cdot M \leqslant M \hat{x}^1,$$

where $\hat{x}^1 = (1 + |\xi_1|) \ldots (1 + |\xi_q|)$. If $k = 0$, then $f = G'$, and condition (1) is satisfied with $m = r = (1, \ldots, 1)$.

Assume that, for some $k \in \boldsymbol{P}^q$, condition (1) is satisfied for suitably chosen m and r. Let $f = D^{k+e_j} F$, $F \in L^2$. Then

$$\begin{aligned} f = D^{e_j} D^k F = D^{e_j} G^{(m)} &= G^{(m+e_j)} + \tfrac{1}{2} \xi_j G^{(m)} \\ &= G^{(m+e_j)} + \tfrac{1}{2} (\xi_j G)^{(m)} - \tfrac{1}{2} \mu_j G^{(m-e_j)} \end{aligned} \tag{2}$$

where μ_j is the jth coordinate of m, and where if $\mu_j = 0$, the product $\mu_j G^{(m-e_j)}$ is to be replaced by 0. Hence

$$f = H^{(m+e_j)},$$

where

$$H(x) = G(x) + \tfrac{1}{2} \int_0^x \tau_j G(t)\, dt^{e_j} - \tfrac{1}{2} \mu_j \int_0^x G(t)\, dt^{2e_j}. \tag{3}$$

By induction hypothesis $\hat{x}^{-r}G$ is bounded, so that $\hat{x}^{-r-2e_j}H$ is also bounded. By induction, the necessity of the condition follows.

Assume now, conversely, that $f = G^{(m)}$, where G is a continuous function such that $\hat{x}^{-r}G$ is bounded. Then, by 7.2.1, there is a continuous function F such that $D^{m+r+1}F = G^{(m)}$ and $|F(x)| < M\hat{x}^{-1}$. F is clearly square integrable, and the proof is thus complete.

7.3.2. THEOREM. *If a distribution f is tempered, the distributions $D^p f$, $f^{(p)}$, $x^p f$ and $d^p f$ are also tempered for every $p \in \boldsymbol{P}^q$.*

PROOF. If f is tempered, there is, by definition, a square integrable function F such that $f = D^k F$ for some $k \in \boldsymbol{P}^q$. Hence $D^p f = D^{k+p}F$, which shows that $D^p f$ is tempered.

There is also a continuous function G satisfying (1). Hence $f^{(p)} = G^{(m+p)}$ which shows that $f^{(p)}$ is tempered, by Theorem 7.3.1.

We have

$$x^p f = x^p G^{(m)} = \sum_s (-1)^s \binom{m}{s} [(x^p)^{(s)} G]^{(m-s)}. \tag{4}$$

Clearly $(x^p)^{(s)} = \dfrac{p!}{(p-s)!} x^{p-s}$, if $s \leqslant p$, and $(x^p)^{(s)} = 0$ otherwise.

Let

$$G_s(x) = \frac{p!}{(p-s)!} \int_0^x t^{p-s} G(t)\, dt^s$$

for $s \leqslant p$ and let $G_s = 0$ otherwise. Then (1) implies that the products $\hat{x}^{-r-p}G_s$ are bounded. Setting $H = \sum_s (-1)^s \binom{m}{s} G_s$ we see that the product $\hat{x}^{-r-p}H$ is also bounded. But $H^{(m)} = x^p f$ by (4), and thus the distribution $x^p f$ is tempered, by 7.3.1. Finally, we have $d^{e_j} f = D^{e_j} f - \xi_j f$, by 7.1(3). Hence, by what we have proved, $d^{e_j} f$ is tempered. Now, if $d^p f$ is tempered for some $p \in \boldsymbol{P}^q$, then $d^{p+e_j} f$ is also tempered, and so $d^p f$ is tempered for all $p \in \boldsymbol{P}^q$, by induction.

7.4. SUBCLASSES OF TEMPERED DISTRIBUTIONS

We say that a distribution f belongs to the class T^k, $k \in \boldsymbol{P}^q$, iff it is of the form $f = D^k F$, $F \in L^2(\boldsymbol{R}^q)$.

7.4.1. THEOREM. $T^k \subset T^m$ *for* $k \leqslant m$.

In the proof we shall make use of the following

7.4.2. LEMMA. *If a function f of a real variable ξ is square integrable in the interval* $-\infty < \xi < \infty$, *then*

$$\int_{-\infty}^{\infty} e^{-\xi^2/2}\left|\int_0^{\xi} e^{\tau^2/4} f(\tau)\,d\tau\right|^2 d\xi \leqslant 5\int_{-\infty}^{\infty} |f(\xi)|^2 d\xi. \tag{1}$$

PROOF OF LEMMA 7.4.2. Let $F(\xi) = \int_0^{\xi} E(\tau)|f(\tau)|\,d\tau$, where $E(\tau) = e^{\tau^2/4}$, and let B be an arbitrary fixed number greater than 1. Then

$$I_B = \int_0^B E^{-2}F^2 = \int_0^1 E^{-2}F^2 + \int_1^B E^{-2}F^2.$$

We have

$$\int_0^1 E^{-2}F^2 \leqslant \int_0^1\Big(\int_0^{\xi}|f(\tau)|\,d\tau\Big)^2 d\xi \leqslant \int_0^1\Big(\int_0^{\xi} d\tau\cdot\int_0^{\xi}|f(\tau)|^2\,d\tau\Big)\,d\xi \leqslant \int_0^1 \xi\cdot A\,d\xi = \tfrac{1}{2}A,$$

where

$$A = \int_0^{\infty} |f(\tau)|^2\,d\tau.$$

On the other hand

$$\int_1^B E^{-2}F^2 \leqslant \int_1^B \xi E^{-2}F^2 \leqslant \int_0^B (-E^{-2})'F^2.$$

Integrating by parts and ignoring negative terms, we find

$$\int_1^B E^{-2}F^2 \leqslant \int_0^B (E^{-2}2FE|f|) = 2\int_0^B (E^{-1}F|f|)$$

$$\leqslant 2\Big(\int_0^B E^{-2}F^2\Big)^{1/2}\Big(\int_0^B |f|^2\Big)^{1/2} \leqslant 2\sqrt{I_B A}.$$

Thus $I_B \leqslant \frac{1}{2}A + 2\sqrt{I_B A}$. Hence if $I_B \geqslant \frac{1}{2}A$, then $(I_B - \frac{1}{2}A)^2 \leqslant 4I_B A$ and hence $I_B^2 - I_B A \leqslant 4I_B A$. Dividing by I_B, we obtain

$$I_B \leqslant 5A. \tag{2}$$

If $I_B \leqslant \frac{1}{2}A$, (2) holds trivially and thus the inequality (2) holds for every $B < \infty$. Hence

$$\int_0^{\infty} e^{-\xi^2/2}\left|\int_0^{x} e^{\tau^2/4} f(\tau)\,d\tau\right|^2 d\xi \leqslant 5\int_0^{\infty} |f(\xi)|^2\,d\xi.$$

Similarly,

$$\int_{-\infty}^{0} e^{-\xi^2/2} \left|\int_0^x e^{\tau^2/4} f(\tau)\,d\tau\right|^2 d\xi \leqslant 5 \int_{-\infty}^{0} |f(\xi)|^2\,d\xi.$$

From these two inequalities, the required inequality (1) now follows.

PROOF OF THEOREM 7.4.1. The assertion is trivially true for $m = k$. Assume that it is true for some $m \geqslant k$, i.e., that there is a function $F_m \in L^2(\boldsymbol{R}^q)$ such that $f = D^m F_m$. Let $F_{m+e_j} = S^{e_j} F_m$. Then

$$D^{m+e_j} F_{m+e_j} = f.$$

We shall show that $F_{m+e_j} \in L^2(\boldsymbol{R}^q)$. In fact, we have

$$F_{m+e_j}(x) = E^{-1}(x) \int_0^x E(t) F_m(t)\,dt^{e_j} = E^{-1}(\xi_j) \int_0^{\xi_j} E(\tau_j) F_m(x, \tau_j)\,d\tau_j,$$

where $F_m(x, \tau_j)$ is the function obtained from $F_m(x)$ on replacing ξ_j by τ_j. Hence

$$\int_{R^q} |F_{m+e_j}(x)|^2 dx = \int_{R^q} dx^{1-e_j} \int_{-\infty}^{\infty} E^{-2}(\xi_j) \left|\int_0^{\xi_j} E(\tau_j) F_m(x, \tau_j)\,d\tau_j\right|^2 d\xi_j$$

and by 7.4.2,

$$\int_{R^q} |F_{m+e_j}(x)|^2\,dx \leqslant 5 \int_{R^q} dx^{1-e_j} \int_{-\infty}^{\infty} |F_m(x, \tau_j)|^2 d\tau_j$$

$$= 5 \int_{R^q} dx^{1-e_j} \int_{R^q} |F_m(x)|^2\,dx^{e_j}.$$

Hence

$$\int_{R^q} |F_{m+e_j}(x)|^2\,dx \leqslant 5 \int_{R^q} |F_m(x)|^2\,dx.$$

This proves that $T^k \subset T^m$ implies $T^k \subset T^{m+e_j}$. Hence Theorem 7.4.1 follows by induction.

Every class T^k ($k \in \boldsymbol{P}^q$) is a linear space, i.e., if $f \in T^k$ and $g \in T^k$, then $\alpha f + \beta g \in T^k$ for any numbers α, β. From 7.4.1 it follows that if $f \in T^k$ and $g \in T^m$, then $\alpha f + \beta g \in T^r$, where $r = \max(k, m)$. The space of tempered distributions is clearly the union of all the classes T^k ($k \in \boldsymbol{P}^q$), so that if f and g are tempered distributions, so is $\alpha f + \beta g$. In other words, the space of tempered distributions is a linear space. This fact can also be deduced from 7.3.1.

7.4.3. COROLLARY. *A distribution f is tempered, iff there exist two finite systems of vectors $m_1, \ldots, m_\nu$ and $k_1, \ldots, k_\nu$ and square integrable functions $f_1, \ldots, f_\nu$*

such that

$$f = \sum_{i=1}^{\nu} x^{m_i} f_i^{(k_i)},$$

where $x^m = \xi_1^{\mu_1} \ldots \xi_q^{\mu_q}$ *if* $x = (\xi_1, \ldots, \xi_q)$ *and* $m = (\mu_1, \ldots, \mu_q)$.

This Corollary follows easily from the definition of tempered distributions and 7.3.2.

7.5. TEMPERED CONVERGENCE OF SEQUENCES

We say that a sequence of tempered distributions f_n is *tempered* to a distribution f and write $f_n \xrightarrow{t} f$, iff there exist square integrable functions F_n, F such that $D^k F_n = f_n$, $D^k F = f$ for some fixed $k \in \boldsymbol{P}^q$ and $F_n \xrightarrow{2} F$, i.e.,

$$\int_{R^q} |F_n - F| \xrightarrow{2} 0, \quad \text{as} \quad n \to \infty.$$

7.5.1. Theorem. *A sequence of tempered distributions f_n is tempered to a distribution f, i.e., $f_n \xrightarrow{t} f$, iff there exist $m \in \boldsymbol{P}^q$, $r \in \boldsymbol{P}^q$, and continuous functions G_n, G such that*

$$G_n^{(m)} = f_n, \qquad G^{(m)} = f \tag{1}$$

and the sequence $\hat{x}^{-r} G_n$ is bounded and converges uniformly in $\boldsymbol{R}^q$ to $\hat{x}^{-r} G$.

Proof. Assume that f_n is tempered to f and let

$$G_n(x) = \int_0^x F_n(t)\,dt, \qquad G(x) = \int_0^x F(t)\,dt.$$

Then the boundedness of $\hat{x}^{-1} G$ follows easily, and we have

$$|G_n - G| \leqslant \left| \int_0^x dt \int_0^x |F_n - F|^2 \right|^{1/2} \leqslant \hat{x}^1 \varepsilon_n \quad \text{with} \quad \varepsilon_n^2 = \int_{R^q} |F_n - F|^2.$$

This implies that $\hat{x}^{-1} G_n$ converges uniformly to $\hat{x}^{-1} G$. If $k = 0$, then $G_n' = f_n$ and $G' = f$. The necessity of our condition for the particular case $k = 0$ thus follows (with $m = r = (1, \ldots, 1)$).

Assume that for some $k \in \boldsymbol{P}^q$, the condition is necessary and holds with suitably chosen m and r. Let $f_n = D^{k+e_j} F_n, f = D^{k+e_j} F$. We then have 7.3(2) and, similarly,

$$f_n = G_n^{(m+e_j)} + \tfrac{1}{2}(\xi_j G_n)^{(m)} - \tfrac{1}{2}\mu_j G_n^{(m-e_j)}.$$

Assuming H is defined by 7.3 (3) and setting

$$H_n(x) = G_n(x) + \tfrac{1}{2}\int_0^x \tau_j G_n(t)\,dt^{e_j} - \tfrac{1}{2}\mu_j \int_0^x G_n(t)\,dt^{2e_j},$$

we have

$$H_n^{(m+e_j)} = f_n \quad \text{and} \quad H^{(m+e_j)} = f.$$

By the induction hypothesis, $\hat{x}^{-r}G_n$ converges uniformly to $\hat{x}^{-r}G$ and is bounded for some $r \in \boldsymbol{P}^q$, i.e., there is a sequence $\varepsilon_n > 0$, $\varepsilon_n \to 0$, such that

$$|G_n - G| \leqslant \varepsilon_n \hat{x}^r.$$

Hence the boundedness of $\hat{x}^{-r-2e_j} H_n$ follows, and we have

$$|H_n - H| \leqslant \varepsilon_n \hat{x}^r + \varepsilon_n \left|\int_0^x \tau_j \hat{t}^r d\hat{t}^{e_j}\right| + \varepsilon_n \mu_j \left|\int_0^x \hat{t}^r dt^{2e_j}\right|$$

$$\leqslant \varepsilon_n \hat{x}^r + \frac{\varepsilon_n}{\varrho_j + 2} \hat{x}^{r+2e_j} + \frac{\varepsilon_n \mu_j}{(\varrho_j+1)(\varrho_j+2)} \hat{x}^{r+2e_j},$$

where ϱ_j is the jth coordinate of r. We also have

$$|H_n - H| \leqslant \varepsilon_n M_j \hat{x}^{r+2e_j}$$

which proves that $\hat{x}^{-r-2e_j} H_n$ converges uniformly to $\hat{x}^{-r-2e_j} H$. The necessity of the condition is therefore proved.

To prove its sufficiency, we assume, conversely, that (1) holds and that $G_n \hat{x}^{-r}$ converges uniformly to $G\hat{x}^{-r}$ and is bounded. There then exist positive numbers ε_n and M such that $\varepsilon_n \to 0$ and

$$|G_n - G| < \varepsilon_n \hat{x}^r, \quad |G| \leqslant M\hat{x}^r.$$

By 7.2.1, there exist continuous functions H_n, F and a number β such that

$$D^{m+r+1} H_n = (G_n - G)^{(m)}, \quad D^{m+r+1} F = G^{(m)},$$

$$|H_n| \leqslant \varepsilon_n \beta \hat{x}^{-1}, \quad |F| \leqslant M\beta \hat{x}^{-1}.$$

It is clear that the functions H_n, F are square integrable in $\boldsymbol{R}^q$ and that the sequence H_n converges in square mean to 0. The functions $F_n = H_n + F$ are thus square integrable in $\boldsymbol{R}^q$ and the sequence F_n converges in square mean to F. Since, clearly, $D^{m+r+1} F_n = f_n$ and $D^{m+r+1} F = f$, the sequence f_n is tempered to f.

From 7.5.1 follows

7.5.2. THEOREM. *If f_n is tempered to f, f_n converges distributionally* (*i.e., in the sense of Section* 3.7, *Part* II) *to f.*

Using 7.5.1 it can easily be proved that if $\alpha_n > 0$, $\alpha_n \to 0$, then the sequence $e^{-\alpha_n x^2}$ is tempered to 1 and the sequence $\left(\sqrt{\alpha_n \pi}\right)^{-q} e^{-x^2/\alpha_n}$ is tempered to the delta distribution δ.

From 7.5.1 it also follows that if a sequence of continuous functions F_n converges almost uniformly to F and is bounded by a slowly increasing function, then $F_n \overset{t}{\to} F$.

7.5.3. THEOREM. *If a sequence f_n is tempered to f, the sequences $D^p f_n, f_n^{(p)}, x^p f_n$ and $d^p f_n$ are tempered to $D^p f, f^{(p)}, x^p f$ and $d^p f$, respectively.*

The proof closely resembles the proof of Theorem 7.3.2.

7.5.4. THEOREM. *If f_n are tempered distributions and $f_n \overset{t}{\to} f$, then $f_n * \delta_n \overset{t}{\to} f$ for any delta sequence δ_n.*

The proof of this theorem is based on the following

7.5.5. LEMMA. *If p is a polynomial in $\mathbf{R}^q$ and δ_n is a delta sequence, the sequence $|p| * |\delta_n|$ is bounded by a polynomial.*

PROOF OF THE LEMMA. There exist an even $r \in \mathbf{P}^q$ (i.e., all the coordinates of r are even numbers) and a number A such that $|p| \leqslant A + x^r$. Clearly

$$|p| * |\delta_n| \leqslant (A + x^r) * |\delta_n| = A \int |\delta_n| + \int (x-t)^r |\delta_n(t)|\, dt.$$

If all the coordinates of t are absolutely $\leqslant 1$, we have $(x-t)^r \leqslant \hat{x}^r$. Hence,

$$\int (x-t)^r |\delta_n(t)|\, dt \leqslant \hat{x}^r \int |\delta_n(t)|\, dt \leqslant M_0 \hat{x}^r,$$

for large n, which implies our assertion.

PROOF OF THEOREM 7.5.4. There exist continuous functions G_n, G satisfying 7.5.1. The sequence G_n converges a.u. (almost uniformly) and is bounded by a polynomial p. The sequence $G_n * \delta_n$ converges to G a.u., by 2.3.1. Moreover, we have $|G_n * \delta_n| \leqslant |p| * |\delta_n| \leqslant M_0 \hat{x}^r$ for some even $r \in \mathbf{P}^q$ and number M_0, by 7.5.5. This implies that $(G_n * \delta_n)\hat{x}^{-r-1}$ converges uniformly in $\mathbf{R}^q$ to $\hat{x}^{-r-1} G$. We thus have $G_n * \delta_n \overset{t}{\to} G$. Since $(G_n * \delta_n)^{(m)} = f_n * \delta_n$ and $G^{(m)} = f$, we finally have $f_n * \delta_n \overset{t}{\to} f$.

7.5.6. COROLLARY. *If f is a tempered distribution and δ_n is a delta sequence, then $f * \delta_n \overset{t}{\to} f$.*

7.5.7. THEOREM. *For any tempered distribution f there exists a sequence of smooth functions f_n with bounded carriers such that $f_n \overset{t}{\to} f$.*

The proof of this theorem is based on the following

7.5.8. LEMMA. *Let Ω be a smooth function of bounded carrier such that $\Omega(0) = 1$. Then, if f is a tempered distribution,*

$$f(x)\Omega\left(\frac{x}{n}\right) \overset{t}{\to} f(x).$$

PROOF OF THE LEMMA. By 7.3.1, there exist a slowly increasing continuous function G and an order $k \in \boldsymbol{P}^q$ such that $f = G^{(k)}$. We then have (see Appendix 16.6 (2))

$$f(x)\Omega\left(\frac{x}{n}\right) = \sum_m (-1)^m \binom{k}{m}\left[G(x)n^{-m}\Omega^{(m)}\left(\frac{x}{n}\right)\right]^{(k-m)}. \qquad (2)$$

For $m = 0$, the product enclosed by the brackets reduces to $G(x)\Omega\left(\frac{x}{n}\right)$; it converges a.u. in $\boldsymbol{R}^q$ and is bounded by the slowly increasing function $|G(x)|\alpha$, where $\alpha = \max_{x \in \boldsymbol{R}^q} \Omega(x)$. Hence $G(x)\Omega\left(\frac{x}{n}\right) \xrightarrow{t} G$ (see the last remark after 7.5.2). Similarly, for $m \neq 0$, the product $G(x)n^{-m}\Omega^{(m)}\left(\frac{x}{n}\right)$ is tempered to 0, as $n \to \infty$. Consequently, (2) is tempered to $G^{(k)} = f$, by 7.5.3 and 7.5.1.

PROOF OF THEOREM 7.5.7. By Lemma 7.5.8, there exists for any tempered distribution f, a sequence of distributions f_n of bounded carriers such that $f_n \xrightarrow{t} f$. If δ_n is a delta sequence, $f_n * \delta_n$ are smooth functions of bounded carries and such that $f_n * \delta_n \xrightarrow{t} f$, by 7.5.4. This implies our assertion.

7.5.9. THEOREM. *If a sequence f_n is tempered in $\boldsymbol{R}^q$ to f and the sequence g_n is tempered in $\boldsymbol{R}^r$ to g, then the sequence $u_n(z) = f_n(x)g_n(y)$, where $x \in \boldsymbol{R}^q$, $y \in \boldsymbol{R}^r$, $z = (x, y) \in \boldsymbol{R}^{q+r}$ is tempered in $\boldsymbol{R}^{q+r}$ to $u(z) = f(x)g(y)$.*

PROOF. There exist functions $F_n, F \in L^2(\boldsymbol{R}^q)$ and $G_n, G \in L^2(\boldsymbol{R}^r)$ such that $D^kF_n = f_n$, $D^kF = f$, $D^lG_n = g_n$ and $D^lG = g$ for some $k \in \boldsymbol{P}^q$, $l \in \boldsymbol{P}^r$, and such that $||F_n - F||_q \to 0$ and $||G_n - G||_r \to 0$, where the symbol $||\ ||$ denotes the square norm and the index indicates the number of dimensions.

Clearly

$$D^{k+l}F_n(x)G_n(y) = u_n(z), \qquad D^{k+l}F(x)G(y) = u(z)$$

and

$$\begin{aligned} ||F_n(x)G_n(y) - F(x)G(y)||_{q+r} &\leqslant ||F_n(x)[G_n(y) - G(y)]||_{q+r} + ||[F_n(x) - F(x)]G(y)||_{q+r} \\ &= ||F_n||_q||G_n - G||_r + ||F_n - F||_q||G||_r \to 0, \end{aligned}$$

as $n \to \infty$.

7.5.10. COROLLARY. *If $f_{1n}, \ldots, f_{qn}$ are sequences tempered in $\boldsymbol{R}^1$ to $f_1, \ldots, f_q$ respectively, then the sequence $u_n(x) = f_{1n}(\xi_1) \ldots f_{qn}(\xi_q)$ is tempered in $\boldsymbol{R}^q$ to $u(x) = f_1(\xi_1) \ldots f_q(\xi_q)$.*

This Corollary follows from 7.5.9 by induction.

7.6. INNER PRODUCT WITH A SMOOTH FUNCTION OF BOUNDED CARRIER

The space of all smooth functions of bounded carriers is denoted, following Schwartz, by $\mathscr{D}$.

Let $O \subset \boldsymbol{R}^q$ be an open set and suppose that the carrier of a function $\psi \in \mathscr{D}$ is inside O. The inner product

$$(\varphi, \psi) = \int_O \varphi\psi$$

is defined for every smooth function φ in O and its value is a number. We shall show that, for each fixed $\psi \in \mathscr{D}$ (whose carrier is inside O), (φ, ψ) is a regular operation on φ. In other words, that if φ_n is a fundamental sequence in O, then the sequence (φ_n, ψ) is fundamental.

However, the sequence (φ_n, ψ) consists of numbers, and we must therefore first explain what a fundamental sequence of numbers is.

7.7. FUNDAMENTAL SEQUENCES AND DISTRIBUTIONS IN $\boldsymbol{R}^0$

We say that a sequence of numbers α_n is *fundamental*, iff it is convergent. It is convenient to link fundamental sequences of numbers with the general theory of regular operations. In order to retain the validity of definitions and theorems without changing their formulation, fundamental sequences of numbers will, equivalently, be called fundamental sequences of smooth functions in $\boldsymbol{R}^0$. Consequently, smooth functions in $\boldsymbol{R}^0$ will be nothing else, but numbers under another name. We do not define $\boldsymbol{R}^0$; the symbol $\boldsymbol{R}^0$ will only have a meaning when placed in conjuction with other words.

According to the former definition, two fundamental sequences φ_n and ψ_n are equivalent, iff the interlaced sequence $\varphi_1, \psi_1, \varphi_2, \psi_2, \ldots$ is fundamental. Hence in the present context, two fundamental sequences of numbers (or smooth functions in $\boldsymbol{R}^0$) are equivalent, iff they converge to the same number f. This implies that there is a one-to-one correspondence between equivalence classes, which will be called distributions in $\boldsymbol{R}^0$, and numbers. Because of this correspondence we are able to identify distributions in $\boldsymbol{R}^0$ with numbers. The only distributions in $\boldsymbol{R}^0$ are thus smooth functions in $\boldsymbol{R}^0$. In this way, the theory of distributions in $\boldsymbol{R}^0$ is seen to be trivial and reduces, in fact, to the theory of real (or complex) numbers.

However, the fact that numbers may be called smooth functions in $\boldsymbol{R}^0$ or distributions in $\boldsymbol{R}^0$ opens to us the possibility of including numbers in the theory of regular operations under a uniform formulation of all definitions and theorems.

7.8. PROOF OF THE REGULARITY OF INNER PRODUCT

Assume first that the carrier of ψ is in an interval I inside O. If φ_n is fundamental in O, there exist smooth functions Φ_n, Φ in I such that $\Phi_n^{(k)} = \varphi_n$ for some $k \in \boldsymbol{P}^q$ and $\Phi_n \rightrightarrows \Phi$. Integrating by parts, we get

$$(\varphi_n, \psi) = \int_I \Phi_n^{(k)}\psi = (-1)^k \int_I \Phi_n \psi^{(k)} \to (-1)^k \int_I \Phi \psi^{(k)}. \tag{1}$$

Hence the sequence (φ_n, ψ) is convergent and, consequently, the operation (φ, ψ) is regular.

We now remove the restriction that the carrier of ψ be in I; we assume only that it is bounded and inside O. However we can then write $\psi = \psi_1 + \ldots + \psi_r$, where the ψ_i are smooth functions whose carriers are in intervals I_i inside O. Since

$$(\varphi_n, \psi) = (\varphi_n, \psi_1) + \ldots + (\varphi_n, \psi_r)$$

and the inner products on the right hand side are regular operations by what has just been proved, the inner product (φ_n, ψ) is also a regular operation, as it is a finite sum of regular operations.

The inner product with the smooth function ψ whose carrier is bounded and inside $O \subset \boldsymbol{R}^q$ is thus defined for every distribution f in O and we have

$$(f, \psi) = \lim_{n \to \infty} (\varphi_n, \psi),$$

where φ_n is a fundamental sequence for f.

In Section 7.10 we shall need the following formula

$$(D^k f, \psi) = (-1)^k (f, d^k \psi), \tag{2}$$

where f is a distribution defined in $\boldsymbol{R}^q$ and ψ is a smooth function of bounded carrier.

In fact, if f is a smooth function in $\boldsymbol{R}^q$, we have

$$(D^k f, \psi) = \int E^{-1}(Ef)^{(k)}\psi = -\int (Ef)^{(k-e_j)}(E^{-1}\psi)^{(e_j)}$$

$$= -\int E^{-1}(Ef)^{(k-e_j)} E(E^{-1}\psi)^{(e_j)} = -(D^{k-e_j} f, d^{e_j}\psi).$$

Hence, if f is a smooth function the equality (2) follows at once by induction. Since all the operations involved in (2) are regular, (2) holds for any distribution f in $\boldsymbol{R}^q$.

The last definition of the inner product (f, ψ) is meaningful for any distribution f in O and any smooth function ψ whose carrier is bounded and inside O. In particular, f can be a tempered distribution. Since tempered distributions are defined in $\boldsymbol{R}^q$, it follows that the inner product (f, ψ) is defined for each tempered distribution f and each $\psi \in \mathscr{D}$. However, if we know that f is a tempered distribution, the inner product (f, ψ) can be defined for a larger class of functions ψ, whose carriers are not necessarily bounded. This class consists of smooth functions which together with all their derivatives decrease rapidly. The following sections are devoted to a more detailed study of this class.

7.9. THE SPACE OF RAPIDLY DECREASING SMOOTH FUNCTIONS

A smooth function ψ in $\boldsymbol{R}^q$ is said to be *rapidly decreasing together with all its derivatives* or, in short, *rapidly decreasing*, iff, for any polynomial p and $k \in \boldsymbol{P}^q$, the product $p\psi^{(k)}$ is bounded. Following Schwartz, the class of all rapidly decreasing smooth functions is denoted by $\mathscr{S}$. Clearly, $\mathscr{S}$ is a linear space. Moreover, if $\psi \in \mathscr{S}$, we have, for each polynomial p and each $k \in \boldsymbol{P}^q$, $p\psi^{(k)} \in \mathscr{S}$ and $(p\psi)^{(k)} \in \mathscr{S}$. We prove that, moreover, $D^k\psi$, $d^k\psi \in \mathscr{S}$ for each $k \in \boldsymbol{P}^q$. In fact, since

$$D^{e_j}\psi = \psi^{(e_j)} + \tfrac{1}{2}\xi_j\psi \quad \text{and} \quad d^{e_j}\psi = \psi^{(e_j)} - \tfrac{1}{2}\xi_j\psi,$$

we have $D^{e_j}\psi$, $d^{e_j}\psi \in \mathscr{S}$. Hence, if $D^k\psi, d^k\psi \in \mathscr{S}$, then $D^{k+e_j}\psi$, $d^{k+e_j}\psi \in \mathscr{S}$, which proves our assertion.

We now show that, if $\psi \in \mathscr{S}$, the following formulae hold

$$\int (d^{e_j}\psi)^2 - \int (D^{e_j}\psi)^2 = \int \psi^2, \tag{1}$$

$$\int (D^{e_j}d^{e_j}\psi)^2 - \int (d^{e_j}D^{e_j}\psi)^2 = \int (d^{e_j}\psi)^2 + \int (D^{e_j}\psi)^2. \tag{2}$$

In fact, from 7.1(3) we get

$$(d^{e_j}\psi)^2 - (D^{e_j}\psi)^2 = -2\xi_j\psi\psi^{(e_j)} = -\xi_j(\psi^2)^{(e_j)}.$$

Integrating by parts, equality (1) now follows. The proof of (2) is left to the reader.

In Section 8.2, we shall show that the class $\mathscr{S}$ coincides with the class of all tempered distributions f such that $D^kf \in L^2(\boldsymbol{R}^q)$ for each $k \in \boldsymbol{P}^q$ and also with the class of all tempered distributions f such that $d^kf \in L^2(\boldsymbol{R}^q)$ for each $k \in \boldsymbol{P}^q$.

7.9.1. THEOREM. *Let $\psi_n, \psi \in \mathscr{S}$. If $D^k\psi_n \overset{2}{\to} D^k\psi$ for every $k \in \boldsymbol{P}^q$, $d^k\psi_n \overset{2}{\to} d^k\psi$, and conversely.*

PROOF. We first show that, if

$$D^k\psi_n \overset{2}{\to} D^k\psi \quad \text{for each } k \in \boldsymbol{P}^q, \tag{3}$$

then

$$d^{e_j}D^k\psi_n \overset{2}{\to} d^{e_j}D^k\psi \quad \text{for each } k \in \boldsymbol{P}^q \tag{4}$$

and

$$D^k d^{e_j}\psi_n \xrightarrow{2} D^k d^{e_j}\psi \quad \text{for each } k \in \boldsymbol{P}^q. \tag{5}$$

In fact, by (1), we have

$$\int [d^{e_j}D^k(\psi_n-\psi)]^2 = \int [D^{k+e_j}(\psi_n-\psi)]^2 + \int [D^k(\psi_n-\psi)]^2$$

and (4) follows. In turn, we have by 7.1(4)

$$D^k d^{e_j}(\psi_n-\psi) = d^{e_j}D^k(\psi_n-\psi) - \varkappa_j D^{k-e_j}(\psi_n-\psi)$$

and (5) follows. Since (3) implies (5), we conclude that $D^k d^m \psi_n \xrightarrow{2} D^k d^m \psi$, by induction with respect to m. In particular, for $k = 0$, we get the first part of our assertion.

The proof of the converse is similar.

We say that a sequence $\psi_n \in \mathscr{S}$ *converges in* $\mathscr{S}$ to $\psi \in \mathscr{S}$ and write $\psi_n \xrightarrow{\mathscr{S}} \psi$, iff, for each $k \in \boldsymbol{P}^q$, we have $D^k\psi_n \xrightarrow{2} D^k\psi$ or, which turns out to be the same, iff $d^k\psi_n \xrightarrow{2} d^k\psi$.

Clearly $\mathscr{D} \subset \mathscr{S}$. Moreover, we show that $\mathscr{D}$ is dense in $\mathscr{S}$, i.e., that we have the following

7.9.2. THEOREM. *For any $\psi \in \mathscr{S}$, there exists a sequence $\varphi_n \in \mathscr{D}$ such that $\varphi_n \xrightarrow{\mathscr{S}} \psi$.*

PROOF. Let $\psi \in \mathscr{S}$ and let Ω be a smooth function of bounded carrier such that $\Omega(0) = 1$. We show that, for each $k \in \boldsymbol{P}^q$,

$$D^k\left(\psi(x)\Omega\left(\frac{x}{n}\right)\right) - D^k\psi(x) \xrightarrow{2} 0.$$

By 7.1(1), we have

$$D^k\left(\psi(x)\Omega\left(\frac{x}{n}\right)\right) - D^k\psi(x) = \sum_m \binom{k}{m} D^{k-m}\psi(x) n^{-m}\Omega^{(m)}\left(\frac{x}{n}\right), \tag{6}$$

where $0 \leqslant m \leqslant k$ and $m \neq 0$. We thus see that

$$\int_{R^q} \left(D^{k-m}\psi(x)n^{-m}\Omega^{(m)}\left(\frac{x}{n}\right)\right)^2 dx \leqslant n^{-2m} \int_{R^q} (D^{k-m}\psi(x)K_m)^2 \to 0 \quad \text{as} \quad n \to \infty,$$

where $K_m = \max\limits_x |\Omega^{(m)}(x)|$. Hence, the sum on the right hand side of (6) converges to 0 in square mean, which implies our assertion.

7.9.3. THEOREM. *If f_n are tempered distributions and $f_n \xrightarrow{t} f$, then for each $\psi \in \mathscr{S}$ we have $f_n\psi \xrightarrow{t} f\psi$ and $f_n * \psi \xrightarrow{t} f * \psi$.*

PROOF. By Theorem 7.5.1, there exist $m \in \boldsymbol{P}^q$, $r \in \boldsymbol{P}^q$ and continuous functions G_n, G such that $G_n^{(m)} = f_n$, $G^{(m)} = f$ and the sequence $\hat{x}^{-r}G_n$ is bounded and converges uniformly to $\hat{x}^{-r}G$. We have

$$f_n\psi = G_n^{(m)}\psi = \sum_{0 \leqslant k \leqslant m} (-1)^k \binom{m}{k} (G_n\psi^{(k)})^{(m-k)}.$$

Since $\psi^{(k)} \in \mathscr{S}$, the sequence $G_n\psi^{(k)}$ is bounded and converges uniformly to $G\psi^{(k)}$, for each fixed k. This implies that $(G_n\psi^{(k)})^{(m-k)}$ is tempered to $(G\psi^{(k)})^{(m-k)}$. Consequently $f_n\psi$ is tempered to

$$\sum_{0 \leqslant k \leqslant m} (-1)^k \binom{m}{k} (G\psi^{(k)})^{(m-k)},$$

i.e., to $f\psi$. This proves the first part of the theorem.

The existence of the convolutions $f_n * \psi$ and $f * \psi$ follows from Section 6.3 because the functions from $\mathscr{S}$ are rapidly decreasing distributions. We have, by Theorem 6.3.4,

$$f_n * \psi = G_n^{(m)} * \psi = G_n * \psi^{(m)} \quad \text{and similarly,} \quad f * \psi = G * \psi^{(m)}.$$

Moreover,

$$|G_n| < M\hat{x}^r, \quad |G_n - G| < \varepsilon_n \hat{x}^r,$$

where the ε_n are positive numbers converging to 0. There is a polynomial P such that $\hat{x}^r < P$ for all $x \in \boldsymbol{R}^q$. Thus,

$$|G_n| < MP, \quad |G_n - G| < \varepsilon_n P.$$

Hence

$$|G_n * \psi^{(m)}| \leqslant MQ, \quad |G_n * \psi^{(m)} - G * \psi^{(m)}| \leqslant \varepsilon_n Q,$$

where $Q = P * |\psi^{(m)}|$.

It is easy to see that Q is a polynomial. In fact,

$$x^k * |\psi^{(m)}| = \int (x-t)^k |\psi^{(m)}(t)|\, dt$$

$$= \int \sum_{0 \leqslant j \leqslant k} \binom{k}{j} x^j (-t)^{k-j} |\psi^{(m)}(t)|\, dt = \sum_{0 \leqslant j \leqslant k} \beta_j x^j$$

with

$$\beta_j = \binom{k}{j} \int (-t)^{k-j} |\psi^{(m)}(t)|\, dt.$$

Thus, if $P = \sum_{0 \leqslant k \leqslant m} \alpha_k x^k$, then $Q = \sum_{0 \leqslant k \leqslant m} \alpha_k x^k * |\psi^{(m)}|$ and we see that Q is a sum of polynomials. Consequently Q is itself a polynomial and there is an $s \in \boldsymbol{P}^q$ such that $Q < K\hat{x}^s$. This proves that $(G_n * \psi^{(m)})\hat{x}^{-s}$ is bounded and converges uniformly to $(G * \psi^{(m)})\hat{x}^{-s}$. This proves that $f_n * \psi \xrightarrow{t} f * \psi$, by Theorem 7.5.1.

7.10. EXTENSION OF THE DEFINITION OF AN INNER PRODUCT

Let ψ be a fixed element in $\mathscr{S}$ and let f be a tempered distribution. We show that all sequences (f, φ_n), where $\varphi_n \in \mathscr{D}$ and $\varphi_n \overset{\mathscr{S}}{\to} \psi$, are convergent to the same limit. This limit is denoted by (f, ψ) and called the *inner product* of f with ψ.

First, there is a function $F \in L^2$ such that $D^k F = f$. By 7.8(2), we have

$$(f, \varphi_n) = (D^k F, \varphi_n) = (-1)^k (F, d^k \varphi_n). \tag{1}$$

From 7.9.1, it follows that $d^k \varphi_n \overset{2}{\to} d^k \psi$. This implies that $(F, d^k \varphi_n) \to (F, d^k \psi)$, and so the limit of (f, φ_n) does exists.

We now show that this limit is independent of the choice of φ_n. Assume that, in addition to φ_n, there is another sequence of functions $\psi_n \in \mathscr{D}$ such that $\psi_n \overset{\mathscr{S}}{\to} \psi$. Then the interlaced sequence

$$\varphi_1, \psi_1, \varphi_2, \psi_2, \ldots$$

also converges in $\mathscr{S}$ to ψ and, consequently, the sequence (f, φ_1), (f, ψ_1), (f, φ_2), (f, ψ_2), ... is convergent. This implies that the sequences (f, φ_n) and (f, ψ_n) are convergent to the same limit.

The definition of the inner product (f, ψ) is therefore consistent. Note that the main idea of the above proof is similar to that used in the case of regular operations. However, (f, ψ) with $\psi \in \mathscr{S}$ cannot be called a regular operation on f, because it cannot be carried out, in general, with arbitrary (non tempered) distributions.

Letting $n \to \infty$, in (1), we get the formula

$$(D^k F, \psi) = (-1)^k (F, d^k \psi) \quad (F \in L^2), \tag{2}$$

which will be useful in the sequel.

The inner product (f, ψ) is clearly a linear functional on $\mathscr{S}$. It is also continuous in the sense of the following

7.10.1. THEOREM. *If f is tempered, $\psi_n, \psi \in \mathscr{S}$ and $\psi_n \overset{\mathscr{S}}{\to} \psi$, then $(f, \psi_n) \to (f, \psi)$.*

PROOF. There exist a function $F \in L^2(\boldsymbol{R}^q)$ and $k \in \boldsymbol{P}^q$ such that $D^k F = f$. Hence, by (2), we have

$$|(f, \psi_n) - (f, \psi)| = |(F, d^k(\psi_n - \psi))| \leqslant ||F|| \cdot ||d^k(\psi_n - \psi)|| \to 0$$

as $n \to \infty$ which proves our assertion.

In Section 11.3 we shall show that the functionals (f, ψ) with tempered f are the only continuous linear functionals on $\mathscr{S}$.

7.10.2. THEOREM. *If* $f_n \xrightarrow{t} f$ *and* $\psi_n \xrightarrow{\mathscr{S}} \psi$, *then* $(f_n, \psi_n) \to (f, \psi)$.

PROOF. There are functions $F_n, F \in L^2(\boldsymbol{R}^q)$ and $k \in \boldsymbol{P}^q$ such that $D^k F_n = f_n$, $D^k F = f$ and $F_n \xrightarrow{2} F$. We therefore have

$$\begin{aligned} |(f_n, \psi_n) - (f, \psi)| &\leqslant |(f_n, \psi_n - \psi)| + |(f_n - f, \psi)| \\ &= |(F_n, d^k(\psi_n - \psi))| + |(F_n - F, d^k\psi)| \\ &\leqslant ||F_n|| \cdot ||d^k(\psi_n - \psi)|| + ||F_n - F|| \cdot ||d^k\psi|| \to 0, \end{aligned}$$

as $n \to \infty$ which implies our assertion.

8. Tempered Hermite series

8.1. HERMITE SERIES AND THEIR DERIVATIVES

We show that every tempered distribution can be expanded into an Hermite series which is tempered to that distribution. We first recall the formulae (see Section 4.6)

$$H_n = (-1)^n E^2 (E^{-2})^{(n)}, \quad k_n = E^{-1} H_n$$

and

$$h_n = (\sqrt{2\pi} n!)^{-1/2} k_n .$$

Simple calculations yield

$$d^k \sqrt{n!}\, h_n = (-1)^k \sqrt{(n+k)!}\, h_{n+k} \tag{1}$$

and, by 4.7(1),

$$D^k \frac{h_n}{\sqrt{n!}} = \frac{1}{\sqrt{(n-k)!}} h_{n-k}, \text{ for } k \leqslant n; \quad D^k h_n = 0, \text{ for } k \nleqslant n, \ (k, n \in \boldsymbol{P}^q). \tag{2}$$

We write

$$f \overset{t}{=} \sum_{n \in \boldsymbol{P}^q} a_n h_n \tag{3}$$

iff, for any sequence A_ν of finite subsets of $\boldsymbol{P}^q$ such that $A_\nu \subset A_{\nu+1}$ and $\lim_{\nu \to \infty} A_\nu = \boldsymbol{P}^q$, the sequence of sums

$$f_\nu = \sum_{n \in A_\nu} a_n h_n$$

is tempered to f. We also say that f_ν is tempered (unconditionally) to the series on the right hand side of (3). It is easy to see that the limit distribution f does not depend on the choice of the subsets A_ν.

We alternatively write

$$f = \sum_{n \in \boldsymbol{P}^q} a_n h_n ,$$

instead of (3), if there is no possibility of misinterpretation.

By 7.5.3, equality (3) implies

$$D^k f = \sum_{n \in P^q} a_n D^k h_n, \qquad d^k f = \sum_{n \in P^q} a_n d^k h_n.$$

Let $F \in L^2(\boldsymbol{R}^q)$. By 4.8.2, we have

$$F = \sum_{n \in P^q} c_n h_n$$

with $\sum_{n \in P^q} |c_n|^2 < \infty$, because square mean convergence implies tempered convergence. Hence, by (1) and (2)

$$D^k F = \sum_{n \geqslant k} \sqrt{\frac{n!}{(n-k)!}}\, c_n h_{n-k},$$

$$d^k F = \sum_{n \in P^q} (-1)^k \sqrt{\frac{(n+k)!}{n!}}\, c_n h_{n+k}.$$

Introducing new coefficients in the first of the above sums, we can write

$$D^k F = \sum_{n \in P^q} a_n h_n \tag{4}$$

where

$$a_n = \sqrt{\frac{(n+k)!}{n!}}\, c_{n+k}.$$

Since $\sum_{n \in P^q} |c_n|^2 < \infty$, we can obtain an estimate involving the coefficients $|a_n|$. To this end, we introduce the symbol $\tilde{n}$: if the jth coordinate of $n \in \boldsymbol{B}^q$ is ν_j, then by $\tilde{n}$ we mean the vector whose jth coordinate is $\max(1, |\nu_j|)$ $(j = 1, \ldots, q)$. We use the inequalities

$$\tilde{n}^k \leqslant \frac{(n+k)!}{n!} \leqslant K\tilde{n}^k \tag{5}$$

which hold for suitably chosen K (e.g., $K = (1+k)^k$), to obtain

$$\sum_{n \in P^q} \tilde{n}^{-k} |a_n|^2 < \infty. \tag{6}$$

Conversely, if (6) holds, the series $\sum_{n \in P^q} a_n h_n$ is tempered to a distribution f of class T^k (see Section 7.4). In fact, the series

$$\sum_{n \in P^q} \sqrt{\frac{n!}{(n+k)!}}\, a_n h_{n+k}$$

then converges in square mean to a square integrable function G. Since $D^kG = \sum_{n\in P^q} a_n h_n$, we have $f = D^kG$.

We have

8.1.1. THEOREM. *If, for some* $k \in \boldsymbol{P}^q$,

$$\sum_{n\in P^q} \tilde{n}^{-k}|a_n|^2 < \infty, \tag{7}$$

then there is a tempered distribution f of class T^k such that

$$f \stackrel{t}{=} \sum_{n\in P^q} a_n h_n. \tag{8}$$

Conversely, if f is a tempered distribution of class T^k, then there are numbers a_n satisfying (7) *such that* (8) *holds. Moreover,*

$$a_n = (f, h_n) \tag{9}$$

which implies that the expansion of the given distribution as an Hermite series is unique.

PROOF. In view of what we have just proved, it only remains to prove equality (9). To this end, we take the inner product with h_p on both sides of (8). Since $(h_n, h_p) = 0$ for $n \neq p$ and $(h_p, h_p) = 1$, we get $(f, h_p) = a_p$, by 7.10.2, which completes the proof.

8.1.2. COROLLARY. *If, for some $k \in \boldsymbol{P}^q$ and a positive number M*

$$|a_n| < M\tilde{n}^k \quad \text{for} \quad n \in \boldsymbol{P}^q, \tag{10}$$

then there is a tempered distribution f such that (8) *holds. Conversely, every tempered distribution f can be expanded into a Hermite series of the form* (8) *so that* (10) *holds for some $k \in \boldsymbol{P}^q$ and some positive number M. Furthermore this expression is unique and its coefficients are given by formula* (9).

PROOF. From (7) it follows that $\tilde{n}^{-k}|a_n|^2 < M^2$ and hence $|a_n| < M\tilde{n}^{k/2}$, which implies (10). Conversely, (10) implies

$$\sum_{n\in P^q} \tilde{n}^{-2k-2}|a_n|^2 < \infty,$$

i.e., that (7) holds with suitably chosen k, which completes the proof of Corollary 8.1.2.

Corollary 8.1.2 tells us that the series (8) represents a tempered distribution, iff its coefficients a_n do not increase more rapidly than some power of n. Theorem 8.1.1 is more precise than Corollary 8.1.2. However, if we are not concerned with the particular class T^k to which the distribution f belongs, condition (10) in Corollary 8.1.2 is more easily checked than condition (7) in Theorem 8.1.1.

8.1.3. THEOREM. *Let f be any tempered distribution and let $\psi \in \mathscr{S}$. If*

$$f = \sum_{n \in P^q} a_n h_n \quad \textit{and} \quad \psi = \sum_{n \in P^q} b_n h_n,$$

then

$$(f, \psi) = \sum_{n \in P^q} a_n b_n.$$

PROOF. Since $\psi \in \mathscr{S}$, we have $D^k \psi \in L^2$ for each $k \in \boldsymbol{P}^q$ and we can therefore write

$$D^k \psi \overset{2}{=} \sum_{n \in P^q} b_n D^k h_n \quad \text{for each } k \in \boldsymbol{P}^q.$$

Let A_ν be an arbitrary sequence of finite subsets of $\boldsymbol{P}^q$ such that $A_\nu \subset A_{\nu+1}$ and $\lim_{\nu \to \infty} A_\nu = \boldsymbol{P}^q$, and let

$$f_\nu = \sum_{n \in A_\nu} a_n h_n \quad \text{and} \quad \psi_\nu = \sum_{n \in A_\nu} b_n h_n.$$

We see that $f_\nu \overset{t}{\to} f$ and $\psi_\nu \overset{\mathscr{S}}{\to} \psi$. Our assertion now follows, by 7.10.2 and the orthonormality of the h_n.

8.2. SQUARE INTEGRABLE FUNCTIONS AND RAPIDLY DECREASING FUNCTIONS

Following Schwartz, we denote, in the sequel, by $\mathscr{S}'$ the space of all tempered distributions.

Substantiating the remark made in Section 7.9, we now prove that the space $\mathscr{S}$ is identical both with the class of all functions f such that $D^k f \in L^2$ for each $k \in \boldsymbol{P}^q$ and with the class of all functions f such that $d^k f \in L^2$ for each $k \in \boldsymbol{P}^q$. We in fact prove a little more, namely:

8.2.1. THEOREM. *Let $f \in \mathscr{S}'$. If $D^k f \in L^2$ or $d^k f \in L^2$ for some $k \in \boldsymbol{P}^q$, then $D^m f \in L^2$ and $d^m f \in L^2$ for each m such that $0 \leqslant m \leqslant k$.*

PROOF. By 8.1.1, the tempered distribution f can be expanded uniquely into a Hermite series

$$f \overset{t}{=} \sum_{n \in P^q} a_n h_n.$$

Applying here the operators D^m and d^m termwise, using the formulae 8.1(1), 8.1(2) and suitably re-indexing the terms in the first equality, we get

$$D^m f = \sum_{n \in P^q} \sqrt{\frac{(n+m)!}{n!}}\, a_{n+m} h_n, \qquad d^m f = (-1)^m \sum_{n \in P^q} \sqrt{\frac{(n+m)!}{n!}}\, a_n h_{n+m}.$$

Now the above series converge in square mean, iff $D^m f \in L^2$ and $d^m f \in L^2$, respectively. The theorem thus reduces to a corresponding theorem about coefficients: If one of the inequalities

$$\sum_{n \in P^q} \frac{(n+m)!}{n!} |a_{n+m}|^2 < \infty \quad \text{or} \quad \sum_{n \in P^q} \frac{(n+m)!}{n!} |a_n|^2 < \infty$$

holds for $m = k$, then they both hold for $0 \leqslant m \leqslant k$. The proof of this is now a routine matter.

8.2.2. THEOREM. *Let $k \in P^q$. Suppose that $D^m f_n, d^m f_n \in L^2$ for $0 \leqslant m \leqslant k$. If $D^k f_n \overset{2}{\to} 0$ or $d^k f_n \overset{2}{\to} 0$, then $D^m f_n \overset{2}{\to} 0$ and $d^m f_n \overset{2}{\to} 0$ for $0 \leqslant m \leqslant k$.*

PROOF. Let

$$f_n = \sum_{p \in P^q} a_{np} h_p.$$

As in the proof of 8.2.1 we get

$$D^m f_n = \sum_{p \in P^q} \sqrt{\frac{(p+m)!}{p!}}\, a_{n,p+m} h_p, \qquad d^m f_n = (-1)^m \sum_{p \in P^q} \sqrt{\frac{(p+m)!}{p!}}\, a_{np} h_{p+m}.$$

Consequently, by 4.8.3, $D^m f_n \overset{2}{\to} 0$ and $d^m f_n \overset{2}{\to} 0$, iff

$$\sum_{p \in P^q} \frac{(p+m)!}{p!} |a_{n,p+m}|^2 \to 0 \quad \text{or} \quad \sum_{p \in P^q} \frac{(p+m)!}{p!} |a_{np}|^2 \to 0 \quad \text{as} \quad n \to \infty.$$

Our assertion now follows.

From 8.2.1 we are able to deduce a few corollaries.

We consider operators of the following types:

$$D^{e_j}, d^{e_j}, \frac{\partial}{\partial \xi_j} \text{ and } \xi_j \ (j = 1, \ldots, q). \tag{1}$$

If A is one of these operators, then Af stands for $D^{e_j} f$, $d^{e_j} f$, $\dfrac{\partial f}{\partial \xi_j}$ or $\xi_j f$ (ordinary product), respectively. For a fixed j, all the operators (1) act on the same variable ξ_j.

8.2.3. COROLLARY. *Let* $k = (\varkappa_1, \ldots, \varkappa_q) \in \boldsymbol{P}^q$, $\varkappa = \varkappa_1 + \ldots + \varkappa_q$. *Let* A_ν $(\nu = 1, \ldots, \varkappa)$ *be operators of the type* (1); *we assume that, among all these* A_ν, *there are exactly* $\varkappa_j$ *operators acting on the variable* ξ_j $(j = 1, \ldots, q)$: *the ordering of the operators is immaterial. We assert that if* $f \in \mathscr{S}'$ *and* $D^k f \in L^2$ *or* $d^k f \in L^2$, *then* $A_\varkappa \ldots A_1 f \in L^2$.

PROOF. Let $B_0 f = f$ and $B_\nu f = A_\nu B_{\nu-1} f$ for $\nu = 1, \ldots, \varkappa$. Moreover, define the jth coordinate of $k_\nu \in \boldsymbol{P}^q$ $(\nu = 0, \ldots, \varkappa)$ to be the number of operators A_μ with $\mu \leqslant \nu$ which act on the variable ξ_j. Thus $k_0 = 0$, $k_\varkappa = k$ and $k_\nu \geqslant k_{\nu-1}$. We show that

$$D^{k-k_\nu} B_\nu f \in L^2 \tag{2}$$

for $\nu = 0, \ldots, \varkappa$. In fact, this is true for $\nu = 0$, by hypothesis. We now prove that if it holds for some $\nu < \varkappa$, then the following also holds

$$D^{k-k_{\nu+1}} B_{\nu+1} f \in L^2. \tag{3}$$

We have to distinguish four cases:

$$1^\circ\ A_{\nu+1} = D^{e_j}; \quad 2^\circ\ A_{\nu+1} = d^{e_j}; \quad 3^\circ\ A_{\nu+1} = \frac{\partial}{\partial \xi_j}; \quad 4^\circ\ A_{\nu+1} = \xi_j.$$

Case 1°. (2) may be rewritten as $D^{k-k_{\nu+1}} A_{\nu+1} B_\nu f \in L^2$ which is (3).

Case 2°. Applying 8.2.1 to the function $B_\nu f$, we have as well as (2), $d^{k-k_\nu} B_\nu f \in L^2$ or $d^{k-k_{\nu+1}} B_{\nu+1} f \in L^2$. Now applying 8.2.1 to the function $B_{\nu+1} f$, we get (3).

Case 3°. As in Case 2°, we get

$$D^{k-k_{\nu+1}} d^{e_j} B_\nu f \in L^2. \tag{4}$$

Relation (3) can be rewritten, for $A_{\nu+1} = D^{e_j}$

$$D^{k-k_{\nu+1}} D^{e_j} B_\nu f \in L^2. \tag{5}$$

From (4) and (5) we now get

$$D^{k-k_{\nu+1}} (\tfrac{1}{2} D^{e_j} B_\nu f + \tfrac{1}{2} d^{e_j} B_\nu f) \in L^2.$$

Hence, using the general formulae (see 7.1(2))

$$D^{e_j} f = f^{(e_j)} + \tfrac{1}{2} \xi_j f, \qquad d^{e_j} f = f^{(e_j)} - \tfrac{1}{2} \xi_j f, \tag{6}$$

we obtain $D^{k-k_{\nu+1}} (B_\nu f)^{(e_j)} \in L^2$, which is just (3) written another way.

Case 4°. From (4) and (5) we get

$$D^{k-k_{\nu+1}} (D^{e_j} B_\nu f - d^{e_j} B_\nu f) \in L^2,$$

i.e., $D^{k-k_{\nu+1}} (\xi_j B_\nu f) \in L^2$.

We have thus proved that in each case (3) follows from (2). This implies, by induction, that (2) holds for each $\nu = 0, \ldots, \varkappa$.

In particular, for $\nu = \varkappa$, relation (2) becomes $B_\varkappa f \in L^2$, i.e., $A_\varkappa \ldots A_1 \in L^2$.

8.2.4. COROLLARY. *Under the conditions of Corollary* 8.2.3, *if* $D^k f_n \in L^2$ *and* $D^k f_n \xrightarrow{2} 0$ *or if* $d^k f_n \in L^2$ *and* $d^k f_n \xrightarrow{2} 0$ $(f_n \in L^2)$, *then* $A_\varkappa \dots A_1 f_n \xrightarrow{2} 0$.

PROOF. By Corollary 8.2.3, we have $A_\varkappa \dots A_1 f_n \in L^2$. Now replacing, in the proof of Corollary 8.2.3, f by f_n, the symbol $\in$ by the symbol $\xrightarrow{2}$ and the symbol L^2 by 0, and applying 8.2.2 instead of 8.2.1 we obtain the proof of our present assertion.

8.2.5. COROLLARY. *Under the conditions of Corollary* 8.2.3, *if* $D^{k+2} f_n \in L^2$ *and* $D^{k+2} f_n \xrightarrow{2} 0$ *or* $d^{k+2} f_n \in L^2$ *and* $d^{k+2} f_n \xrightarrow{2} 0$ $(f_n \in L^2)$, *then* $A_\varkappa \dots A_1 f_n \rightrightarrows 0$ *in* $\mathbf{R}^q$, *i.e.*, $A_\varkappa \dots A_1 f_n$ *converges uniformly to* 0 *in* $\mathbf{R}^q$.

PROOF. By Corollary 8.2.3, we have $[(x-y)^1 A_\varkappa \dots A_1 f_n(x)]' \in L^2$ for each fixed $y \in \mathbf{R}^q$ and then, by Corollary 8.2.4, we obtain

$$[(x-y)^1 A_\varkappa \dots A_1 f_n(x)]' \xrightarrow{2} 0. \tag{7}$$

Moreover, square integrable functions are locally integrable, and we can therefore write

$$\begin{aligned}
|(x-y)^1 A_\varkappa \dots A_1 f_n(x)| &= \Big|\int_y^x [(t-y)^1 A_\varkappa \dots A_1 f_n(t)]' dt\Big| \\
&\leqslant \Big|\int_y^x dt\Big|^{1/2} \Big(\int_{\mathbf{R}^q} |[(t-y)^1 A_\varkappa \dots A_1 f_n(t)]'|^2 dt\Big)^{1/2} \\
&= \sqrt{|(x-y)^1|} \cdot \varepsilon_n,
\end{aligned}$$

where $\varepsilon_n = \Big(\int_{\mathbf{R}^q} |[(t-y)^1 A_\varkappa \dots A_1 f_n(t)]'|^2 dt\Big)^{1/2}$. Hence

$$|A_\varkappa \dots A_1 f_n(x)| \leqslant |(x-y)^1|^{-1/2} \cdot \varepsilon_n,$$

which implies our assertion by (7), since the point y can be chosen arbitrarily.

We recall that the convergence $f_n \xrightarrow{\mathscr{S}} f$ is defined equally by either of the conditions: $D^k f_n \xrightarrow{2} D^k f$ for each $k \in \mathbf{P}^q$, or $d^k f_n \xrightarrow{2} d^k f$ for each $k \in \mathbf{P}^q$. The final theorem of this section tells us that this convergence can equally well be defined in other ways. First, the following theorem gives three characteristics of the space $\mathscr{S}$ (the meaning of $\tilde{n}$ is that given in Section 8.1).

8.2.6. THEOREM. *If* $f \in \mathscr{S}'$, *then each of the conditions*

(i) $D^k f \in L^2$ *for each* $k \in \mathbf{P}^q$;

(ii) $d^k f \in L^2$ *for each* $k \in \mathbf{P}^q$;

(iii) *the Hermite coefficients* a_n *of* f *satisfy*

$$\lim_{n\to\infty} \tilde{n}^k a_n = 0 \quad \textit{for each } k \in \boldsymbol{P}^q,$$

is necessary and sufficient in order that $f \in \mathscr{S}$.

PROOF. If $f \in \mathscr{S}$, then also $f^{(e_j)} \in \mathscr{S}$ and $\xi_j f \in \mathscr{S}$. Hence $D^{e_j} f \in \mathscr{S}$ by (6). By induction we thus have $D^k f \in \mathscr{S}$ for each k. Since $\mathscr{S} \subset L^2$, it follows that $D^k f \in L^2$. This proves the necessity of condition (i).

Assume, now, that $D^k f \in L^2$ for a given $f \in \mathscr{S}'$ and all k. Then, by 8.2.3

$$[(x-y)^1 f^{(k)}(x)]' \in L^2.$$

However, square integrable functions are locally integrable, and so we have,

$$\int_y^x [(t-y)^1 f^{(k)}(t)]' dt = (x-y)^1 f^{(k)}(x)$$

or, more explicitly,

$$\int_{\eta_1}^{\xi_1} \cdots \int_{\eta_q}^{\xi_q} \frac{\partial^q}{\partial \xi_1 \dots \partial \xi_q} [(t_1-\eta_1) \dots (t_q-\eta_q) f^{(k)}(t)] d\tau_1 \dots d\tau_q$$
$$= (\xi_1-\eta_1) \dots (\xi_q-\eta_q) f^{(k)}(x).$$

It follows that $f^{(k)}$ is a continuous function for every k.

By 8.2.1, it also follows that the function $[x^{m+1} f^{(k)}]'$ is square integrable. Since it is also continuous, we have

$$|x^{m+1} f^{(k)}| = \left| \int_0^x [t^{m+1} f^{(k)}(t)]' dt \right| \leqslant \sqrt{|x^1|} \cdot M,$$

and so the functions $x^m f^{(k)}$ are bounded. Since m can be arbitrarily large, it follows that $f^{(k)}$ decreases faster than any polynomial. This proves the sufficiency of condition (i).

The proof for (ii) is similar. We therefore move on to condition (iii).

By condition (i), $f \in \mathscr{S}$ iff $D^k f \in L^2(\boldsymbol{R}^q)$ for each $k \in \boldsymbol{P}^q$. Since

$$D^k f = \sum_{n \in \boldsymbol{P}^q} \sqrt{\frac{(n+k)!}{n!}}\, a_{n+k} h_n,$$

we have $D^k f \in L^2$, iff

$$\sum_{n \in \boldsymbol{P}^q} \frac{(n+k)!}{n!} |a_{n+k}|^2 < \infty \qquad \text{for each } k \in \boldsymbol{P}^q.$$

This last condition is equivalent to the condition in (iii), which follows from the inequalities 8.1(5). The proof is now complete.

8.2.7. THEOREM. *If $f \in \mathscr{S}'$ and $D^{e_j} f \in L^2$ or $d^{e_j} f \in L^2$, then*

$$\int (d^{e_j} f)^2 - \int (D^{e_j} f)^2 = \int f^2.$$

PROOF. From (6) we get

$$\begin{aligned}(d^{e_j} f)^2 - (D^{e_j} f)^2 &= (d^{e_j} f - D^{e_j} f)(d^{e_j} f + D^{e_j} f)\\ &= -\xi_j f \cdot 2 f^{(e_j)} = -\xi_j (f^2)^{(e_j)}.\end{aligned}$$

We can generally write

$$\int_{R^q} G = \int_{-\infty}^{\infty} d\xi_j \int_{R^{q-1}} G(x)\, dx^{1-e_j},$$

where $1 - e_j$ is the vector in $\boldsymbol{R}^q$ each of whose coordinates is 1, except for the jth coordinate which is 0. Thus

$$\begin{aligned}\int_{R^q} (d^{e_j} f)^2 - \int_{R^q} (D^{e_j} f)^2 &= \lim_{a,b \to \infty} \int_{-a}^{b} (-\xi_j F^{(e_j)}(\xi_j))\, d\xi_j\\ &= \lim_{a,b \to \infty} \left(\int_{-a}^{b} d\xi_j \int_{R^{q-1}} f^2 dx^{1-e_j} - aF(-a) - gF(b) \right)\\ &= \int_{R^q} f^2 dx - \lim_{a \to \infty} aF(-a) - \lim_{b \to \infty} bF(b),\end{aligned}$$

where the function

$$F(\xi_j) = \int_{R^{q-1}} f^2 dx^{1-e_j}$$

is integrable over the interval $-\infty < \xi_j < \infty$. This implies the existence of the limits of $\xi F(\xi)$, as $\xi \to -\infty$ and as $\xi \to +\infty$. Since F is integrable, both these limits are 0. The stated formula now follows.

8.2.8. THEOREM. *If $f \in \mathscr{S}'$ and $D^{e_j} d^{e_j} f \in L^2$ or $d^{e_j} D^{e_j} f \in L^2$, then*

$$\int (D^{e_j} d^{e_j} f)^2 - \int (d^{e_j} D^{e_j} f)^2 = \int (D^{e_j} f)^2 + \int (d^{e_j} f)^2.$$

PROOF. From (6) we get

$$\begin{aligned}D^{e_j} d^{e_j} f &= f^{(2e_j)} - \tfrac{1}{2} f - \tfrac{1}{4} \xi_j^2 f,\\ d^{e_j} D^{e_j} f &= f^{(2e_j)} + \tfrac{1}{2} f - \tfrac{1}{4} \xi_j^2 f.\end{aligned}$$

Hence

$$\begin{aligned}(D^{e_j} d^{e_j} f)^2 - (d^{e_j} D^{e_j} f)^2 &= (D^{e_j} d^{e_j} f - d^{e_j} D^{e_j} f)(D^{e_j} d^{e_j} f + d^{e_j} D^{e_j} f)\\ &= -2 f f^{(2e_j)} + \tfrac{1}{2} \xi_j^2 f^2\end{aligned}$$

and, integrating by parts,

$$\int (D^{e_j} d^{e_j} f)^2 - \int (d^{e_j} D^{e_j} f)^2 = 2\int (f^{(e_j)})^2 + \tfrac{1}{2}\int (\xi_j f)^2.$$

Transforming the right hand side by means of formulae (6) we get the required equality.

8.2.9. THEOREM. *If $f_n, f \in \mathscr{S}$, the statement $f_n \xrightarrow{\mathscr{S}} f$ is equivalent to each of the following six conditions:*

(i) $D^k f_n \rightrightarrows D^k f$,

(ii) $d^k f_n \rightrightarrows d^k f$,

(iii) $p f_n^{(k)} \rightrightarrows p f_n^{(k)}$,

(iv) $(p f_n)^{(k)} \rightrightarrows (pf)^{(k)}$,

(v) $p f_n^{(k)} \xrightarrow{2} p f^{(k)}$,

(vi) $(p f_n)^{(k)} \xrightarrow{2} (pf)^{(k)}$,

where the p are polynomials and $k \in \boldsymbol{P}^q$ and it is understood that these conditions hold for all polynomials and all k.

PROOF. If $f_n \xrightarrow{\mathscr{S}} f$, it follows from Corollary 8.2.5 that (i), (ii), (iii) and (iv) hold. Now the equivalence of (iii) and (iv) follows easily from formulae 16.6 (1) and 16.6 (2) of the Appendix. The equivalence of (i), (ii) and (iii) follows, in the case $k = e_j$, from formulae 7.1 (3), and the equivalence for arbitrary k then follows by induction. All the statements (i), (ii), (iii) and (iv) are thus equivalent. Now, if (iii) holds for all p and k, then clearly (v) holds for all p and k. Similarly, (iv) implies (vi). The equivalence of (v) and (vi) follows from formulae 16.6 (1) and 16.6 (2) of the Appendix. To complete the proof, it thus remains to show that (v) implies $f_n \xrightarrow{\mathscr{S}} f$.

In fact, it follows from 7.1 (2) by induction, that $D^k f_n$ and $D^k f$ are linear combinations of $x^r f_n^{(k)}$ and $x^r f^{(k)}$ respectively with the same coefficients. Hence (v) implies $D^k f_n \xrightarrow{2} D^k f$ for each $k \in \boldsymbol{P}^q$, i.e., $f_n \xrightarrow{\mathscr{S}} f$.

8.3. EXAMPLES AND REMARKS

The function $f(x) = 1$ can clearly be regarded as a tempered distribution. It therefore expands into a Hermite series with coefficients

$$a_n = \int h_n.$$

We evaluate them in the case $q = 1$. From 4.6 (8), 4.6 (5) and 4.6 (4) follows

$$\sqrt{n+1}\,h_{n+1} = -2h_n' + \sqrt{n}\,h_{n-1}. \tag{1}$$

Integrating this, we find

$$\int_{-\infty}^{\infty} h_{n+1} = \sqrt{\frac{n}{n+1}} \int_{-\infty}^{\infty} h_{n-1}.$$

Since $\int_{-\infty}^{\infty} h_0 = \sqrt[4]{8\pi}$ and $\int_{-\infty}^{\infty} h_1 = 0$, we obtain

$$\int_{-\infty}^{\infty} h_n = \begin{cases} \sqrt[4]{8\pi}\sqrt{\frac{1}{2}\frac{3}{4}\cdots\frac{n-1}{n}} & \text{for } n \text{ even}, \\ 0 & \text{for } n \text{ odd}. \end{cases} \tag{2}$$

We can thus write

$$1 = \sqrt[4]{8\pi}\left(h_0 + \sqrt{\tfrac{1}{2}}\,h_2 + \sqrt{\tfrac{1}{2}\tfrac{3}{4}}\,h_4 + \ldots\right). \tag{3}$$

It is easy to see that the coefficients in the series (3) decrease to 0. However, this series does not converge in square mean, because the constant function 1 is not square integrable. Since 1 can be regarded as a tempered distribution, viz., we have $1 = DF$, where $F = S1 \in L^2$, which follows from 7.2 (2). It is interesting to remark that 1 cannot be represented in the form $1 = dG$ nor, more generally, in the form $1 = d^kG$ with $G \in L^2$, $k \in \boldsymbol{P}^q$. In fact, if $G \in L^2$, the first coefficient in the Hermite expansion of d^kG ($k \neq 0$) vanishes (see Section 8.1). On the other hand the first coefficient in the expansion (3) of 1 is $\neq 0$.

This example shows that not every tempered distribution can be represented in the form d^kF ($F \in L^2$). The roles of the derivatives d and D are thus not symmetric. In our definition of tempered distributions (see Section 7.3), the derivative D cannot be replaced by d: in fact, the class of distributions of the form d^kF ($F \in L^2$) is strictly smaller than the class of distributions D^kF ($F \in L^2$), i.e., the class of tempered distributions.

Another example of a tempered distribution is the Dirac delta-distribution. We evaluate its Hermite coefficients, again for the one-dimensional case $q = 1$. They are

$$a_n = \int \delta h_n = h_n(0).$$

From

$$\sqrt{n+1}\,h_{n+1}(x) = xh_n(x) - \sqrt{n}\,h_{n-1}(x) \tag{4}$$

we obtain

$$h_{n+1}(0) = -\sqrt{\frac{n}{n+1}}\,h_{n-1}(0).$$

Since $h_0(0) = 1/\sqrt{2\pi}$ and $h_1(0) = 0$, we get

$$h_n(0) = \begin{cases} \dfrac{(-1)^{n/2}}{\sqrt[4]{2\pi}} \sqrt{\dfrac{1}{2}\dfrac{3}{4}\cdots\dfrac{n-1}{n}} & \text{for } n \text{ even,} \\ 0 & \text{for } n \text{ odd.} \end{cases} \tag{5}$$

We can therefore write

$$\delta = \frac{1}{\sqrt[4]{2\pi}}\left(h_0 - \sqrt{\tfrac{1}{2}}\,h_2 + \sqrt{\tfrac{1}{2}\tfrac{3}{4}}\,h_4 - \sqrt{\tfrac{1}{2}\tfrac{3}{4}\tfrac{5}{6}}\,h_6 + \ldots\right). \tag{6}$$

Note that in the expansions (3) and (6) only even terms appear. This could have been foreseen prior to any calculation because 1 and δ are even distributions. Generally, a distribution f is *even*, iff $f(-x) = f(x)$. Thus if f is an even tempered distribution and $f = a_0 + a_1h_1 + a_2h_2 + a_3h_3 + \ldots$, then $0 = f(x) - f(-x) = 2a_1h_1 + 2a_3h_3 + \ldots$ Hence $0 = a_1 = a_3 = \ldots$

Similarly, a distribution f is *odd*, iff $f(-x) = -f(x)$. If f is an odd tempered distribution and $f = a_0 + a_1h_1 + a_2h_2 + a_3h_3 + \ldots$, then we have $0 = f(x) + f(-x) = 2a_0h_0 + 2a_2h_2 + \ldots$, which implies $0 = a_0 = a_2 = \ldots$ The Hermite expansion of an odd tempered distribution thus consists only of odd terms.

Substituting $h_n = (\sqrt{2\pi}\,n!)^{-1/2}\, e^{-x^2/4}H_n$ in (6) and then multiplying the equality throughout by $e^{x^2/4}$, we get

$$\delta = \frac{1}{\sqrt{2\pi}}\left(H_0 - \frac{1}{2}H_2 + \frac{1}{2\cdot 4}H_4 - \ldots\right)$$

since $e^{x^2/4}\delta = \delta$ (see Section 3.1, Part II).

However, the last series is not in fact tempered, although it is convergent in the distributional sense. Note that the partial sums of the series are polynomials: in this way, we obtain an explicit approximation to the delta distribution by polynomials. Such an approximation is of theoretical interest only and is of no apparent use in numerical calculations.

8.4. MULTIDIMENSIONAL EXPANSIONS

The multidimensional case can sometimes be reduced to the one-dimensional one. We first prove

8.4.1. THEOREM. *If f is a tempered distribution in $\boldsymbol{R}^q$ and g is a tempered distribution in $\boldsymbol{R}^q$, then the distribution*

$$h(x, y) = f(x)g(y),$$

where $x \in \boldsymbol{R}^q$, $y \in \boldsymbol{R}^q$ is tempered in $\boldsymbol{R}^{q+r}$. Moreover the Hermite coefficient c_p where $p = (m, n) \in \boldsymbol{P}^{q+r}$ is equal to the product $a_m b_n$ of the Hermite coefficients a_m and b_n of f and g.

PROOF. Let $f = D^k F$, $g = D^l G$, $k \in \boldsymbol{P}^q$, $l \in \boldsymbol{P}^r$, $F \in L^2(\boldsymbol{R}^q)$, $G \in L^2(\boldsymbol{R}^r)$. If $s = (k, l)$, then

$$D^s\big(F(x)G(y)\big) = D^k F(x) D^l G(y) = h(x, y).$$

Hence

$$c_p = \sqrt{\frac{(p+s)!}{p!}} \int_{\boldsymbol{R}^{q+r}} F(x)G(y)h_{m+k}(x)h_{n+l}(y)\,dx\,dy$$

$$= \sqrt{\frac{(m+k)!}{m!}} \int_{\boldsymbol{R}^q} F h_{m+k} \sqrt{\frac{(n+l)!}{n!}} \int_{\boldsymbol{R}^r} G h_{n+l} = a_m b_n.$$

8.4.2. COROLLARY. *If $f_1, \ldots, f_q$ are tempered distributions in $\boldsymbol{R}^1$, then the product $f(x) = f_1(\xi_1) \ldots f_q(\xi_q)$ is a tempered distribution in $\boldsymbol{R}^q$. Moreover, if $n = (\nu_1, \ldots, \nu_q) \in \boldsymbol{P}^q$, the Hermite coefficient a_n of f is equal to the product*

$$a_n = a_{1\nu_1} \ldots a_{q\nu_q},$$

where $a_{j\nu_j}$ denotes the ν_jth Hermite coefficient of $f_j(\xi_j)$.

PROOF. The assertion follows by induction from 8.4.1.

The function which takes the value 1 everywhere in $\boldsymbol{R}^q$ can plainly be regarded as the product of q similar functions, each of whose domains is $\boldsymbol{R}^1$. Hence the required expansion follows immediately from 8.3 (3) and the only difficulty lies with the notation. To clarify it we introduce the symbol $\dot{n}$ where

$$\dot{n} = \begin{cases} \sqrt{\dfrac{1}{2}\dfrac{3}{4}\cdots\dfrac{n-1}{n}} & \text{for } n \text{ an even positive integer,} \\ 0 & \text{for } n \text{ an odd positive integer,} \\ 1 & \text{for } n = 0. \end{cases}$$

If $n = (\nu_1, \ldots, \nu_q) \in \boldsymbol{P}^q$, we put by definition

$$\dot{n} = \dot{\nu}_1 \ldots \dot{\nu}_q.$$

The expansion of the unit function 1 in $\boldsymbol{R}^q$ can now be written in the form

$$1 = (8\pi)^{q/4} \sum_{n \in \boldsymbol{P}^q} \dot{n} h_n \quad \text{in } \boldsymbol{R}^q.$$

Since $\delta(x) = \delta_1(\xi_1) \ldots \delta_q(\xi_q)$ for $x = (\xi_1, \ldots, \xi_q) \in \boldsymbol{R}^q$, we can write, by (6),

$$\delta = (2\pi)^{-q/4} \sum_{n \in \boldsymbol{P}^q} i^n \dot{n} h_n \quad \text{in } \boldsymbol{R}^q.$$

8.5. SOME PARTICULAR EXPANSIONS

The coefficients of the Hermite expansion of $\frac{1}{x}$ are $a_n = \int \frac{h_n}{x}$. They all vanish, for n even (see also Section 13.1). Moreover, we have

$$a_1 = \int \frac{h_1}{x} = \frac{1}{\sqrt[4]{2\pi}} \int e^{-x^2/4} dx = \sqrt[4]{8\pi}.$$

The odd coefficients can now be evaluated successively by means of a recursive formula. In fact, from 8.3 (4), we get

$$\sqrt{n+1} \int \frac{h_{n+1}}{x} = \int h_n - \sqrt{n} \int \frac{h_{n-1}}{x},$$

which is the required formula, since the values of $\int h_n$ are known and are given by 8.3 (2). Hence, for even n,

$$\sqrt{n+1}\, a_{n+1} = \sqrt[4]{8\pi} \sqrt{\frac{1}{2}\,\frac{3}{4} \cdots \frac{n-1}{n}} - \sqrt{n}\, a_{n-1}.$$

Instead of using this formula as it stands, it helps to simplify it by setting

$$a_n = \frac{\sqrt[4]{8\pi}}{\sqrt{n!}} u_n \qquad \text{(for odd } n\text{)}.$$

Then $u_1 = 1$, and the recursive formula for u_n is

$$u_{n+1} = 1\cdot 3 \ldots (n-1) - n u_{n-1} \qquad \text{(for even } n \geqslant 2\text{)}. \tag{1}$$

Using this we arrive at the following table for the values of u_n:

n	1	3	5	7	9	11	13	15	17
u_n	1	−1	7	−27	321	−2265	37575	−390915	8281665

It is easy to prove that the sign of u_n always alternates. In fact, if $u_{n-3} < 0$, then by (1), $u_{n-1} > 1\cdot 3 \ldots (n-3)$. Hence, again by (1), $u_{n+1} < 1\cdot 3 \ldots (n-1) - n\cdot 1\cdot 3 \ldots (n-3) = -1\cdot 3 \ldots (n-3)$. Noting a few initial values of u_n, this implies by induction that the sign alternates.

The Hermite expansion of $\frac{1}{x}$ can be written in the form

$$\frac{1}{x} = \sqrt[4]{8\pi} \left(\frac{\omega_1}{\sqrt{1!}} h_1 - \frac{\omega_3}{\sqrt{3!}} h_3 + \frac{\omega_5}{\sqrt{5!}} h_5 - \ldots \right),$$

where $\omega_n = |u_n|$.

The function $\operatorname{sgn} x$ is defined as follows

$$\operatorname{sgn} x = \begin{cases} -1 & \text{for} \quad x < 0, \\ 0 & \text{for} \quad x = 0, \\ 1 & \text{for} \quad x > 0. \end{cases}$$

Its Hermite coefficients are

$$b_n = \int_{-\infty}^{\infty} \operatorname{sgn} x\, h_n(x)\, dx = \int_0^{\infty} h_n - \int_{-\infty}^{0} h_n.$$

Hence $b_n = 0$, if n is even, and

$$b_n = 2\int_0^{\infty} h_n \qquad \text{for odd } n.$$

In particular

$$b_1 = \frac{2}{\sqrt[4]{2\pi}} \int_0^{\infty} x e^{-x^2/4}\, dx = \frac{4}{\sqrt[4]{2\pi}}.$$

As in the preceding case, the remaining coefficients can be found using a recursive formula. Integrating 8.3(1) we find

$$\sqrt{n+1} \int_0^{\infty} h_{n+1} = 2h_n(0) + \sqrt{n} \int_0^{\infty} h_{n-1}$$

and, by 8.3(5),

$$\sqrt{n+1}\, b_{n+1} = \frac{4(-1)^{n/2}}{\sqrt[4]{2\pi}} \sqrt{\frac{1}{2}\frac{3}{4} \cdots \frac{n-1}{n}} + \sqrt{n}\, b_{n-1} \qquad \text{(for even } n\text{)}.$$

Setting

$$b_n = \frac{4}{\sqrt[4]{2\pi}} \frac{(-1)^{(n-1)/2}}{\sqrt{n!}} v_n \qquad \text{(for odd } n\text{)}$$

we find $v_1 = 1$ and

$$v_{n+1} = 1 \cdot 3 \ldots (n-1) - n v_{n-1} \qquad \text{(for even } n \geqslant 2\text{)}.$$

The recursive formulae for u_n and v_n are thus the same, which implies that $v_n = u_n$. Hence the required expansion is

$$\operatorname{sgn} x = \frac{4}{\sqrt[4]{2\pi}} \left(\frac{\omega_1}{\sqrt{1!}} h_1 + \frac{\omega_3}{\sqrt{3!}} h_3 + \frac{\omega_5}{\sqrt{5!}} h_5 + \ldots \right);$$

note that all the coefficients in this expansion are positive.

We can incidentally establish an interesting integral equality for the odd Hermite functions. The equality $u_n = v_n$ implies that, for a_n and b_n,

$$a_n = (-1)^{(n-1)/2} \frac{\sqrt{\pi}}{2} b_n.$$

But

$$a_n = \int_{-\infty}^{\infty} \frac{h_n}{x} = 2 \int_0^{\infty} \frac{h_n}{x}, \qquad \text{as } n \text{ is odd.}$$

Hence

$$\frac{2}{\sqrt{\pi}} \int_0^{\infty} \frac{h_n}{x} = (-1)^{(n-1)/2} \int_0^{\infty} h_n \qquad \text{for odd } n.$$

We finally remark that, just as for the function 1 and the distribution δ, the above results can be extended to any number of dimensions.

8.6. THE FOURIER TRANSFORM

If f is any tempered distribution and $f = \sum_{n \in P^q} a_n h_n$, then the Fourier transform of f is defined by

$$\mathscr{F}(f) = \sum_{n \in P^q} i^n a_n h_n.$$

Clearly, the Fourier transform of a tempered distribution is itself a tempered distribution. By 4.9.1, the above definition is compatible with the definition for square integrable functions.

It follows from the definition that the Fourier transform of an even distribution is an even distribution and the Fourier transform of an odd distribution is an odd distribution.

We also have

8.6.1. THEOREM. *If $f \in \mathscr{S}$, then $\mathscr{F}(f) \in \mathscr{S}$.*

PROOF. This follows from the estimate of the Hermite coefficients, given by 8.2.6.

From the expansions of the preceding sections, we see at once that, in $\boldsymbol{R}^q$,

$$\mathscr{F}(1) = (2\sqrt{\pi})^q \delta, \qquad \mathscr{F}(\delta) = (2\sqrt{\pi})^{-q},$$

$$\mathscr{F}\left(\frac{1}{x}\right) = (i\sqrt{\pi})^q \cdot 2^{-q} \operatorname{sgn} x, \qquad \mathscr{F}(2^{-q} \operatorname{sgn} x) = (-i\sqrt{\pi})^{-q} \frac{1}{x}.$$

(The formulae in the last line follow in the first place for the one dimensional case, but they can easily be extended to any number of dimensions, using the appropriate definitions of $\frac{1}{x}$ and $\operatorname{sgn} x$).

If $f = \sum_{n \in \boldsymbol{P}^q} a_n h_n$, we can write

$$\mathscr{F}(D^{e_j} f) = \sum_{n \in \boldsymbol{P}^q} i^{n-e_j} a_n D^{e_j} h_n = -iD^{e_j} \sum_{n \in \boldsymbol{P}^q} i^n a_n h_n = -iD^{e_j} \mathscr{F}(f),$$

by definition of the Fourier transform and by 8.1(2). Similarly we have

$$\mathscr{F}(d^{e_j} f) = id^{e_j} \mathscr{F}(f).$$

By induction, we obtain

$$\mathscr{F}(D^k f) = (-iD)^k \mathscr{F}(f) \tag{1}$$

and similarly

$$\mathscr{F}(d^k f) = (id)^k \mathscr{F}(f) \tag{2}$$

for any $k \in \boldsymbol{P}^q$ and any tempered distribution f. Hence, using 7.1(3), we obtain the formulae

$$\xi_j \mathscr{F}(f) = 2i\mathscr{F}(f^{(e_j)}) \quad \text{and} \quad \mathscr{F}(\xi_j f) = -2i[\mathscr{F}(f)]^{(e_j)}$$

for any tempered distribution f. More generally, we find, by induction,

$$x^k \mathscr{F}(f) = (2i)^k \mathscr{F}(f^{(k)}) \quad \text{and} \quad \mathscr{F}(x^k f) = (-2i)^k [\mathscr{F}(f)]^{(k)} \tag{3}$$

for any tempered distribution f.

Formula (1) is particularly important. It shows that the Fourier transform of a distribution of class T^k is again a distribution of class T^k. Moreover, it implies

8.6.2. THEOREM. *If a sequence of distributions f_n is tempered to f, then the sequence of Fourier transforms $\mathscr{F}(f_n)$ is tempered to $\mathscr{F}(f)$.*

PROOF. In fact, there is a sequence of square integrable functions F_n which converges in mean to F and such that $D^k F_n = f_n$, $D^k F = f$ for some $k \in \boldsymbol{P}^q$. Hence, by (1),

$$\mathscr{F}(f_n) = (-iD)^k \mathscr{F}(F_n).$$

Since F_n converges in square mean to F the sequence $\mathscr{F}(F_n)$ converges in square mean to $\mathscr{F}(F)$. Hence $\mathscr{F}(f_n)$ is tempered to

$$(-iD)^k \mathscr{F}(F) = \mathscr{F}(f),$$

again by (1), and the proof is complete.

The above theorem enables us to prove the equalities

$$\mathscr{F}\big(f(x+a)\big) = e^{-iax/2} g(x) \quad \text{and} \quad \mathscr{F}\big(e^{iax/2} f(x)\big) = g(x+a) \quad (a \in \boldsymbol{R}^q) \tag{4}$$

for any tempered distribution f and its Fourier transform $\mathscr{F}(f) = g$.

In fact, there is a sequence of square integrable functions f_n which is tempered to f, e.g., $f_n = \sum_{p \in N_n} a_p h_p$, where the N_n are finite subsets of $\boldsymbol{P}^q$ such that $N_n \to \boldsymbol{P}^q$, as $n \to \infty$. By 4.3 (7) and 4.3 (8) we have

$$\mathscr{F}\big(f_n(x+a)\big) = e^{-iax/2} g_n(x) \quad \text{and} \quad \mathscr{F}\big(e^{iax/2} f_n(x)\big) = g_n(x+a),$$

where $g_n = F(f_n)$. Hence (4) follows, by 8.6.2.

If $\alpha_n > 0$, then

$$\mathscr{F}(e^{-\alpha_n x^2}) = (2\sqrt{\pi})^q (\sqrt{\pi\beta_n})^{-q} e^{-x^2/\beta_n},$$

where $\beta_n = 16\alpha_n$ (see Section 4.3). Letting $\alpha_n \to 0$, we hence obtain

$$\mathscr{F}(1) = (2\sqrt{\pi})^q \delta. \tag{5}$$

Applying the inverse Fourier transform, we get

$$\mathscr{F}(\delta) = (2\sqrt{\pi})^{-q}.$$

Both the formulae were found at the beginning of this section, but by quite different methods.

If $g = \mathscr{F}\left(\frac{1}{x}\right)$ and $x \in \boldsymbol{R}^1$, we have by (3)

$$g' = \frac{1}{-2i} \mathscr{F}\left(x \cdot \frac{1}{x}\right) = \tfrac{1}{2} i \mathscr{F}(1).$$

(The equality $x \cdot \frac{1}{x} = 1$ is trivial in the classical sense; the proof that it also holds when $\frac{1}{x}$ is regarded as a distribution is given in Section 12.6.) By what we have just proved, $g' = i\sqrt{\pi}\delta$. Hence $g = i\sqrt{\pi}H + c$, where c is a constant. To make g odd, we have to put $g = i\sqrt{\pi}(H - \frac{1}{2})$. Hence $\mathscr{F}\left(\frac{1}{x}\right) = i\sqrt{\pi}2^{-1} \operatorname{sgn} x$. To generalize this result to any number of dimensions, we use

8.6.3. THEOREM. *If $f_1(\xi_1), \ldots, f_q(\xi_q)$ are tempered distributions in $\boldsymbol{R}^1$ and $x = (\xi_1, \ldots, \xi_q)$, then the product $f(x) = f_1(\xi_1) \ldots f_q(\xi_q)$ is a tempered distribution in $\boldsymbol{R}^q$ and its Fourier transform is equal to the product $\mathscr{F}(f) = g_1(\xi_1) \ldots g_q(\xi_q)$, where $g_j(\xi_j)$ denotes the Fourier transform of $f_j(\xi_j)$.*

PROOF. The fact that f is a tempered distribution was stated in Corollary 8.4.2. By that Corollary it also follows that

$$\sum_{0 \leq n \leq A} i^n a_n h_n(x) = \sum_{\nu_1=0}^{A} i^{\nu_1} a_{1\nu_1} h_{\nu_1}(\xi_1) \ldots \sum_{\nu_q=0}^{A} i^{\nu_q} a_{q\nu_q} h_{\nu_q}(\xi_q).$$

Letting $A \to \infty$, we obtain the required result.

Applying 8.6.3 to the distribution $\frac{1}{x} = \frac{1}{\xi_1} \dots \frac{1}{\xi_q}$ we obtain, by the previous result

$$\mathscr{F}\left(\frac{1}{x}\right) = (i\sqrt{\pi})^q 2^{-q} \operatorname{sgn} x,$$

where $\operatorname{sgn} x = \operatorname{sgn} \xi_1 \dots \operatorname{sgn} \xi_q$. Hence, applying the inverse transform, we have

$$\mathscr{F}(2^{-q} \operatorname{sgn} x) = (-i\sqrt{\pi})^{-q} \frac{1}{x}.$$

Note that these equalities were previously obtained at the beginning of this section, directly from the appropriate Hermite expansions.

8.7. AN ANALOGY WITH POWER SERIES

Formula 8.1(2) recalls the ordinary differentiation of a power

$$\left(\frac{d}{dz}\right)^k \frac{z^n}{n!} = \begin{cases} \dfrac{z^{n-k}}{(n-k)!} & \text{for} \quad k \leqslant n, \\ 0 & \text{for} \quad k \nleq n. \end{cases}$$

There is also an analogy between the Hermite series $f = \sum_{n \in P^q} a_n h_n$ and the power series $\psi(z) = \sum_{n \in P^q} a_n z^n$, in which, to every tempered distribution f, there corresponds, one-to-one, an analytic function $\psi(z)$ such that $\sum_{n \in P^q} n^{-k} |a_n|^2 < \infty$ for some k. The study of tempered distributions is thus equivalent to the study of a particular class A of analytic functions: to every property of the class A, there corresponds a property of tempered distributions, and conversely. We discuss this using an example in the one dimensional case. In this case A is a class of analytic functions of a single complex variable z, which are holomorphic in the circle $|z|<1$. Rotation around the origin acts as a linear operator from A onto A. In particular, so does the rotation T through the angle $\pi/2$, which can be expressed analytically either by $T(\psi(z)) = \psi(iz)$ or by $T(\sum_{n=0}^{\infty} a_n z^n) = \sum_{n=0}^{\infty} a_n i^n z^n$. Clearly, to this particular rotation there corresponds the Fourier transform $\mathscr{F}(\sum_{n=0}^{\infty} a_n h_n) = \sum_{n=0}^{\infty} a_n i^n z^n$. The inverse transformation T^{-1} is the rotation through the angle $-\pi/2$, and so $T^{-1}(\psi(z)) = \psi(-iz)$. Hence we obtain the inverse Fourier transform $\mathscr{F}^{-1}(\sum_{n=0}^{\infty} a_n h_n)$

$= \sum_{n=0}^{\infty} a_n(-i)^n h_n$. For the rotation T, it is also plain that $T^4 = I$, i.e., that the operation consisting of four successive rotations through $\pi/2$ is the identity transformation. Consequently, we also have the corresponding property for the Fourier transform, that is $\mathscr{F}^4 = I$.

Given any positive integer n, we thus see that there is no problem in finding a transformation G of the space of tempered distribution such that $G^n = I$ (Fourier–Mehler Transformation).

8.8. THE FOURIER TRANSFORM OF A CONVOLUTION

The Fourier transform $\mathscr{F}(f)$ of a function $f \in L^2$ was defined, in Section 4.3, as the limit in L^2 of

$$F_A(x) = (2\sqrt{\pi})^{-q} \int_{-A}^{A} e^{ixt/2} f(t)\,dt,$$

as $A \to \infty$. In particular, if $f \in \mathscr{S}$, the above integral converges absolutely and we have, more simply

$$\mathscr{F}(f) = (2\sqrt{\pi})^{-q} \int_{R^q} e^{ixt/2} f(t)\,dt.$$

8.8.1. THEOREM. *If f, $g \in \mathscr{S}$, then*

$$\mathscr{F}(f * g) = (2\sqrt{\pi})^q \mathscr{F}(f) \cdot \mathscr{F}(g). \tag{1}$$

In other words, the Fourier transform of a convolution is equal up to a constant factor to the product of the Fourier transforms of its factors.

PROOF.

$$\mathscr{F}(f*g) = (2\sqrt{\pi})^{-q} \int_{R^q} e^{ixt/2}\,dt \int_{R^q} f(t-\tau)g(\tau)\,d\tau$$

$$= (2\sqrt{\pi})^{-q} \int_{R^q} e^{ix\tau/2} g(\tau)\,d\tau \int_{R^q} e^{ix(t-\tau)/2} f(t-\tau)\,dt.$$

Making the substitution $t-\tau = u$ in the innermost integral above we get

$$\mathscr{F}(f*g) = (2\sqrt{\pi})^{-q} \int_{R^q} e^{ixt/2} g(\tau)\,d\tau \int_{R^q} e^{ixu/2} f(u)\,du,$$

and the required equality follows.

From $f, g \in \mathscr{S}$, it follows that $\mathscr{F}(f)$, $\mathscr{F}(g) \in \mathscr{S}$, by Theorem 8.2.2. Hence $\mathscr{F}(f) \cdot \mathscr{F}(g) \in \mathscr{S}$ and consequently, $\mathscr{F}(f*g) \in \mathscr{S}$. Taking the inverse Fourier transform, we get $f*g$. We thus have

8.8.2. THEOREM. *If* $f,\ g \in \mathscr{S}$, *then* $f{*}g \in \mathscr{S}$.

This theorem can also be proved directly, by estimating integrals.

From 8.8.1 we deduce the more general

8.8.3. THEOREM. *If* $f, g \in L^2$, *then* (1) *holds.*

PROOF. Let $f_n, g_n \in \mathscr{S}$, $f_n \overset{2}{\to} f$, $g_n \overset{2}{\to} g$. By 8.8.1 we have

$$\mathscr{F}(f_n{*}g_n) = (\sqrt{2\pi})^q \mathscr{F}(f_n) \cdot \mathscr{F}(g_n).$$

We have $\mathscr{F}(f_n) \overset{2}{\to} \mathscr{F}(f)$ and $\mathscr{F}(g_n) \overset{2}{\to} \mathscr{F}(g)$, by 4.3 (3). Hence

$$\mathscr{F}(f_n) \cdot \mathscr{F}(g_n) \to \mathscr{F}(f) \cdot \mathscr{F}(g) \quad \text{in } L^1.$$

Now, $f_n{*}g_n \to f{*}g$ uniformly, by 4.2.5. Hence

$$\mathscr{F}(f_n{*}g_n) \overset{2}{\to} \mathscr{F}(f{*}g),$$

by 7.2.3. Thus $f_n{*}g_n$ is tempered to $f{*}g$. Consequently, $\mathscr{F}(f_n{*}g_n)$ is tempered to $\mathscr{F}(f{*}g)$. Hence (1) follows for $f, g \in L^2$, and the proof is complete.

8.8.4. THEOREM. *If* f *is tempered and* $g \in \mathscr{S}$, *then* (1) *holds.*

PROOF. Let $f_n \in \mathscr{S}$, $f_n \overset{t}{\to} f$. By 8.8.1 we have

$$\mathscr{F}(f_n{*}g) = (2\sqrt{\pi})^q \mathscr{F}(f_n) \cdot \mathscr{F}(g).$$

If $n \to \infty$, we have $f_n{*}g \overset{t}{\to} f{*}g$, by 8.8.3 and hence $\mathscr{F}(f_n{*}g) \overset{t}{\to} \mathscr{F}(f{*}g)$, by 8.6.2. Again by 8.6.2, we have $\mathscr{F}(f_n) \overset{t}{\to} \mathscr{F}(f)$. Since $\mathscr{F}(f) \in \mathscr{S}$, this implies $\mathscr{F}(f_n) \cdot \mathscr{F}(g) \overset{t}{\to} \mathscr{F}(f) \cdot \mathscr{F}(g)$, by 7.9.3.

9. Periodic distributions

9.1. SMOOTH INTEGRAL

For the study of periodic distributions it is useful to introduce the concept of a *smooth integral*. We first discuss the case of a single variable. We assume that ω is of bounded carrier and that $\int_{-\infty}^{\infty} \omega = 1$. By the *smooth integral* of φ from ω to x we mean

$$\int_{\omega}^{x} \varphi = \int_{-\infty}^{\infty} \omega(\eta)\, d\eta \int_{\eta}^{x} \varphi(t)\, dt. \tag{1}$$

Clearly,

$$\Big(\int_{\omega}^{x} \varphi\Big)' = \varphi(x).$$

$\int_{\omega}^{x} f$ is thus an inverse operation for differentiation, in addition to the ordinary integral

$$\int_{a}^{x} \varphi = \int_{a}^{x} \varphi(t)\, dt, \tag{2}$$

where a is a fixed point. Notwithstanding this similarity, there is a striking difference between the integrals (1) and (2). The smooth integral (1) is a regular operation and the ordinary integral (2) is not a regular operation. The smooth integral can therefore be extended to arbitrary distributions φ, whereas this is not feasible for the ordinary integral.

The reader can verify that, if δ_n is a delta sequence, then

$$\int_{\delta_n}^{x} \varphi \to \int_{0}^{x} \varphi.$$

More generally, we also consider partial smooth integrals

$$\int_{\omega, j}^{x} \varphi = \int_{-\infty}^{\infty} \omega(\eta)\, d\eta \int_{\eta}^{\xi_j} \varphi\big(x + (\xi - \xi_j) e_j\big)\, d\xi, \tag{3}$$

where φ is a smooth function in $\boldsymbol{R}^q$, and j indicates the coordinate with respect to which the integration is carried out.

Note that (1) is just a particular case of (3), namely when $q = 1$.

We show that $\int_{\omega,j}^{x} \varphi$ is a regular operation on φ, where ω is assumed fixed. To this end, we first introduce the following definition. A fundamental sequence φ_n is said to be *of order* k in an open set $I \subset \boldsymbol{R}^q$, iff there is a uniformly convergent sequence of smooth functions Φ_n in I such that $\Phi_n^{(k)} = \varphi_n$ in I. If a fundamental sequence is of order k in I, then it is also of order l, for any $l \geqslant k$.

9.1.1. THEOREM. *If a sequence φ_n is fundamental in $\boldsymbol{R}^q$, the sequence $\int_{\omega,j}^{x} \varphi_n$ is also fundamental in $\boldsymbol{R}^q$. Moreover, if φ_n is of order $k \geqslant e_j$ in the interval I, then $\int_{\omega,j}^{x} \varphi_n$ is of order $k - e_j$ in I.*

PROOF. We have, in I,

$$\int_{\eta}^{\xi_j} \varphi_n\big(x + (\xi - \xi_j)e_j\big)d\xi = \Phi_n^{(k-e_j)}(x) - \Phi_n^{(k-e_j)}\big(x + (\eta - \xi_j)e_j\big).$$

Multiplying both sides by $\omega(\eta)$ and then integrating with respect to η, we get

$$\int_{\omega,j}^{x} \varphi_n = \Phi_n^{(k-e_j)}(x) - \Psi_n(x), \tag{4}$$

where

$$\Psi_n(x) = \int_{-\infty}^{\infty} \omega(\eta)\Phi_n^{(k-e_j)}\big(x + (\eta - \xi_j)e_j\big)d\eta$$

$$= \Big(\int_{-\infty}^{\infty} \omega(\eta)\Phi_n^{((\varkappa_j - 1)e_j)}\big(x + (\eta - \xi_j)e_j\big)d\eta\Big)^{(k - \varkappa_j e_j)}$$

and $\varkappa_j$ is the jth coordinate of k. Hence, integrating by parts $(\varkappa_j - 1)$ times

$$\Psi_n(x) = (-1)^{\varkappa_j - 1}\Big(\int_{-\infty}^{\infty} \omega^{(\varkappa_j - 1)}(\eta)\Phi_n\big(x + (\eta - \xi_j)e_j\big)d\eta\Big)^{(k - \varkappa_j e_j)}.$$

Now, since Φ_n converges uniformly in I, it follows that the sequence of integrals in parentheses also converges uniformly in I. The sequence Ψ_n is therefore fundamental and of order $k - \varkappa_j e_j$ in I. Consequently it is also of order $k - e_j$ in I. This implies that the sequence (4) is fundamental and of order $k - e_j$ in I.

From 9.1.1 it follows that the integral $\int_{\omega,j}^{x} \varphi$, for fixed ω and j, is a regular operation on φ, and therefore extends to arbitrary distributions.

We say that a distribution f is of order $k \in \boldsymbol{P}^q$ in an interval $I \subset \boldsymbol{R}^q$, if it is, in I, the derivative of order k of a continuous function. From Theorem 9.1.1, we at once obtain

9.1.2. THEOREM. *If a distribution f is of order $k \geqslant e_j$ in an open interval I, then the distribution $\int_{\omega,j}^{x} f$ is of order $k-e_j$ in I.*

A sequence of distributions f_n is said to be of order k in an open set O iff, for each interval I inside O there is a uniformly convergent sequence of continuous functions Φ_n in I such that $\Phi_n^{(k)} = f_n$ in I.

9.1.3. THEOREM. *If a sequence of distributions f_n is distributionally convergent in $\boldsymbol{R}^q$, then the sequence $\int_{\omega,j}^{x} f_n$ is also distributionally convergent in $\boldsymbol{R}^q$. Moreover, if f_n is of order $k \geqslant e_j$ in an interval I, then $\int_{\omega,j}^{x} f_n$ is of order $k-e_j$ in I.*

PROOF. Since all the operations which are used in the proof of 9.1.1 are regular, the proof of 9.1.3 is simply the proof of 9.1.1 repeated step for step using f_n instead of φ_n.

From 9.1.3 it follows that if f_n converges distributionally to f, then $\int_{\omega,j}^{x} f_n$ converges distributionally to $\int_{\omega,j}^{x} f$.

9.2. INTEGRAL OVER THE PERIOD

A distribution f in $\boldsymbol{R}^q$ is said to be *periodic*, iff

$$f(x+e_j) = f(x) \quad \text{for} \quad j = 1, \ldots, q.$$

(We thus restrict ourselves to periods of length 1. This is not a real restriction and we can pass without difficulty, if necessary, to the period 2π or to any other period by means of a simple substitution.)

By induction it follows that f is periodic, iff $f(x+p) = f(x)$ for $p \in \boldsymbol{B}^q$ (we recall that $\boldsymbol{B}^q$ is the set of all integer points of $\boldsymbol{R}^q$).

By the *integral over the period* e_j we mean

$$\int_{0,j}^{1} f = F_j(x+e_j) - F_j(x), \tag{1}$$

where F_j is the jth primitive of f, i.e., a distribution such that $F_j^{(e_j)} = f$. For example, we can take $F_j = \int_{\omega, j}^{x} f$. To show that the definition is consistent, we have to prove that the expression (1) does not depend on the choice of F_j. In fact, if G_j is another jth primitive of f, then $G_j^{(e_j)} = f$. Hence $(F_j - G_j)^{(e_j)} = 0$. This implies that the distribution $K_j = F_j - G_j$ is independent of ξ_j. Consequently

$$[F_j(x+e_j) - F_j(x)] - [G_j(x+e_j) - G_j(x)] = K_j(x+e_j) - K_j(x) = 0,$$

which proves the consistency of the definition.

Note that the distribution (1) is independent of ξ_j. In fact, this follows immediately from the equalities

$$\Big(\int_{0,j}^{1} f\Big)^{(e_j)} = f(x+e_j) - f(x) = 0.$$

Since $\int_{0,j}^{1} f$ is constant with respect to ξ_j, it follows immediately from (1) that $\int_{0,j}^{1} f$ is a periodic distribution in $\boldsymbol{R}^q$.

From (1) and 9.1.3 it follows that if a sequence of periodic distributions f_n is distributionally convergent to f in $\boldsymbol{R}^q$, then sequence $\int_{0,j}^{1} f_n$ is distributionally convergent to $\int_{0,j}^{1} f$ in $\boldsymbol{R}^q$.

9.3. DECOMPOSITION THEOREM FOR PERIODIC DISTRIBUTIONS

Our aim is to prove

9.3.1. THEOREM. *A sequence of periodic distributions is distributionally convergent, iff it can be represented as a finite sum of derivatives of uniformly convergent sequences of continuous periodic functions. In other words:*

A sequence of periodic distributions f_n in $\boldsymbol{R}^q$ is distributionally convergent, iff, for some $r \in \boldsymbol{P}^q$, there exist continuous periodic functions F_{mn} $(0 \leqslant m \leqslant r$, $n = 1, 2, \ldots)$ such that

$$f_n = \sum_{0 \leqslant m \leqslant r} F_{mn}^{(m)}$$

and $F_{mn} \rightrightarrows$ in $\boldsymbol{R}^q$ for each $0 \leqslant m \leqslant r$. In particular, if the f_n are smooth periodic functions, then the F_{mn} can be chosen smooth.

PROOF. Let I be an interval in $\mathbf{R}^q$ containing the origin and the point $(1, \ldots, 1)$. Assume that a sequence f_n of periodic distributions is of order $k \geqslant e_j$ in I. Since the f_n are periodic it follows that the sequence f_n is of order k in $\mathbf{R}^q$. Let

$$g_n(x) = \int_{\omega,j}^{x} f_n - \xi_j \int_{0,j}^{x} f_n \tag{1}$$

and $F_n(x) = \int_{\omega,j}^{x} f_n$. We can then write

$$g_n(x) = F_n(x) - \xi_j\big(F_n(x+e_j) - F_n(x)\big).$$

It now follows by 9.1.3 that g_n is of order $k-e_j$. Moreover, since the difference $F_n(x+e_j)-F_n(x)$ is constant with respect to ξ_j, we obtain $g_n(x+e_j) = g_n(x)$. From the periodicity of f_n it follows that $g_n(x+e_i) = g_n(x)$ for $i \neq j$ as well. The sequence g_n thus consists of periodic distributions and is of order $k-e_j$. Moreover, if the f_n are smooth, so are g_n and $\int_{0,j}^{1} f_n$.

Now, differentiating (1) with respect to ξ_j, we get

$$f_n = \int_{0,j}^{1} f_n + g_n^{(e_j)}.$$

We have therefore proved that if a sequence of periodic distributions (or smooth functions) f_n is of order $k \geqslant e_j$, it can be expressed as the sum of two sequences which are of order $k-e_j$ and whose terms are periodic distributions of order $k-e_j$.

By induction, it follows that the sequence f_n can be expressed as a finite sum of derivatives of sequences of order 0 whose terms are periodic distributions of order 0. In other words, the sequences converge uniformly in $\mathbf{R}^q$ and their terms are continuous functions. In particular, if the functions f_n are smooth, the corresponding uniformly convergent sequences also consist of smooth functions.

9.3.2. COROLLARY. *A sequence of periodic distributions is convergent, iff it is tempered.*

This Corollary follows from the preceding theorem and 7.4.1.

9.3.3. COROLLARY. *A distribution f is periodic, iff it is a finite sum of distributional derivatives of continuous periodic functions, i.e., if there is an $r \in \mathbf{R}^q$ such that*

$$f = \sum_{0 \leqslant m \leqslant r} f_m^{(m)}$$

where the f_m are continuous periodic functions.

9.3.4. Corollary. *Periodic distributions are tempered.*

The class of periodic distributions is clearly a linear space. It is therefore a subspace of the space of tempered distributions.

9.4. PERIODIC INNER PRODUCT

The ordinary inner product $(\varphi, \psi) = \int_{R^q} \varphi\psi$ does not exist for periodic functions φ and ψ, because the integral is in general divergent for such functions, the exceptional case being when the product $\varphi\psi$ vanishes almost everywhere. For periodic smooth functions φ, ψ we define the *periodic inner product* by

$$(\varphi, \psi) = \int_0^1 \varphi\psi .$$

It should be emphasized that this definition of the inner product of periodic smooth functions differs from our former definition (see Section 5.1). However the use of the same symbol for it does not give rise to any confusion.

It is clear that all the equalities 5.1(1) remain true for periodic inner products. Moreover, it is easy to check, that, for each $k \in \boldsymbol{P}^q$, we have

$$(\varphi^{(k)}, \psi) = (-1)^k (\varphi, \psi^{(k)}). \tag{1}$$

9.4.1. Lemma. *If φ_n is a fundamental sequence of periodic smooth functions, the sequence (φ_n, ψ) is convergent for each periodic smooth function ψ.*

Proof. By 9.3.1 there exist periodic smooth functions Φ_{mn} $(0 \leqslant m \leqslant r \in \boldsymbol{P}^q)$ such that

$$\varphi_n = \sum_{0 \leqslant m \leqslant r} \Phi_{mn}^{(m)}$$

and $\Phi_{mn} \rightrightarrows$ in $\boldsymbol{R}^q$ as $n \to \infty$. Hence, by (1), we can write

$$(\varphi_n, \psi) = \sum_{0 \leqslant m \leqslant r} (-1)^m (\Phi_{mn}, \psi^{(m)}).$$

Since the sequence $(\Phi_{mn}, \psi^{(m)})$ converges for each $0 \leqslant m \leqslant r$, this implies our assertion.

If f is a periodic distribution and ψ is a periodic smooth function, the periodic inner product (f, ψ) is defined by the equality

$$(f, \psi) = \lim_{n \to \infty} (\varphi_n, \psi), \tag{2}$$

where the φ_n are periodic smooth functions such that the sequence φ_n is distributionally convergent to f.

By Lemma 9.4.1, the limit on the right hand side of (2) always exists and does not depend on the choice of the sequence φ_n.

9.4.2. THEOREM. *If f is a periodic distribution and ψ is a periodic smooth function, we have, for each $k \in \boldsymbol{P}^q$,*

$$(f^{(k)}, \psi) = (-1)^k (f, \psi^{(k)}). \tag{3}$$

PROOF. By (2) and (1), we have

$$(f^{(k)}, \psi) = \lim_{n\to\infty} (\varphi_n^{(k)}, \psi) = (-1)^k \lim_{n\to\infty} (\varphi_n, \psi^{(k)}) = (-1)^k (f, \psi^{(k)}).$$

9.4.3. THEOREM. *If a sequence of periodic distributions f_n converges distributionally to f, then*

$$(f_n, \psi) \to (f, \psi),$$

for any periodic smooth function ψ.

PROOF. By 9.3.1, there exist periodic continuous functions F_{mn}, F_m, where $0 \leqslant m \leqslant r \in \boldsymbol{P}^q$, such that

$$f_n = \sum_{0\leqslant m\leqslant r} F_{mn}^{(m)} \quad \text{and} \quad f = \sum_{0\leqslant m\leqslant r} F_m^{(m)}$$

and $F_{mn} \rightrightarrows F_m$ in $\boldsymbol{R}^q$ as $n \to \infty$. Hence, by (3), we have

$$(f_n, \psi) = \sum_{0\leqslant m\leqslant r} (-1)^m (F_{mn}, \psi^{(m)}) \quad \text{and} \quad (f, \psi) = \sum_{0\leqslant m\leqslant r} (-1)^m (F_m, \psi^{(m)}).$$

This implies our assertion since $(F_{mn}, \psi^{(m)}) \to (F_m, \psi^{(m)})$ for $0 \leqslant m \leqslant r$.

REMARK. Since each distributionally convergent sequence of smooth functions is fundamental, Lemma 9.4.1 is merely a special case of Theorem 9.4.3.

We say that a sequence of smooth periodic functions in $\boldsymbol{R}^q$ is *smoothly convergent* to φ iff, for each $k \in \boldsymbol{P}^q$, the sequence $\varphi_n^{(k)}$ is uniformly convergent to $\varphi^{(k)}$ in $\boldsymbol{R}^q$.

9.4.4. THEOREM. *If a sequence of periodic distributions f_n converges distributionally to f and a sequence of periodic smooth functions ψ_n converges smoothly to ψ, then*

$$\lim_{n\to\infty} (f_n, \psi_n) = (f, \psi).$$

PROOF. In fact, by 9.4.3, it is enough to show that $(f_n, \psi_n - \psi) \to 0$ as $n \to \infty$. Let F_{mn} be the functions used in the proof of 9.4.1. By (3), we then have

$$(f_n, \psi_n - \psi) = \sum_{0\leqslant m\leqslant r} (-1)^m (F_{mn}, \psi_n^{(m)} - \psi^{(m)}).$$

Since $F_{mn} \rightrightarrows$ in $\boldsymbol{R}^q$ and $\psi_n^{(m)} \rightrightarrows \psi^{(m)}$, for each $0 \leqslant m \leqslant r$, our assertion follows.

9.5. PERIODIC CONVOLUTION

Let φ and ψ be two periodic smooth functions. By the *periodic convolution* $\varphi*\psi$ we mean the integral

$$\int_0^1 \varphi(x-t)\psi(t)\,dt. \tag{1}$$

It should be emphasized, just as in the case of the inner product, that the definition of the convolution of periodic functions differs from our former definition, although the use of the same symbol for it (asterisk) should not give rise to any confusion. A close connection exists between the inner product (φ, ψ) and the convolution $\varphi*\psi$. In fact, if $\tilde{\varphi}_x(t) = \varphi(x-t)$, we have

$$\varphi*\psi = (\tilde{\varphi}_x, \psi).$$

For periodic functions, convolution can thus be defined by means of the inner product. Conversely, the inner product (φ, ψ) can be defined as the value at 0 of the convolution $\overline{\varphi}*\psi$, where $\overline{\varphi}(t) = \varphi(-t)$; in other words

$$(\varphi, \psi) = (\overline{\varphi}*\psi)(0).$$

Since φ and ψ are periodic we can write

$$\int_0^1 \varphi(x-t)\psi(t)\,dt = \int_a^{a+1} \varphi(x-t)\psi(t)\,dt \tag{2}$$

for each $a \in \boldsymbol{R}^q$. A simple substitution yields the commutativity law

$$\varphi*\psi = \psi*\varphi. \tag{3}$$

Moreover, we have

$$(\lambda\varphi)*\psi = \varphi*(\lambda\psi) = \lambda(\varphi*\psi) \tag{4}$$

for any real number λ and

$$\varphi*(\psi+\chi) = \varphi*\psi+\varphi*\chi. \tag{5}$$

Making a simple substitution and using the Fubini theorem, we get the associativity law

$$(\varphi*\psi)*\chi = \varphi*(\psi*\chi). \tag{6}$$

Since the integral (1) exists for arbitrary smooth periodic functions φ and ψ it follows that, for any $k, l \in \boldsymbol{P}^q$, the convolution $\varphi^{(k)}*\psi^{(l)}$ exists in $\boldsymbol{R}^q$. We have

$$(\varphi*\psi)^{(k)} = \varphi^{(k)}*\psi = \varphi*\psi^{(k)} \tag{7}$$

for every $k \in \boldsymbol{P}^q$. In fact, if $k = e_j$, the first equality in (7) follows from differentiating

$$\int_0^1 \varphi(x-t)\psi(t)\,dt$$

under the integral sign. The equality for arbitrary k follows by q-dimensional induction. The second equality follows from the first one by commutativity.

9.5.1. LEMMA. *If φ_n and ψ_n are fundamental sequences of smooth periodic functions in $\boldsymbol{R}^q$, then the sequence $\varphi_n * \psi_n$ is also fundamental in $\boldsymbol{R}^q$.*

PROOF. By 9.3.1, there exist periodic continuous functions Φ_{mn} and Ψ_{kn} $(0 \leqslant m \leqslant r \in \boldsymbol{P}^q, 0 \leqslant k \leqslant r_1 \in \boldsymbol{P}^q)$ such that

$$\varphi_n = \sum_{0 \leqslant m \leqslant r} \Phi_{mn}^{(m)}, \qquad \psi_n = \sum_{0 \leqslant k \leqslant r_1} \Psi_{kn}^{(k)}$$

and $\Phi_{mn} \rightrightarrows$ in $\boldsymbol{R}^q$ and $\Psi_{kn} \rightrightarrows$ in $\boldsymbol{R}^q$ as $n \to \infty$ $(0 \leqslant m \leqslant r, 0 \leqslant k \leqslant r_1)$. Hence, by (5) and (7), we can write

$$\varphi_n * \psi_n = \sum_{0 \leqslant m \leqslant r} \sum_{0 \leqslant k \leqslant r_1} (\Phi_{mn} * \Psi_{kn})^{(k+m)},$$

which implies our assertion.

Let f and g be periodic distributions in $\boldsymbol{R}^q$ and let $f_n = f * \delta_n$, $g_n = g * \delta_n$ be their regular sequences. It is easy to check that the f_n and g_n are periodic smooth functions.

If f and g are arbitrary periodic distributions, their convolution $f * g$ is given by the fundamental sequence $\varphi_n * \psi_n$, where φ_n and ψ_n are fundamental sequences of periodic smooth functions corresponding to f and g respectively.

It follows from Lemma 9.5.1 that the above definition is consistent (by similar argument to that in Section 6.3).

If the periodic distributions f and g reduce to smooth functions, their convolution under the present definition coincides with that given by the former definition.

By simple verification we can check that all the equalities from (3) to (7) remain true for distributions.

9.5.2. THEOREM. *If the sequences f_n and g_n of periodic distributions are distributionally convergent, then the sequence $f_n * g_n$ of their periodic convolutions is also distributionally convergent.*

PROOF. By 9.3.1, there exist periodic continuous functions Φ_{mn} and Ψ_{kn} $(0 \leqslant m \leqslant r \in \boldsymbol{P}^q, 0 \leqslant k \leqslant r_1 \in \boldsymbol{P}^q)$ such that

$$f_n = \sum_{0 \leqslant m \leqslant r} \Phi_{mn}^{(m)}, \qquad g_n = \sum_{0 \leqslant k \leqslant r_1} \Psi_{kn}^{(k)}$$

and $\Phi_{mn} \rightrightarrows$ in R^q and $\Psi_{kn} \rightrightarrows$ in R^q as $n \to \infty$ $(0 \leqslant m \leqslant r, 0 \leqslant k \leqslant r_1)$. Hence, by (5) and (7), we can write

$$f_n * g_n = \sum_{0 \leqslant m \leqslant r} \sum_{0 \leqslant k \leqslant r_1} (\Phi_{mn} * \Psi_{kn})^{(k+m)},$$

which implies our assertion.

If f_n and g_n converge distributionally to f and g respectively, the sequence $f_n * g_n$ clearly converges distributionally to $f * g$.

EXAMPLE. As an example we consider the periodic convolution $f * \sum_{m \in B^q} \delta(x-m)$, where f is a periodic distribution. Let δ_n be an arbitrary delta sequence and let $f_n = f * \delta_n$. Clearly,

$$g_n = \delta_n * \sum_{m \in B^q} \delta(x-m) = \sum_{m \in B^q} \delta_n(x-m).$$

Hence we have, by (2),

$$f_n * g_n = \int_a^{a+1} f_n(x-t) g_n(t)\, dt = \int_a^{a+1} f_n(x-t)\, \delta_n(t)\, dt$$

$$= \int_{-\infty}^{\infty} f_n(x-t) \delta_n(t)\, dt = f_n * \delta_n$$

for sufficiently large n and $-1 < a < 0$. Hence it follows that

$$f * \sum_{m \in B^q} \delta(x-m) = f. \tag{8}$$

REMARK. Lemma 9.5.1 is a special case of Theorem 9.5.2; this follows from two facts: 1° distributionally convergent sequences of periodic smooth functions are fundamental, and conversely; 2° the convolution of two periodic smooth functions is a smooth function.

9.6. EXPANSIONS IN FOURIER SERIES

It is a well known fact of Classical Analysis that the saw-function $g(\xi)$ (Fig. 9.1), which is periodic (of period 1) and equal to $1-\xi$ in (0, 1), can be expanded as the series

$$g(\xi) = \sum_{\substack{n=-\infty \\ n \neq 0}}^{\infty} (2\pi n i)^{-1} e^{2\pi i n \xi};$$

the sequence of partial sums of this series is bounded and converges a.e. (at non integer points to be exact).

The series can thus be regarded as distributionally convergent. Differentiating it distributionally, we get

$$-1+\sum_{n=-\infty}^{\infty}\delta(\xi-n)=\sum_{\substack{n=-\infty\\ n\neq 0}}^{\infty}e^{2\pi in\xi}$$

and hence

$$\sum_{n=-\infty}^{\infty}\delta(\xi-n)=\sum_{n=-\infty}^{\infty}e^{2\pi in\xi}.$$

Successively putting $\xi=\xi_1,\ldots,\xi_q$ in the above and multiplying the resulting equalities together we get

$$\sum_{n\in B^q}\delta(x-n)=\sum_{n\in B^q}e^{2\pi inx} \tag{1}$$

for $x=(\xi_1,\ldots,\xi_q)\in R^q$. (The equality between the sums on the two sides can be rigorously justified, by first considering the products of partial sums.)

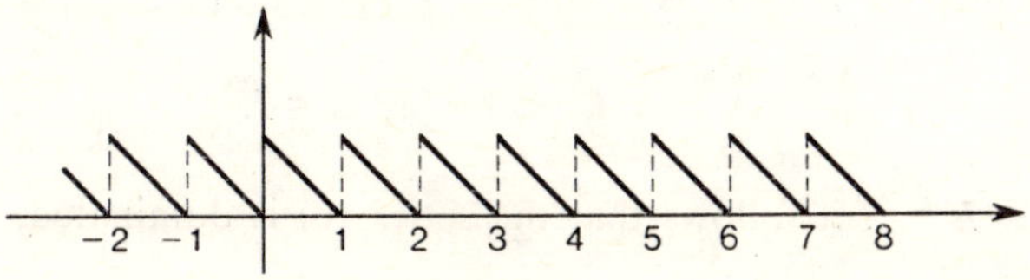

FIG. 9.1

It is interesting to note that the particular expansion (1) can be used to obtain a Fourier expansion for any periodic distribution. This method is a distributional one and turns out to be simpler than any classical method for expanding functions. Since we are working in an arbitrary number of dimensions, we first need to introduce, as in Section 8.1, the symbol $\tilde{n}$. Namely, we recall that if the jth coordinate of the vector n is ν_j, $\tilde{n}$ denotes the vector whose jth coordinate is $\max(1,|\nu_j|)$ $(j=1,\ldots,q)$.

For the sake of brevity we write E^n instead of $e^{2\pi inx}$.

9.6.1. THEOREM. *The series*

$$f=\sum_{n\in B^q}a_nE^n \tag{2}$$

converges distributionally, iff there exist $k\in P^q$ and $M\in R^1$ such that

$$|a_n|\leqslant M\tilde{n}^k \quad \text{for} \quad n\in B^q, \tag{3}$$

in which case, the sum f is a periodic distribution.

Conversely, if f is a periodic distribution, there exist coefficients a_n satisfying (3) *such that* (2) *holds.*

Moreover,

$$a_n = (f, E^{-n}), \tag{4}$$

which implies that the expansion (2), *for a given periodic distribution f, is unique.*

PROOF. Assume that (2) holds for some fixed $k = (\varkappa_1, \ldots, \varkappa_q)$ and $M \in \boldsymbol{R}^1$. We introduce vectors $l_n = (\lambda_{n1}, \ldots, \lambda_{nq})$ such that $\lambda_{nj} = \varkappa_j + 2$, if $\nu_j = 0$, and $\lambda_{nj} = 0$, if $\nu_j \neq 0$. We then have

$$\left(\frac{x^{l_n}}{l_n!} E^n\right)^{(k+2)} = \acute{n}^{k+2} E^n, \tag{5}$$

where $\acute{n} = (\nu_1', \ldots, \nu_q')$ with $\nu_j' = 2\pi i \nu_j$, if $\nu_j \neq 0$, and $\nu_j' = 1$, if $\nu_j = 0$. Let

$$F = \sum_{n \in B^q} a_n \acute{n}^{-k-2} \frac{x^{l_n}}{l_n!} E^n. \tag{6}$$

For any fixed interval I, there is a number K such that

$$\frac{|x^{l_n}|}{l_n!} < K, \quad \text{for all } x \in I.$$

Since $|\acute{n}^{-k-2}| \leqslant \tilde{n}^{-k-2}$, it follows that the series (6) is dominated, in I, by each of the series

$$K \sum_{n \in B^q} \tilde{n}^{-k-2} |a_n| \leqslant KM \sum_{n \in B^q} \tilde{n}^{-2}.$$

Now this last series is convergent, and this implies that the series (6) is uniformly convergent in I. Since I can be chosen arbitrarily, it follows that the series (6) is almost uniformly convergent in $\boldsymbol{R}^q$, and thus distributionally convergent. Differentiating (6) $k+2$ times, we get, by (5),

$$F^{(k+2)} = \sum_{n \in B^q} a_n E^n.$$

The distribution $f = F^{(k+2)}$ is periodic, since all the E^n are periodic.

Assume, now, that f is a periodic distribution. Then we have, by 9.5 (8) and (1),

$$f = f * \sum_{n \in B^q} \delta(x-n) = f * \sum_{n \in B^q} E^n.$$

Hence, by 9.5.2, we have

$$f = \sum_{n \in B^q} f * E^n.$$

Since

$$f * E^n = \int_0^1 f(t) E^n(x-t)\,dt = (f, E^{-n}) E^n,$$

equalities (2) and (4) follow.

By Corollary 9.3.3, we can write

$$f = \sum_{0 \leqslant m \leqslant r} f_m^{(m)}$$

where the f_m are continuous periodic functions. We have

$$f_m^{(m)} = \sum_{n \in B^q} a_{mn} E^n,$$

where

$$a_{mn} = (f_m^{(m)}, E^{-n}) = (2\pi i n)^m (f_m, E^{-n}).$$

Hence

$$|a_{mn}| \leqslant (2\pi)^m \int_0^1 |f_m| \tilde{n}^m$$

and

$$|a_n| \leqslant \sum_{0 \leqslant m \leqslant r} |a_{mn}| \leqslant M \tilde{n}^r,$$

where

$$M = \sum_{0 \leqslant m \leqslant r} (2\pi)^m \int_0^1 |f_m|.$$

It remains to show that if a series of the form (2) converges distributionally then there exist $k \in \boldsymbol{P}^q$ and $M \in \boldsymbol{R}^1$ such that (3) holds. In fact, since the E^n are periodic, so is the limit distribution f. By what we have just proved, the expansion (2) is unique and (3) holds.

We denote the class of all periodic smooth functions by $\mathscr{P}$. Clearly, $\mathscr{P}$ is a linear space. In fact we have

9.6.2. THEOREM. *The space $\mathscr{P}$ coincides with the class of all periodic functions f such that*

$$|a_n| < M_k \tilde{n}^{-k} \quad \textit{for each } k \in \boldsymbol{P}^q \tag{7}$$

where $a_n = (f_n, E^{-n})$ and M_k is a number depending on k only.

PROOF. If (7) holds, then, plainly, $\sum_{n \in B^q} |n^k a_n| < \infty$ for any $k \in \boldsymbol{P}^q$. This implies that, for each $k \in \boldsymbol{P}^q$, the series $\sum_{n \in B^q} (2\pi i n)^k a_n E^n$ is uniformly convergent. But this is the expansion of

$$\Big(\sum_{n\in B^q} a_n E^n\Big)^{(k)}$$

and the function $f = \sum_{n\in B^q} a_n E^n$ is thus smooth and periodic.

Assume, conversely, that f is a smooth periodic function and let k be an arbitrary fixed vector of P^q. Then the coefficients a_n for f satisfy

$$a_n = (f, E^{-n}) = (2\pi i)^{-k_n}(f^{(k_n)}, E^{-n}),$$

where the coordinate of k_n is $\varkappa_j$, if $\nu_j \neq 0$, and 0, if $\nu_j = 0$. Hence, $|a_n| \leqslant \tilde{n}^{-k}M$, where $M = \max_{n\in B^q} |(f^{(k_n)}, E^{-n})|$ (note that, for each fixed k, there is only a finite number of distinct k_n).

9.6.3. THEOREM. *Let f be a periodic distribution and let ψ be a periodic smooth function. If*

$$f = \sum_{n\in B^q} a_n E^n \quad \textit{and} \quad \psi = \sum_{n\in B^q} b_n E^n,$$

then

$$(f, \psi) = \sum_{n\in B^q} a_n b_n.$$

PROOF. Let A_ν be an arbitrary sequence of finite subsets of B^q such that $A_\nu \subset A_{\nu+1}$ and $\lim_{\nu\to\infty} A_\nu = B^q$, and let

$$f_\nu = \sum_{n\in A_\nu} a_n E^n \quad \text{and} \quad \psi_\nu = \sum_{n\in A_\nu} b_n E^n.$$

The sequence f_ν is distributionally convergent to f and $\psi_\nu^{(k)}$ is uniformly convergent to $\psi^{(k)}$. Hence, our assertion follows by 9.4.4.

9.7. THE FOURIER TRANSFORM OF PERIODIC DISTRIBUTIONS

From 8.6 (4) and 8.6 (5) we obtain

$$\mathscr{F}(E^n) = (2\sqrt{\pi})^q \cdot \delta(x+4\pi n).$$

Hence, by Theorem 8.6.2, we see that the Fourier transform $\mathscr{F}(f)$ of a periodic function f is

$$\mathscr{F}(f) = (2\sqrt{\pi})^q \sum_{n\in B^q} a_n \cdot \delta(x+4\pi n),$$

where the a_n are the Fourier coefficients of f.

10. The Köthe spaces

10.1. GENERAL REMARKS

As we have shown, every tempered distribution can be expanded uniquely into an Hermite series and every periodic distribution can be expanded into a Fourier series. There is thus a one-to-one correspondence between such distributions and certain matrices (matrices of coefficients). The properties of the distributions are in some way reflected by corresponding properties of the matrices. E.g., if we differentiate the distribution in the appropriate way, the effect on the corresponding matrix is that its elements are multiplied by certain coefficients. It is therefore possible to reduce the study of certain classes of distributions to the corresponding study of certain classes of matrices. In addition, we can order the elements of the matrices in ordinary sequences, which is a further simplification, and clearly, the concept of a sequence of numbers is much simpler than the concept of a distribution.

An elegant theory of spaces of sequences has been developopped by G. Köthe (see [11]).

In this chapter, we will present a little of that theory. However, there is a difference between Köthe's approach and ours. Köthe's work is based on deep theorems of modern topology, whereas our method is quite elementary and, in spite of this, gives the results in a slightly more general form. More precisely, Köthe restricts himself to sequences of numbers; in contrast, our method works, without any alteration, for sequences of vectors (elements of normed spaces). In this detail, the present chapter differs from the rest of the book, which is devoted, essentially, to the real or complex domain. Of course, the reader may always substitute real or complex numbers for elements of the normed space under consideration, if this leads to a greater degree of understanding.

10.2. SPACES OF SEQUENCES

$\mathscr{X}$ denotes the set of all sequences $A = (a_1, a_2, \ldots)$ whose terms a_j belong to a normed space X over the field of complex numbers. The norm of $x \in X$ is denoted by $|x|$. In the sequel, sequences A will be called *vectors* and the terms

a_j their *coordinates*. The vector A is said to be *real* or *complex*, iff all its coordinates a_j are real or complex respectively. We introduce the notation:

$$\lambda A = (\lambda a_1, \lambda a_2, \ldots)$$

for any complex number λ. If $B = (b_1, b_2, \ldots)$ is antoher vector in $\mathscr{X}$, we write

$$A+B = (a_1+b_1, a_2+b_2, \ldots).$$

If one of the vectors A, B is complex and the other belongs to $\mathscr{X}$, we write

$$AB = (a_1 b_1, a_2 b_2, \ldots).$$

If all the coordinates of A are real and $\neq 0$, then

$$A^{-1} = \left(\frac{1}{a_1}, \frac{1}{a_2}, \ldots\right).$$

We make use of the notation:

$$|A| = \sup_j |a_j| \quad \text{and} \quad ||A|| = |a_1|+|a_2|+ \ldots$$

Clearly $|A| \leqslant ||A||$. Moreover

$$|\lambda A| = |\lambda| \cdot |A|, \quad |A+B| \leqslant |A|+|B|, \quad |AB| \leqslant |A| \cdot |B|$$

$$||\lambda A|| = |\lambda| \cdot ||A||, \quad ||A+B|| \leqslant ||A||+||B||, \quad ||AB|| \leqslant ||A|| \cdot ||B||.$$

The last inequality follows from the stronger inequality

$$||AB|| \leqslant |A| \cdot ||B||.$$

By the *inner product* (A, B), where one of the factors is a complex vector, we mean the sum of the series $a_1 b_1 + a_2 b_2 + \ldots$; the inner product exists, iff the series converges. Note that $||AB|| = |a_1 b_1| + |a_2 b_2| + \ldots$ The inner product (A, B) therefore exists, whenever $||AB|| < \infty$, and we have $|(A, B)| \leqslant ||AB||$.

10.3. KÖTHE'S ECHELON SPACE AND CO-ECHELON SPACE

Assume that a sequence of vectors $T_1, T_2, \ldots$ all of whose coordinates are positive is given and that the following property holds:

$$|T_k T_{k+1}^{-1}| < \infty.$$

Any vector A such that $|T_k^{-1} A| < \infty$ for some k is said to be *tempered*. Any real vector A such that $||T_k A|| < \infty$ for $k = 1, 2, \ldots$ is said to be *rapidly decreasing*. In the case when the coordinates of A are numbers, the set of all tempered vectors is called a Köthe *co-echelon space* (Stufenraum) and the set of all rapidly decreasing vectors is called a Köthe *echelon space* (gestufter Raum).

If A is rapidly decreasing and B is tempered, the inner product (A, B) exists. This follows from the inequality $||AB|| \leqslant ||T_k A|| \cdot |T_k^{-1} B|$.

If the sequence $T_1, T_2, \ldots$ is given, then both the space $\mathscr{T}$ of all tempered vectors and the space $\mathscr{R}$ of all real rapidly decreasing vectors are determined uniquely. However the converse is not true. Viz., even if we know all the vectors belonging to $\mathscr{T}$ and to $\mathscr{R}$, the sequence $T_1, T_2, \ldots$ inducing them is not determined uniquely. For any given positive number α we can always choose a sequence $\tilde{T}_1, \tilde{T}_2, \ldots$ which induces precisely the given spaces $\mathscr{T}$ and $\mathscr{R}$ and, moreover, satisfies

$$|\tilde{T}_k \tilde{T}_{k+1}^{-1}| < \alpha .$$

In fact, we can put $\tilde{T}_k = \alpha^{-k} m_1 \ldots m_k T_k$ with $m_1 = 1$ and $|T_k T_{k+1}^{-1}| \leqslant m_{k+1}$.

10.4. STRONG AND WEAK BOUNDEDNESS

We say that a sequence of tempered vectors A_n is *strongly bounded,* iff there is an index k and a number M such that $|T_k^{-1} A_n| < M$. We say that a sequence of tempered vectors A_n is *weakly bounded,* iff the sequence (A_n, R) is bounded for every fixed rapidly decreasing real vector R.

REMARK. Note that both the above concepts are somewhat different from similar concepts considered in functional analysis. In fact, strong boundedness is usually defined by means of the dual space. Weak boundedness is usually defined by means of functionals, whose values are numbers, whereas our "inner product" can take its values in any normed space.

The following theorem plays an important role throughout the theory of this chapter.

10.4.1. THEOREM (Boundedness Theorem). *A sequence of tempered vectors is weakly bounded, iff it is strongly bounded.*

Before giving the proof we first establish a general theorem on matrices. This theorem will be called the *Diagonal Theorem.*

10.5. DIAGONAL THEOREM

As before, let X be any normed space and let $|x|$ be the norm of $x \in X$.

10.5.1. THEOREM (Diagonal Theorem). *If* $x_{ij} \in X$ *and* $\lim_{j\to\infty} x_{ij} = 0$ *for* $i = 1,2,\ldots,$ *then there exist an infinite set I of positive integers and a subset J (finite or infinite) of I such that, for almost all $i \in I$, we have*

$$\sum_{j\in J} |x_{ij}| < \infty \tag{1}$$

and

$$\Big|\sum_{j\in J} x_{ij}\Big| \geqslant \tfrac{1}{2}\,|x_{ii}|. \tag{2}$$

If X is complete, the meaning of this theorem is clear, since the inequality (1) implies the existence of the sum $\sum_{j\in J} x_{ij}$. Further, if X is not complete, but J is finite, the last sum is meaningful. However, if X is not complete and J is infinite, it could happen that the sum does not exist. In this case we understand that

$$\Big|\sum_{j\in J} x_{ij}\Big| = \lim_{n\to\infty}\Big|\sum_{j\in J_n} x_{ij}\Big|,$$

where the J_n are finite subsets of J such that $\lim_{n\to\infty} J_n = J$. The existence of the limit on the right hand side follows from (1). (Clearly, the above equality also holds, when J is finite or X is complete.)

Proof of the Diagonal Theorem. We may assume, that if J is a finite set satisfying

$$\Big|\sum_{j\in J} x_{ij}\Big| \geqslant \tfrac{1}{2}\,|x_{ii}|, \quad \text{for each } i\in J, \tag{3}$$

then there exists a positive integer r greater than each element of J and such that

$$\Big|\sum_{j\in J} x_{ij}\Big| < \tfrac{1}{2}\,|x_{ii}| \quad \text{for} \quad i > r. \tag{4}$$

Otherwise, the theorem is trivially true. Under this additional assumption, we select an increasing sequence $i_1, i_2, \ldots$ of positive integers and construct a sequence $\varepsilon_1, \varepsilon_2, \ldots$ of positive numbers such that:

$$\Big|\sum_{k=1}^{n-1} x_{i_n i_k}\Big| = (\tfrac{1}{2} - \varepsilon_n)\,|x_{i_n i_n}| \tag{5}$$

where, if $n = 1$, we take $\sum_{k=1}^{n-1} x_{i_n i_k} = 0$, and

$$|x_{i_n i_{n+q}}| \leqslant 2^{-q}\varepsilon_n |x_{i_n i_n}| \tag{6}$$

for any positive integers n and q.

From the additional assumption it follows that there exists a positive integer r such that $|x_{ii}| > 0$ for $i \geqslant r$. Let $i_1 = r$ and $\varepsilon_1 = \frac{1}{2}$. Also by the additional assumption, there is an index $s > i_1$ such that $|x_{ii_1}| < \frac{1}{2}|x_{ii}|$ for $i > s$. We choose $i_2 \geqslant s$ so that $|x_{i_1 i_2}| < 2^{-1}\varepsilon_1 |x_{i_1 i_1}|$ and we take ε_2 so that $|x_{i_2 i_1}| = (\frac{1}{2}-\varepsilon_2)|x_{i_2 i_2}|$. Since $i_2 \geqslant s$, we have $\varepsilon_2 > 0$. It is now easy to verify that (5) and (6) hold for $n = q = 1$.

Assume that we have already found $i_1, \ldots, i_p$ ($p \geqslant 2$) and $\varepsilon_1, \ldots, \varepsilon_p$ so that (5) and (6) hold for $1 \leqslant n \leqslant p-1$ and $n+q \leqslant p$. By the additional assumption, there is a positive integer $u > i_p$ such that

$$\left|\sum_{k=1}^{p} x_{ii_k}\right| < \tfrac{1}{2}|x_{ii}| \tag{7}$$

for $i > u$. We choose $i_{p+1} > u$ so that

$$|x_{i_n i_{p+1}}| \leqslant 2^{n-p-1}\varepsilon_n |x_{i_n i_n}|$$

holds for $1 \leqslant n \leqslant p$. This is possible, because $|x_{ii}| > 0$ for $i \geqslant r$ and $\lim_{j\to\infty} x_{ij} = 0$ for every fixed i. Next, we take ε_{p+1} so that

$$\left|\sum_{k=1}^{p} x_{i_{p+1} i_k}\right| = (\tfrac{1}{2}-\varepsilon_{p+1})|x_{i_{p+1} i_{p+1}}|.$$

Since $|x_{i_{p+1} i_{p+1}}| > 0$ and $i_{p+1} > u$, we have $\varepsilon_{p+1} > 0$ by (7), and it is then easy to check that (5) and (6) hold for $1 \leqslant n \leqslant p$ and $n+q \leqslant p+1$. Hence, the existence of an increasing sequence $i_1, i_2, \ldots$ such that (5) and (6) hold follows by induction. By (5) and (6), we have

$$\left|\sum_{k=1}^{p} x_{i_n i_k}\right| \geqslant |x_{i_n i_n}| - \left|\sum_{k=1}^{n-1} x_{i_n i_k}\right| - \sum_{k=n+1}^{\infty} |x_{i_n i_k}| \geqslant \tfrac{1}{2}|x_{i_n i_n}| \tag{8}$$

for $n = 1, 2, \ldots$ Let I be the set of all $i_1, i_2, \ldots$ and let $J = I$. Then (1) follows from (6) and (2) follows from (8). The proof of the Diagonal Theorem is now complete.

REMARK. The number $\frac{1}{2}$ in (2) cannot be replaced, in the Diagonal Theorem, by any greater number. To see this, consider the matrix x_{ij} in which $x_{ii} = 2$, and $x_{ij} = 0$ for all $j > i$ and $x_{ij} = -1$ for $j < i$ in each line i. Then, for each I and each $J \subset I$, we have $\sum_{j\in J} x_{ij} = 1$ for some $i \in I$.

10.6. THE PROOF OF THE BOUNDEDNESS THEOREM

If A_n is a strongly bounded sequence of tempered vectors, there exists an index k and a number M such that $|A_n T_k^{-1}| < M$ for all $n = 1, 2, \ldots$ For each rapidly decreasing vector R, we have $\|T_k R\| < \infty$. Hence, from the inequalities

$$|(A_n, R)| \leqslant \|A_n R\| \leqslant |A_n T_n^{-1}| \cdot \|T_k R\|,$$

it follows that every strongly bounded sequence is weakly bounded.

It remains to prove that every weakly bounded sequence is strongly bounded. Assume on the contrary, that this is not true, i.e., that there is a weakly bounded sequence A_n which is not strongly bounded. Moreover, we may assume that $|T_k T_{k+1}^{-1}| < \frac{1}{2}$, and it therefore follows that

$$|T_i T_j^{-1}| < 2^{i-j} \tag{1}$$

for $i, j = 1, 2, \ldots$ and $i < j$. Since A_n is not strongly bounded, the sequence $|A_n T_i^{-1}|$ is unbounded for each fixed i. Hence there exists an increasing sequence n_i of positive integers such that

$$|A_{n_i} T_i^{-1}| \to \infty \quad \text{as} \quad i \to \infty. \tag{2}$$

There thus exists a sequence r_i of positive integers such that

$$|A_{n_i} T_i^{-1}| - 1 < |e_{r_i} A_{n_i} T_i^{-1}| = |(A_{n_i}, e_{r_i} T_i^{-1})| \quad \text{for} \quad i = 1, 2, \ldots, \tag{3}$$

where e_{r_i} denotes a vector whose r_ith coordinate is 1 and whose remaining coordinates are 0. Setting $x_{ij} = (A_{n_i}, e_{r_j} T_j^{-1})$ we have $x_{ij} \in X$ and $x_{ij} \to 0$ as $j \to \infty$, by (1). Hence, by the Diagonal Theorem, there exists an infinite set I of positive integers and a subset J (finite or infinite) of I such that, for all $i \in I$, we have

$$\sum_{j \in J} |(A_{n_i}, e_{r_j} T_j^{-1})| < \infty \tag{4}$$

and

$$\left|\sum_{j \in J} (A_{n_i}, e_{r_j} T_j^{-1})\right| > \tfrac{1}{2} |(A_{n_i}, e_{r_i} T_i^{-1})|. \tag{5}$$

Let R be a vector such that

$$e_i R = \sum_{j \in J} e_i e_{r_j} T_j^{-1}, \quad \text{for each } i = 1, 2, \ldots$$

By (1), the series on the right hand side of this equality is convergent for each $i = 1, 2, \ldots$ We prove that R is rapidly decreasing. In fact, for any fixed k, we have

$$\|T_k R\| = \Big\|T_k \sum_{j\in J} e_{r_j} T_j^{-1}\Big\| \leqslant \Big\|T_k \sum{}' e_{r_j} T_j^{-1}\Big\| + \Big\|T_k \sum{}'' e_{r_j} T_j^{-1}\Big\|$$
$$\leqslant \Big\|T_k \sum{}' e_{r_j} T_j^{-1}\Big\| + \sum{}'' 2^{k-j} < \infty,$$

where $\sum'$ indicates summation over all $j \in J$ such that $j \leqslant k$ and $\sum''$ indicates summation over all $j \in J$ such that $j > k$.

Moreover, we see that

$$\sum_{j\in J} (A_{n_i}, e_{r_j} T_j^{-1}) = \Big(A_{n_i}, \sum_{j\in J} e_{r_j} T_j^{-1}\Big) = (A_{n_i}, R).$$

Hence, by (5), (4), (3) and (2) we have $|(A_{n_i}, R)| \to \infty$ as $i \to \infty$, which contradicts the weak boundedness of the sequence A_n. This completes the proof of the Boundedness Theorem.

10.7. STRONG CONVERGENCE AND WEAK CONVERGENCE

We say that a sequence of vectors A_n *converges coordinatewise*, iff for every fixed index j the sequence (e_j, A_n) is convergent. We say that a sequence of tempered vectors A_n *converges strongly*, iff it converges coordinatewise and, moreover, is strongly bounded, i.e., there is an index k and a number M such that $|T_k^{-1} A_n| < M$ for all $n = 1, 2, \ldots$ There is then a tempered vector A to which A_n converges coordinatewise and such that $|T_k^{-1} A| < M$. In fact, the jth coordinate of the limit vector A is given by $a_j = (e_j, A) = \lim_{n\to\infty} (e_j, A_n)$. We can therefore say that every strongly convergent sequence of tempered vectors A_n converges strongly to some tempered vector A.

We may say that a sequence of tempered vectors A_n *converges weakly*, iff for every rapidly decreasing real vector R, the sequence of inner products (R, A_n) is convergent. Clearly, every weakly convergent sequence is weakly bounded.

10.7.1. THEOREM. *A sequence of tempered vectors A_n converges strongly, iff it converges weakly.*

PROOF. Assume that A_n converges strongly to A. Then there is an index k and a number M such that $|T_k^{-1} A_n| < M$ for all n and $|T_k^{-1} A| < M$. Let $R = (r_1, r_2, \ldots)$ be any rapidly decreasing vector. Then $\|T_k R\| < \infty$. For any given number $\varepsilon > 0$, there is therefore an index j_k such that

$$\|T_k R_k'\| < \varepsilon/4M,$$

where R_k' is the vector all of whose coordinates up to the j_kth coordinate are 0, and all of whose subsequent coordinates are equal to the corresponding coordinates of R. Setting $R_k = R - R'_k$, we have

$$|(R, A_n - A)| \leqslant ||(R_k + R'_k)(A_n - A)|| \leqslant ||R_k(A_n - A)|| + ||T_k R'_k \cdot T_k^{-1}(A_n - A)||$$
$$\leqslant ||R_k(A_n - A)|| + ||T_k R'_k|| \cdot |T_k^{-1}(A_n - A)|$$
$$\leqslant ||R_k(A_n - A)|| + \varepsilon/2.$$

The vector $R_k(A_n - A)$ tends to 0 coordinatewise, as $n \to \infty$, and all its coordinates with index $> j_k$ are 0. This implies that

$$||R_k(A_n - A)|| \to 0, \quad \text{as} \quad n \to \infty.$$

Hence $|(R, A_n - A)| < \varepsilon$ for sufficiently large values of n. Hence

$$(R, A_n) \to (R, A), \tag{1}$$

which shows that A_n converges weakly.

Assume now, conversely, that A_n converges weakly. Then the sequence (e_j, A_n) is convergent for every fixed j, since the vector e_j is rapidly decreasing. Moreover, since A_n is weakly bounded, it is strongly bounded, and A_n thus converges strongly and the proof is complete.

We say that a sequence A_n *converges weakly* to A, iff (1) holds for every rapidly decreasing real vector R. We can now say that if a sequence of tempered vectors A_n converges strongly or weakly, then there is a tempered vector A to which A_n converges strongly and weakly. Clearly, the limit vector A is determined uniquely.

10.8. A MORE GENERAL FORMULATION OF THE THEORY

The whole of the preceding theory continues to apply when we replace the vectors $A = (a_1, a_2, \ldots)$ by matrices of any arbitrary (but fixed) number of dimensions. Still more generally, let K be any countable set of indices p, and consider functions $A = \{a_p\}$ defined in K and taking values in a given normed space X over the field of complex numbers; the functions in K will be called matrices and the set of all such functions will be denoted by X, for the sake of convenience in the sequel, although traditionally, this definition is reserved for certain special sets K. As in Section 10.2 we put $\lambda A = \{\lambda a_p\}$ and $A + B = \{a_p + b_p\}$ if $A = \{a_p\}$, $B = \{b_p\}$ and A, B belong to X. We write $AB = \{a_p b_p\}$, provided one of the factors is a complex or real matrix. Moreover, we make use of the following notation

$$|A| = \sup_{p \in K} |a_p|, \qquad ||A|| = \sum_{p \in K} |a_p|.$$

By the inner product (A, B) we mean the sum $\sum_p a_p b_p$ provided that this sum is independent of the order of summation and that one of the factors is a complex or real matrix. Clearly (A, B) exists whenever $||AB|| < \infty$.

We assume that a sequence of positive matrices $T_1, T_2, \ldots$ defined in K is given and that the following property holds:

$$|T_k T_{k+1}^{-1}| < \infty.$$

Any matrix A such that $|T_k^{-1} A| < \infty$ for some k is said to be *tempered*. Any real matrix A such that $||T_k A|| < \infty$ for $k = 1, 2, \ldots$ is said to be *rapidly decreasing*. If A is rapidly decreasing and B is tempered, their inner product always exists.

We say that a sequence of tempered matrices A_n is *weakly bounded*, iff, for each rapidly decreasing matrix R the sequence of inner products is bounded. We say that a sequence of tempered matrices A_n is *strongly bounded*, iff there exist an index k and a number M such that $|T_k^{-1} A_n| < M$ for all $n = 1, 2, \ldots$

10.8.1. THEOREM. *A sequence of tempered matrices A_n is weakly bounded, iff it is strongly bounded.*

This theorem immediately reduces to 10.4.1, when we arrange the set K as a sequence $k_1, k_2, \ldots$ However one minor point should be noted: all the inner products involved in the proof of 10.4.1 converge absolutely, and so their value does not depend on the ordering of the summands.

We say that a sequence of matrices A_n *converges strongly*, iff it converges pointwise on K and, moreover, is strongly bounded, i.e, there exist an index k and a number M such that $|T_k^{-1} A_n| < M$ for all $n = 1, 2, \ldots$ Clearly, the pointwise limit A satisfies the same inequality $|T_k^{-1} A| < M$. Every strongly convergent sequence thus converges to some limit which, like the terms of the sequence, is a tempered matrix.

We say that a sequence of tempered matrices A_n *converges weakly*, iff, for each rapidly decreasing matrix R, the sequence of inner products (R, A_n) is convergent.

10.8.2. THEOREM. *A sequence of tempered matrices A_n converges strongly, iff it converges weakly.*

This theorem reduces to 10.7.1, if we arrange the set K as a sequence. Further, the remark at the end of Section 10.7 continues to hold, i.e., if a sequence of tempered matrices A_n converges strongly or weakly, then there exists a tempered matrix to which it converges strongly and weakly.

10.9. FUNCTIONALS ON THE SPACE OF RAPIDLY DECREASING MATRICES

We say that $U(R)$ is a functional on the space $\mathcal{R}$ of rapidly decreasing matrices, iff for each $R \in \mathcal{R}$ the value of $U(R)$ is a number. The functional $U(R)$ is linear, iff, for any numbers α, β and $R, S \in \mathcal{R}$ we have

$$U(\alpha R+\beta S) = \alpha U(R)+\beta U(S).$$

The functional $U(R)$ is continuous, iff

$$\lim_{n\to\infty} U(R_n) = U(R)$$

for each sequence of matrices $R_n \in \mathcal{R}$ such that $R_n \overset{\mathcal{R}}{\to} R$, i.e., such that, for each fixed k, $||T_k(R_n-R)|| \to 0$, as $n \to \infty$.

By the remark at the end of Section 10.3, we can assume that $|T_k T_{k+1}^{-1}| \leqslant 1$. Hence, $|T_k^{-1}T_l| \leqslant 1$ provided $l \leqslant k$.

10.9.1. Theorem. *For each continuous linear functional U on $\mathcal{R}$ there exists a unique tempered matrix A such that*

$$U(R) = (A, R) \quad \textit{for each } R \in \mathcal{R}. \tag{1}$$

Conversely, for each tempered matrix A, (1) *represents a continuous linear functional. The correspondence* (1) *between continuous linear functionals on $\mathcal{R}$ and tempered matrices is thus one-to-one.*

Proof. Let $a_p = U(e_p)$, where e_p is the matrix which takes the value 1 at the index p and the value 0 elsewhere. We show that the matrix $A = \{a_p\}$ $(p \in K)$ is tempered. Assume it is not. Then, for every integer k, the matrix $T_k^{-1}A$ is unbounded, and so, for each positive integer k, there is an index $p_k \in K$ such that $|a_{p_k}e_{p_k}T_k^{-1}| > k^2$. Let $r_{p_k} = a_{p_k}^{-1}$ and let $r_p = 0$ for $p \neq p_k$. Then $R = \{r_p\} \in \mathcal{R}$. In fact, given any positive integer l, we have

$$|r_{p_k}e_{p_k}T_l| \leqslant |a_{p_k}^{-1}e_{p_k}T_k| < \frac{1}{k^2} \quad \text{for} \quad k \geqslant l.$$

This implies that

$$||RT_l|| = \sum_{k=1}^{\infty} |r_{p_k}e_{p_k}T_k| < \infty,$$

which proves that $R \in \mathcal{R}$.

If the sets S_n are finite subsets of K such that $S_n \to K$, then

$$\sum_{p\in S_n} r_p e_p \overset{\mathcal{R}}{\to} R, \quad \text{as} \quad n \to \infty.$$

By the linearity of U, we have

$$U\Big(\sum_{p \in S_n} r_p e_p\Big) = \sum_{p \in S_n} a_p r_p \to \infty, \quad \text{as} \quad n \to \infty.$$

This contradicts the continuity of U. Hence the matrix A is tempered and we have

$$U(R) = \lim_{n \to \infty} \sum_{p \in S_n} a_p r_p = \sum_{p \in K} a_p r_p = (A, R).$$

We still have to show that the matrix A in (1) is uniquely determined by U. Assume that there is another tempered matrix $B = \{b_n\}$ such that $U(R) = (B, R)$ for each $R \in \mathcal{R}$. Then $(A, R) = (B, R)$ and, in particular, $(A, e_p) = (B, e_p)$, i.e., $a_p = b_p$ for each $p \in K$. This completes the proof of the first part of the theorem.

The proof of the second part is simpler. The linearity of $U(R)$ follows from

$$(A, \alpha Q + \beta R) = \alpha(A, Q) + \beta(A, R).$$

To prove the continuity of $U(R)$ we first write

$$U(R_n) - U(R) = U(R_n - R) = (A, R_n - R) = \big(T_k^{-1} A, T_k(R_n - R)\big).$$

Hence

$$|U(R_n) - U(R)| \leqslant |T_k^{-1} A| \cdot \|T_k(R_n - R)\|.$$

If A is tempered, then $|T_k^{-1} A| < \infty$ for some k. If $R_n \xrightarrow{\mathcal{R}} R$, the second factor on the right hand side tends to 0. This implies $|U(R_n) - U(R)| \to 0$, which proves that U is continuous.

11. Applications of the Köthe spaces

11.1. APPLICATIONS TO TEMPERED DISTRIBUTIONS

We say that a sequence of tempered distributions f_n $(n = 1, 2, \ldots)$ is *tempered of order* $k \in \boldsymbol{P}^q$ *to* f, iff there exist square integrable functions F_n, F such that

$$D^k F_n = f_n, \quad D^k F = f \tag{1}$$

and the sequence F_n converges in square mean to F, as $n \to \infty$.

11.1.1. THEOREM. *A sequence of tempered distributions* $f_n = \sum_{p \in \boldsymbol{P}^q} a_{np} h_p$ *is tempered of order* k *to* $f = \sum_{p \in \boldsymbol{P}^q} a_p h_p$, *iff*

$$\sum_{p \in \boldsymbol{P}^q} \tilde{p}^{-k} |a_{np} - a_p|^2 \to 0, \quad \text{as} \quad n \to \infty, \tag{2}$$

where $\tilde{p}$ *denotes the vector whose jth coordinate is* $\max(1, \pi_j)$ *if the jth coordinate of* p *is* π_j $(j = 1, \ldots, q)$.

PROOF. Assume that F_n, $F \in L^2$ and that (1) holds. By the Parseval formula we have

$$\int |F_n - F|^2 = \sum_{p \in \boldsymbol{P}^q} |c_{np} - c_p|^2,$$

where c_{np} and c_p are the Hermite coefficients of F_n and F, respectively. Since

$$a_{np} - a_p = \sqrt{\frac{(p+k)!}{p!}}\,(c_{n,p+k} - c_{p+k}),$$

we have, by 8.1(5),

$$|c_{n,p+k} - c_{p+k}|^2 \leqslant \tilde{p}^{-k} |a_{np} - a_p|^2 \leqslant K |c_{n,p+k} - c_{p+k}|^2,$$

where K depends only on k. Hence our assertion follows.

According to the definition in Section 7.5, a sequence of tempered distributions f_n is tempered to f, iff it is tempered to f of some order $k \in \boldsymbol{P}^q$.

11.1.2. THEOREM. *A sequence of tempered distributions f_n is tempered to f, iff $a_{np} \to a_p$ as $n \to \infty$, and, moreover, there exist an index $k \in \boldsymbol{P}^q$ and a number M such that*

$$\tilde{p}^{-k}|a_{np}| < M \quad \text{for all } n = 1, 2, \dots \text{ and } p \in \boldsymbol{P}^q. \tag{3}$$

PROOF. Assume that f_n is tempered to f. By 11.1.1, there then exists an index $k \in \boldsymbol{P}^q$ such that (3) holds for a suitably chosen M.

Assume now, conversely, that $a_{np} \to a_p$, as $n \to \infty$, and that (3) holds for some k and M. Then also

$$\tilde{p}^{-k}|a_p| \leqslant M \quad \text{for all } p \in \boldsymbol{P}^q$$

and

$$\tilde{p}^{-2k}|a_{np} - a_p|^2 < 4M^2.$$

Given any number $\varepsilon > 0$ there is a positive integer ν_0 such that

$$\sum_{\nu_0 \not> p \in \boldsymbol{P}^q} \tilde{p}^{-2} < \frac{\varepsilon}{8} M^{-2}.$$

Hence

$$\sum_{p \in \boldsymbol{P}^q} \tilde{p}^{-2k-2}|a_{np} - a_p|^2 \leqslant \sum_{\nu_0 > p \in \boldsymbol{P}^q} \tilde{p}^{-2k-2}|a_{np} - a_p|^2 + \varepsilon/2.$$

Since $a_{np} \to a_p$, as $n \to \infty$, the sum on the right hand side is less than $\varepsilon/2$ for sufficiently large n, and so the sum on the left hand side is less than ε for such n. This proves, by 11.1.1, that the sequence f_n is tempered to f of order $2k+2$, and the proof is complete.

Let $T_k = \{\tilde{p}^k\}$ for $p \in \boldsymbol{P}^q$ and $k = 1, 2, \dots$ We denote by $\mathscr{T}$ the class of all complex valued matrices $A = \{a_p\}$ $(p \in \boldsymbol{P}^q)$ such that $|AT_k^{-1}| < \infty$ for some k. The elements of $\mathscr{T}$ are clearly tempered matrices in the sense of Section 10.8. We say that $\{a_p\}$ is the coefficient matrix of a tempered distribution f, iff $f = \sum_{p \in \boldsymbol{P}^q} a_p h_p$. From Corollary 8.1.2 it follows that the correspondence between $\mathscr{S}'$ and $\mathscr{T}$ is one-to-one. Note that for every $k \in \boldsymbol{P}^q$ there is a vector $k' \geqslant k$ all of whose coordinates are equal. This implies that in 11.1.2, we can specify that all the coordinates of k be equal. According to our convention of Section 4.3, we can also state that k is a positive integer.

Using now the language of Section 10.8, Theorem 11.1.2 can be formulated in the following way:

11.1.3. THEOREM. *A sequence of tempered distributions f_n is tempered to f, iff the sequence of coefficient matrices of f_n converges strongly to the coefficient matrix of f.*

Let A_n, A and R be the coefficient matrices of $f_n, f \in \mathscr{S}'$ and $\psi \in \mathscr{S}$. Then, by 8.1.3, we have

$$(f_n, \psi) = (A_n, R) \quad \text{and} \quad (f, \psi) = (A, R). \tag{4}$$

If (A_n, R) converges to (A, R) for every rapidly decreasing R, then A_n converges weakly to A according to the definition in Section 10.7. Similarly, we say that the sequence f_n is tempered weakly to f iff (f_n, ψ) converges to (f, ψ) for each $\psi \in \mathscr{S}$. Further, it will be convenient in what follows to use the terms "is strongly convergent" and "is tempered" synonymously. The equalities (4) imply

11.1.4. THEOREM. *A sequence of tempered distributions f_n is tempered weakly to f, iff the sequence of coefficient matrices of f_n converges weakly to the coefficient matrix of f.*

PROOF. In fact, in addition to (4), we only need the fact that the coefficient matrix is rapidly decreasing iff it corresponds to a function from $\mathscr{S}$, and this follows from 8.2.6.

Theorem 11.1.3 and 11.1.4 are similar. They say that strong and weak tempered convergence of tempered distributions are equivalent, respectively, to strong and weak convergence of coefficient matrices. Now, 10.8.2 states that strong and weak convergence of a sequence of matrices are equivalent. This implies the equivalence of strong and weak tempered convergence of a sequence of tempered distributions. More precisely, we have the following important

11.1.5. THEOREM. *A sequence of tempered distributions f_n is tempered strongly to f, iff it is tempered weakly to f.*

REMARK. Just as in Section 10.8, we should distinguish between *being tempered* and *being tempered to some distribution*. E.g., we say that a sequence f_n is tempered weakly, iff the sequence (f_n, ψ) is convergent for every $\psi \in \mathscr{S}$. It is obvious that if f_n is tempered weakly to f, then f_n is tempered weakly. However it is not obvious that if f_n is tempered weakly, then there exists an f to which f_n is tempered weakly. This fact follows from the corresponding properties of matrices stated at the end of Section 10.8. The same can also be said of strong tempered convergence, but here the situation is rather trivial and it is not necessary to go to the theory of sequence spaces.

11.2. CONVERGENCE IN $\mathscr{S}$ AND $\mathscr{R}$

We recall that $\psi_n \overset{\mathscr{S}}{\to} \psi$ $(\psi_n, \psi \in \mathscr{S})$ means that $D^k\psi_n \overset{2}{\to} D^k\psi$ for each $k \in \boldsymbol{P}^q$ or equivalently $d^k\psi_n \overset{2}{\to} d^k\psi$ for each $k \in \boldsymbol{P}^q$. We show that, in this definition, we can restrict ourselves to vectors $k = (\varkappa, \ldots, \varkappa)$ with $\varkappa \in \boldsymbol{P}^1$, and that this modification does not affect the scope of the definition. In fact, this follows immediately from the following

11.2.1. THEOREM. *Let $f_n, f \in \mathscr{S}'$. If $d^k f_n, d^k f \in L^2$ and $d^k f_n \overset{2}{\to} d^k f$ for some $k \in \boldsymbol{P}^q$, then $d^m f_n, d^m f \in L^2$ and $d^m f_n \overset{2}{\to} d^m f$ for each m such that $0 \leqslant m \leqslant k$. The theorem remains true, if we replace d by D troughout.*

PROOF. By 8.1.1, we can write

$$f_n \overset{t}{=} \sum_{p \in \boldsymbol{P}^q} a_{np} h_p \quad \text{and} \quad f \overset{t}{=} \sum_{p \in \boldsymbol{P}^q} a_p h_p .$$

Applying the operator d^m term by term and using formula 8.1(1), we get

$$d^m f_n = (-1)^m \sum_{p \in \boldsymbol{P}^q} \sqrt{\frac{(p+m)!}{p!}}\, a_{np} h_{p+m}$$

and

$$d^m f = (-1)^m \sum_{p \in \boldsymbol{P}^q} \sqrt{\frac{(p+m)!}{p!}}\, a_p h_{p+m}.$$

By 8.2.1, $d^m f_n, d^m f \in L^2$. By the Parseval formula, we get, for $m \leqslant k$,

$$\int |d^m f_n - d^m f|^2 = \sum_{p \in \boldsymbol{P}^q} \frac{(p+m)!}{p!} |a_{np} - a_p|^2 \leqslant \sum_{p \in \boldsymbol{P}^q} \frac{(p+k)!}{p!} |a_{np} - a_p|^2$$

$$= \int |d^k f_n - d^k f|^2$$

which proves our assertion.

The proof with D instead of d is similar (see also the proof of 8.2.1).

We recall that $R_n \overset{\mathscr{R}}{\to} R$ $(R_n, R \in \mathscr{R})$ means that $||T_k(R_n - R)|| \to 0$ for each $k \in \boldsymbol{P}^1$, where $T_k = \{\tilde{p}^k\}$. From the definition of $\tilde{p}$ it follows that the element of T_k with index p is the kth power of the product of the non vanishing coordinates of p. We now show that $R_n \overset{\mathscr{R}}{\to} R$, iff $||T_k(R_n - R)^2|| \to 0$. This will be proved only for the special case when $R = 0$. In fact, let $R_n = \{r_{np}\}$. We have

$$||T_k R_n^2|| = \sum_{p \in \boldsymbol{P}^q} \tilde{p}^k |r_{np}|^2 \quad \text{and} \quad ||T_k R_n|| = \sum_{p \in \boldsymbol{P}^q} \tilde{p}^k |r_{np}| .$$

Now if $||T_k R_n|| \to 0$ as $n \to \infty$, we see that there is a positive integer n_0, independent of k, such that $|r_{np}| \leqslant 1$ for $n > n_0$ and all $p \in \boldsymbol{P}^q$. Hence $||T_k R_n^2|| \leqslant ||T_k R_n||$ for sufficiently large n, and consequently, $||T_k R_n|| \to 0$ implies $||T_k R_n^2|| \to 0$. Conversely, assume that $||T_k R_n^2|| \to 0$ for each fixed positive integer k.

Let ε be an arbitrary positive number and let S be a finite subset of $\boldsymbol{P}^q$ such that

$$\sum_{p\in S'} \tilde{p}^{-2} < \varepsilon/2,$$

where S' is the set of all $p \in \boldsymbol{P}^q$ which do not belong to S. It is easy to check that

$$\sum_{p\in \boldsymbol{P}^q} \tilde{p}^k|r_{np}| \leqslant \sum_{p\in \boldsymbol{P}^q} \tilde{p}^{2k+2}|r_{np}|^2 + \sum_{p\in S} \tilde{p}^k|r_{np}| + \sum_{p\in S'} \tilde{p}^{-2}. \tag{1}$$

In fact, if for some p, $\tilde{p}^k|r_{np}| > \tilde{p}^{2k+2}|r_{np}|^2$, then $\tilde{p}^k|r_{np}| < \tilde{p}^{-2}$. Now if $||T_k R_n^2|| \to 0$ for each $k \in \boldsymbol{P}^1$, the first sum on the right hand side of (1) tends to 0, as $n \to \infty$. Hence, for each $p \in \boldsymbol{P}^q$, $|r_{np}| \to 0$ as $n \to \infty$. Since the set S is finite, the second sum on the right hand side also tends to 0, as $n \to \infty$. Consequently, the left hand side of (1) is less than ε for sufficiently large n, i.e., $||T_k R_n|| < \varepsilon$.

The general case follows, on setting $R_n = R_n' - R'$.

11.2.2. THEOREM. *If* $\psi_n \overset{\mathscr{S}}{\to} \psi$ $(\psi_n,\ \psi \in \mathscr{S})$, *then* $R_n \overset{\mathscr{R}}{\to} R$ *for the corresponding matrices, and conversely.*

PROOF. By the Parseval equality and by 8.1(1) we have

$$\int |d^k\psi_n - d^k\psi|^2 = \sum_{p\in \boldsymbol{P}^q} \frac{(p+k)!}{p!} |r_{np} - r_p|^2.$$

Using the general inequalities

$$\tilde{p}^k \leqslant \frac{(p+k)!}{p!} \leqslant K\tilde{p}^k$$

(see Section 8.1) we get

$$||T_k(R_n - R)^2|| \leqslant \int |d^k\psi_n - d^k\psi|^2 \leqslant K||T_k(R_n - R)^2||,$$

and our assertion now follows.

11.3. TEMPERED DISTRIBUTIONS AS FUNCTIONALS

L. Schwartz has defined tempered distributions as continuous linear functionals on the space $\mathscr{S}$. We say that $T(\psi)$ is a *functional on* $\mathscr{S}$, iff the value of $T(\psi)$ is a number for each $\psi \in \mathscr{S}$. We say that the functional $T(\psi)$ is *linear*, iff

$$T(\alpha\varphi + \beta\psi) = \alpha T(\varphi) + \beta T(\psi)$$

for any number α, β and $\varphi, \psi \in \mathscr{S}$. Finally, we say that the functional $T(\psi)$ is *continuous*, iff

$$\lim_{n\to\infty} T(\psi_n) = T(\psi)$$

for every sequence of functions $\psi_n \in \mathscr{S}$ converging to ψ in $\mathscr{S}$ (see Section 8.2).

Let Mf denote the matrix corresponding to the tempered distribution f and mA the distribution corresponding to the complex (or real) matrix A, so that $mMA = A$ and $Mmf = f$. Clearly

$$M(\alpha f+\beta g) = \alpha Mf+\beta Mg,$$
$$m(\alpha A+\beta B) = \alpha mA+\beta mB.$$

This implies that if $T(\psi)$ is a linear functional on $\mathscr{S}$, then $U(R) = T(mR)$ is a linear functional on the space $\mathscr{R}$ of rapidly decreasing matrices. Conversely, if U is a linear functional on $\mathscr{R}$, then $T(\psi) = U(M\psi)$ is a linear functional on $\mathscr{S}$. This correspondence is one-to-one, because $U(R) = T(mR)$ implies, on substituting $R = M\psi$, that $U(M\psi) = T(mM\psi) = T(\psi)$.

Moreover, if U is continuous, then T is continuous, and conversely. In fact, if U is continuous and $\psi_n \overset{\mathscr{S}}{\to} \psi$, then $M\psi_n \overset{\mathscr{R}}{\to} M\psi$, by 11.2.2 and therefore $T(\psi_n) = U(M\psi_n) \to U(M\psi) = T(\psi)$. The converse is proved similarly.

We have thus proved that there is a one-to-one correspondence between continuous linear fimctionals on $\mathscr{S}$ and on $\mathscr{R}$. By means of this correspondence, 10.9.1 can be translated into the language of tempered distributions.

11.3.1. THEOREM. *For each continuous linear functional T on $\mathscr{S}$ there exists a unique tempered distribution f such that*

$$T(\psi) = (f, \psi) \quad \textit{for each } \psi \in \mathscr{S}. \tag{1}$$

Conversely, for each tempered distribution f, (1) *represents a continuous linear functional on $\mathscr{S}$. The correspondence* (1) *between continuous linear functionals on $\mathscr{S}$ and tempered distributions is thus one-to-one.*

PROOF. If $T(\psi)$ is a continuous linear functional on $\mathscr{S}$, then $T(mR)$ is a continuous linear functional on $\mathscr{R}$. By 10.9.1, there is a tempered matrix A such that $T(mR) = (A, R)$ for each $R \in \mathscr{R}$. By 11.1(4), we have $(A, R) = (mA, mR)$. Let $f = mA$. Then $T(mR) = (f, mR)$. Now each $\psi \in \mathscr{S}$ is of the form mR for some $R \in \mathscr{R}$, and so $T(\psi) = (f, \psi)$ for each $\psi \in \mathscr{S}$. The distribution f is uniquely determined by T. In fact, if $(f, \psi) = (g, \psi)$ for each $\psi \in \mathscr{S}$, then also $(Mf, M\psi) = (Mg, M\psi)$, by 11.1(4). Since every matrix $R \in \mathscr{R}$ is of the form $R = M\psi$ with $\psi \in \mathscr{S}$, we have $(Mf, R) = (Mg, R)$ for each $R \in \mathscr{R}$. This implies $Mf = Mg$, and hence $f = g$. This proves the first part of the theorem. The proof of the second part is a matter of simple verification.

By 11.3.1, tempered distributions can be identified with continuous linear functionals on $\mathscr{S}$. It is therefore possible to start from functionals and develop the theory from this standpoint. However, some knowledge of functional analysis is needed for such an approach.

11.4. APPLICATION TO ARBITRARY DISTRIBUTIONS

We say that a sequence of distributions f_n *converges weakly in an open set* $O \subset \boldsymbol{R}^q$, iff for every fixed smooth function φ whose carrier is bounded and inside O, the sequence of inner products (f_n, φ) (see Section 7.8) is convergent; the inner products need not exist for all n, we require only that they exist for large values of n.

We also recall the definition of ordinary convergence. A sequence of distributions f_n is convergent in O, iff for every fixed interval I inside O there exist continuous functions F_n and an index $k \in \boldsymbol{P}^q$ such that $F_n^{(k)} = f_n$ and the sequence F_n converges uniformly in I. The functions F_n need not exist for all n, we require only that they exist for large values of n. To distinguish this type of convergence from weak convergence we also call it *strong convergence.*

11.4.1. THEOREM. *A sequence of distributions converges weakly in O, iff it converges strongly in O.*

PROOF. Assume that a sequence of distributions f_n converges weakly in O. Let I be any given interval inside O and let J be another interval inside O such that I is inside J. Let ω be a smooth function in $\boldsymbol{R}^q$ such that $\omega = 1$ on I and $\omega = 0$ outside J. It is easy to prove that

$$(f_n\omega, \psi) = (f_n, \omega\psi).$$

Therefore, $(f_n\omega, \psi)$ converges for every $\psi \in \mathcal{S}$, i.e., $f_n\omega$ is tempered weakly. Hence, by 11.1.5, $f_n\omega$ is tempered strongly, and consequently converges strongly in $\boldsymbol{R}^q$, by 7.5.2. Since $\omega f_n = f_n$ in I, the sequence f_n converges strongly in I. But this implies that f_n converges strongly in O, for the interval I inside O can be chosen arbitrarily.

Assume now, conversely, that f_n converges strongly in O. Let φ be any smooth function whose carrier is inside an interval I inside O. There is then a sequence of continuous functions F_n, uniformly convergent in I, such that $F_n^{(k)} = f_n$ for some $k \in \boldsymbol{P}^q$ and $n \geqslant n_0$. Hence the sequence $(f_n, \varphi) = (F_n, \varphi^{(k)})$ is convergent. Assume now that the carrier of φ is inside O but is not inside any interval I inside O. It is always possible to represent such a φ as a finite sum of smooth functions $\varphi_1, \ldots, \varphi_r$ such that the carrier of φ_j is inside an interval I_j inside O, and by the preceding result, the sequence (f_n, φ_j) is convergent as $n \to \infty$ for each j. Hence the sequence

$$(f_n, \varphi) = (f_n, \varphi_1) + \ldots + (f_n, \varphi_r) \tag{1}$$

is also convergent, which means that f_n converges weakly in O.

Let $\mathcal{D}_O$ denote the space of all smooth functions φ whose carriers are inside O.

11.4.2. THEOREM. *If f is a distribution in O and $(f, \varphi) = 0$ for each $\varphi \in \mathcal{D}_O$, then $f = 0$.*

PROOF. Let f_n be a fundamental sequence for f. Then $(f_n, \varphi) \to 0$. If h_n is the interlaced sequence formed from f_n and $g_n = 0$, then $(h_n, \varphi) \to 0$ for each $\varphi \in \mathcal{D}_O$. Thus h_n converges weakly in O and therefore strongly by 11.3.1. But the h_n are smooth functions, and so the sequence h_n is fundamental, by 3.9.3, Part II. This implies that the subsequences f_n and g_n are equivalent. Since $g_n = 0$, it follows that $f = 0$.

In this section, we have until now considered weak and strong convergence with no mention of the limits to which the sequences converge.

We now say that a sequence of distributions f_n *converges weakly to* f in $O \subset \boldsymbol{R}^q$, iff for any fixed smooth function φ whose carrier is inside O, we have

$$(f_n, \varphi) \to (f, \varphi).$$

We say that a sequence of distributions f_n *converges strongly to* f in O, iff for each fixed interval I inside O there exist continuous functions F_n and F and an index $k \in \boldsymbol{P}^q$ such that $F_n^{(k)} = f_n$, $F^{(k)} = f$ and the sequence F_n converges uniformly to F in I.

If we denote by F the limit of the F_n in the last part of the proof of 11.3.1, then $(f, \varphi) = (F, \varphi^{(k)})$ which implies that

$$(f_n, \varphi) \to (f, \varphi), \tag{2}$$

provided the carrier of φ is in I: if it is not, we use the decomposition (1) to arrive at the same equality (2). Thus, if f_n converges strongly to f, then f_n also converges weakly to f. We know that if f_n converges strongly, then f_n converges strongly to some limit f. This implies, now, that if f_n converges weakly, then f_n converges weakly to some limit f. The strong limit and the weak limit are uniquely defined and are the same.

The preceding argument shows that, in fact, strong convergence and weak convergence are exactly the same concept, differing only in the way they are defined. In the sequel, it is therefore unnecessary to distinguish between strong and weak convergence: both can be known simply as "convergence". We can use either definition, according to which is the more convenient.

11.5. DISTRIBUTIONS AS FUNCTIONALS

L. Schwartz has defined distributions in any open set $O \subset \boldsymbol{R}^q$ as functionals on the space $\mathcal{D}_O$ of smooth functions in O whose carriers are bounded and inside O. In particular, the set O can be the whole space $\boldsymbol{R}^q$. We say that $T(\varphi)$ is a functional on $\mathscr{D}_O$, iff the value of $T(\varphi)$ is a number for each $\varphi \in \mathscr{D}_O$. The functional T is linear, iff

$$T(\alpha\varphi+\beta\psi) = \alpha T(\varphi)+\beta T(\psi)$$

for $\alpha, \beta \in \boldsymbol{R}^1$ and $\varphi, \psi \in \mathscr{D}_O$. Finally, the functional $T(\varphi)$ is continuous, iff

$$\lim_{n\to\infty} T(\varphi_n) \to T(\varphi)$$

for every sequence φ_n which converges to φ in $\mathscr{D}_O$, i.e., for every sequence of functions $\varphi_n \in \mathscr{D}_O$ such that the carriers of all the φ_n are contained in a common interval I inside O and

$$\varphi_n^{(k)} \rightrightarrows \varphi^{(k)} \quad \text{in } O \qquad \text{for every } k \in \boldsymbol{P}^q.$$

The continuity of T is thus relative to the kind of convergence (or topology) introduced in $\mathscr{D}_O$.

11.5.1. THEOREM. *For each continuous linear functional T on $\mathscr{D}_O$, there is a unique distribution f in O such that*

$$T(\varphi) = (f, \varphi) \quad \textit{for each } \varphi \in \mathscr{D}_O. \tag{1}$$

Conversely, for each distribution f in O, (1) *represents a continuous linear functional. The correspondence between continuous linear functionals on $\mathscr{D}_O$ and distributions in O is thus one-to-one.*

PROOF. Let Θ be any given bounded open set inside O and let ω be a smooth function in $\boldsymbol{R}^q$ such that $\omega = 1$ in Θ and $\omega = 0$ outside a bounded set Ω inside O such that Θ is inside Ω. If T is a continuous linear functional on $\mathscr{D}_O$, then

$$T_\Theta(\psi) = T(\omega\psi)$$

is a continuous linear functional on $\mathscr{S}$ such that $T_\Theta(\varphi) = T(\varphi)$ for $\varphi \in \mathscr{D}_\Theta$, i.e., for smooth functions $\varphi \in \mathscr{D}_O$ whose carriers are in Θ. By 11.3.1, there is a tempered distribution f_Θ such that $T_\Theta(\psi) = (f_\Theta, \psi)$ for $\psi \in \mathscr{S}$. Consequently, we have $T(\varphi) = (f_\Theta, \varphi)$ for those $\varphi \in \mathscr{D}_O$ with carrier inside Θ.

Assume that Θ and Ξ are two bounded open sets in O. There exist tempered distributions f_Θ and f_Ξ such that $T(\varphi) = (f_\Theta, \varphi)$ or $T(\varphi) = (f_\Xi, \varphi)$ if the carrier of φ is in Θ or Ξ, respectively. Thus $(f_\Theta - f_\Xi, \varphi) = 0$ for each $\varphi \in \mathscr{D}_A$, i.e., whose carrier is inside $A = \Theta \cap \Xi$. Hence $f_\Theta - f_\Xi = 0$ in A, by 11.4.2.

We have shown that there is a family of (tempered) distributions f_Θ corresponding to the family $\mathscr{F}$ of bounded open sets Θ inside O such that $f_\Theta = f_\Xi$ in $\Theta \cap \Xi$. By 3.10.5, Part II, there is a distribution f in O such that $f = f_\Theta$ in each $\Theta \in \mathscr{F}$. It is easy to see that (1) then holds with this f. In fact, if $\varphi \in \mathscr{D}_O$, there is a bounded open set Θ inside O such that the carrier of φ is inside Θ. We thus have

$T(\varphi)=(f_\Theta, \varphi)=(f, \varphi)$, since $f_\Theta = f$ in Θ. The distribution f satisfying (1) is unique: this follows from 11.4.2.

It remains to show, conversely, that if f is a distribution in O, then (1) is a continuous linear functional. This is a matter of simple verification, which is left to the reader.

The last theorem enables us to identify distributions in O with continuous linear functionals on $\mathscr{D}_O$. Following L. Schwartz, we can equally define distributions as functionals and develop the theory from this point of view. By 11.4.1, the two approaches are completely equivalent as indeed one would expect. In the sequential approach sequence and series are fundamental tools, while in the functional analytic method, topology plays the fundamental role.

Furthermore each method suggests its own area of investigation: a mathematician who is familiar with the functional analytic way of thinking will be inclined to investigate connections between different subclasses of distributions. On the other hand, the sequential approach, which is more akin to Classical Analysis, will suggest calculations with distributions, generalizing algorithms. From this point of view, it is more closely linked with applications.

11.6. APPLICATION TO PERIODIC DISTRIBUTIONS

The theory of Köthe spaces also applies to periodic distributions. In the following theorem distributional convergence can be read as weak convergence, since the two notions are equivalent, by 11.4.1.

11.6.1. THEOREM. *A sequence of periodic distributions* $f_n = \sum_{p \in B^q} a_{np} E^p$ *converges distributionally to* $f = \sum_{p \in B^q} a_p E^p$, *iff* $a_{np} \to a_p$ *as* $n \to \infty$ *and, moreover, there exist an index* $k \in \boldsymbol{P}^q$ *and a number* M *such that*

$$\tilde{p}^{-k}|a_{np}| < M \quad \text{for all } n \in \boldsymbol{P}^1 \text{ and } p \in B^q. \tag{1}$$

PROOF. Assume that f_n converges distributionally to f. Then, by 9.3.1, there exist periodic continuous functions F_{mn}, F_m $(0 \leqslant m \leqslant r \in \boldsymbol{P}^q)$ such that

$$f_n = \sum_{0 \leqslant m \leqslant r} F_{mn}^{(m)} \quad (n = 1, 2, \ldots) \quad \text{and} \quad f = \sum_{0 \leqslant m \leqslant r} F_m^{(m)}$$

and $F_{mn} \rightrightarrows F_m$ in $\boldsymbol{R}^q$. Hence, by Theorem 9.6.1 and 9.4 (3), we can write

$$a_{np} - a_p = \sum_{0 \leqslant m \leqslant r} (2\pi i p)^m (F_{mn} - F_m, E^{-p}) \quad \text{and} \quad a_p = \sum_{0 \leqslant m \leqslant r} (2\pi i p)^m (F_m, E^{-p}).$$

But

$$|(F_{mn}-F_m, E^{-p})| \leqslant \int_0^1 |F_{mn}-F_m| \to 0 \quad \text{as } n \to \infty.$$

This implies $a_{np_1^*} \to a_p$. Moreover, for each m there is a number K_m such that

$$\int_0^1 |F_{mn}-F_m| < K_m \quad \text{for all } n \quad \text{and} \quad \int_0^1 |F_m| < K_m.$$

Let $k \geqslant r$. Since $|(2\pi ip)^m| \leqslant (2\pi)^k \tilde{p}^k$ for $m \leqslant k$, we get

$$|a_{np}-a_p| \leqslant M_1 \tilde{p}^k \quad \text{and} \quad |a_p| \leqslant M_1 \tilde{p}^k$$

with $M_1 = (2\pi)^k \sum_{0 \leqslant m \leqslant r} K_m$. Hence we get (1) with $M = 2M_1$.

Assume now, conversely, that $a_{np} \to a_p$ as $n \to \infty$ and that (1) holds for some k and M. Then

$$\tilde{p}^{-k}|a_p| \leqslant M \quad \text{and} \quad \tilde{p}^{-k}|a_{np}-a_p| < 2M. \tag{2}$$

Given any number $\varepsilon > 0$, there is a positive integer ν_0 such that

$$\sum_{\nu_0 \not> p \in B^q} \tilde{p}^{-2} < \frac{\varepsilon}{4} M^{-1}.$$

Hence

$$\sum_{p \in B^q} \tilde{p}^{-k-2}|a_{np}-a_p| \leqslant \sum_{\nu_0 > p \in B^q} \tilde{p}^{-k-2}|a_{np}-a_p| + \varepsilon/2.$$

Since $a_{np} \to a_p$ as $n \to \infty$, the sum on the right hand side is less than $\varepsilon/2$ for sufficiently large n, and so for such n, then sum on the left hand side is less than ε.

For each $p \in B^q$ we can choose a vector $l_p \in P^q$ and a number $\gamma_p \neq 0$ such that

$$(\gamma_p x^{l_p} E^p)^{(k+2)} = E^p. \tag{3}$$

It is a matter of routine calculation to show that

$$|\gamma_p|\, l_p! < \tilde{p}^{-k-2}. \tag{4}$$

Let

$$F_n = \sum_{p \in B^q} a_{np}\gamma_p x^{l_p} E^p \quad \text{and} \quad F = \sum_{p \in B^q} a_p \gamma_p x^{l_p} E^p. \tag{5}$$

For a fixed bounded I, there is a number K such that $|x^{l_p}/l_p!| < K$ for all x in I. It thus follows, by (1) and (2), that both the series (5) are dominated by the convergent series

$$KM \sum_{p \in B^q} \tilde{p}^{-2},$$

and thus they are uniformly convergent in I. Since I can be chosen arbitrarily, it follows that both the series (5) are almost uniformly convergent in $\boldsymbol{R}^q$. Hence F_n, F are continuous functions. Moreover, by (4) we have, in the interval I,

$$|F_n - F| \leqslant K \sum_{p \in B^q} \tilde{p}^{-k-2} |a_{np} - a_p| \to 0, \quad \text{as} \quad n \to \infty,$$

which implies the almost uniform convergence of F_n to F in $\boldsymbol{R}^q$. Hence, differentiating both the equalities (5) $k+2$ times and applying (3), we see that the sequence $f_n = \sum_{p \in B^q} a_{np} E^p$ converges distributionally to $f = \sum_{p \in B^q} a_p E^p$ in $\boldsymbol{R}^q$, and the proof of our assertion is complete.

Using the language of Section 10.8 we can alternatively formulate Theorem 11.6.1 in the following way:

11.6.2. THEOREM. *A sequence of periodic distributions f_n is distributionally convergent to f, iff the sequence of coefficient matrices of f_n converges strongly to the coefficient matrix of f.*

Note that 11.6.2 becomes false, when the word *periodic* is replaced by *tempered.* In fact, the sequence $f_n(x) = \exp\left(x^2 - \dfrac{x^4}{n}\right)$ consists of continuous bounded functions and converges almost uniformly to e^{x^2}. It can therefore be regarded as a sequence of tempered distributions which converges distributionally. However the limit e^{x^2} is not a tempered distribution, and so the sequence f_n is not tempered. Hence, by 11.1.3, the corresponding sequence of coefficient matrices does not converge strongly. On the other hand, 11.6.2 does remain true, when we replace in it the words *distributionally convergent* by *tempered.* This follows from the fact that every tempered sequence is distributionally convergent.

However, the following result should be noted

11.6.3. THEOREM. *A sequence of periodic distributions f_n converges distributionally to f, iff*

$$(f_n, \varphi) \to (f, \varphi) \tag{6}$$

for each smooth periodic function φ.

PROOF. If A_n, A and R are the coefficient matrices of f_n, f and φ, then $(f_n, \varphi) = (A_n, R)$ and $(f, \varphi) = (A, R)$ by 9.6.3. Thus (6) implies that $(A_n, R) \to (A, R)$, and conversely. Hence it follows by Theorem 9.4.3, that (6) holds for each periodic smooth φ, iff the sequence of matrices A_n converges weakly to A, or, by 10.8.2, iff A_n converges strongly to A. Our assertion now follows by 11.6.2.

11.7. PERIODIC DISTRIBUTIONS AS FUNCTIONALS

Periodic distributions can be defined as continuous linear functionals on the space $\mathscr{P}$ of all periodic smooth functions. A functional $T(\varphi)$ on $\mathscr{P}$ is said to be continuous, iff

$$\lim_{n\to\infty} T(\varphi_n) = T(\varphi)$$

whenever $\varphi_n \overset{\mathscr{P}}{\to} \varphi$, i.e., whenever $\varphi_n^{(k)} \rightrightarrows \varphi^{(k)}$ in $\boldsymbol{R}^q$, for every fixed $k \in \boldsymbol{B}^q$.

11.7.1. THEOREM. *For each continuous linear functional on $\mathscr{P}$, there is a periodic distribution f such that*

$$T(\varphi) = (f, \varphi) \quad \textit{for each } \varphi \in \mathscr{P}. \tag{1}$$

Conversely, for each periodic distribution f, (1) *represents a continuous linear functional on $\mathscr{P}$. The correspondence between continuous linear functionals on $\mathscr{P}$ and periodic distributions is thus one-to-one.*

The proof closely resembles the proof of 11.3.1 on tempered distributions.

Theorem 11.7.1 allows us to identify periodic distributions with continuous linear functionals on $\mathscr{P}$.

12. Applications of the equivalence of weak and strong convergence

12.1. CONVERGENCE AND REGULAR OPERATIONS

In Section 3.8, Part II, it is proved that passage to the limit of sequences of distributions commutes with certain regular operations. We here prove the following general result (see Section 2.3 and 3.8, Part II):

12.1.1. THEOREM. *Passage to the limit commutes with every regular operation. In other words, if for any fundamental sequences $\varphi_{1n}, \ldots, \varphi_{rn}$ in $O_1, \ldots, O_r$ the sequence $A(\varphi_{1n}, \ldots, \varphi_{rn})$ is fundamental in a set O, then for any sequences of distributions $f_{1n}, \ldots, f_{rn}$ which converge to $f_1, \ldots, f_r$ in $O_1, \ldots, O_r$ respectively, the sequence $A(f_{1n}, \ldots, f_{rn})$ converges to $A(f_1, \ldots, f_r)$ in O.*

PROOF. Let δ_n be an arbitrary delta-sequence (see Section 2.1) and let $f_{inm} = f_{in} * \delta_m$ for $i = 1, \ldots, r$ and $n, m = 1, 2, \ldots$ By 6.2.1, the sequence f_{inm} is fundamental for f_{in}. Since A is a regular operation we can write

$$A(f_{1nm}, \ldots, f_{rnm}) \to A(f_{1n}, \ldots, f_{rn}) \quad \text{in } O \quad \text{as } m \to \infty. \tag{1}$$

Let ε be an arbitrary positive number and let φ be any smooth function whose carrier is in O. By 11.4.1 and by (1), we have

$$(A(f_{1nm}, \ldots, f_{rnm}), \varphi) \to (A(f_{1n}, \ldots, f_{rn}), \varphi) \quad \text{as } m \to \infty$$

for any $n = 1, 2, \ldots$ Hence, there is a sequence of integers m_n such that $m_n < m_{n+1}$ and

$$|(A(f_{1nm_n}, \ldots, f_{rnm_n}), \varphi) - (A(f_{1n}, \ldots, f_{rn}), \varphi)| < \varepsilon/2 \quad (n = 1, 2, \ldots). \tag{2}$$

Let $\varphi_{in} = f_{inm_n} = f_{in} * \bar{\delta}_n$, where $\bar{\delta}_n = \delta_{m_n}$. Since $\bar{\delta}_n$ is a delta-sequence, the sequence φ_{in} is a fundamental sequence for f_i, by 6.2.2. Now A is a regular operation, and so we can write

$$A(\varphi_{1n}, \ldots, \varphi_{rn}) \to A(f_1, \ldots, f_r) \quad \text{in } O \quad \text{as } n \to \infty.$$

Consequently, by 11.4.1, there is an n_0 such that

$$|(A(f_{1nm_n}, \ldots, f_{rnm_n}), \varphi) - (A(f_1, \ldots, f_r), \varphi)| < \varepsilon/2$$

for $n > n_0$. Hence, by (2), we have

$$|(A(f_1, \ldots, f_r), \varphi) - (A(f_{1n}, \ldots, f_{rn}), \varphi)| < \varepsilon$$

for $n > n_0$. This implies that the sequence $A(f_{1n}, \ldots, f_{rn})$ converges to $A(f_1, \ldots, f_r)$ in O, which completes the proof.

REMARK. Note that we have established the weak convergence of $A(f_{1n}, \ldots, f_{rn})$. The equivalence of weak and strong convergence is thus essential to the proof.

12.2. THE VALUE OF A DISTRIBUTION AT A POINT

We say that a distribution f takes the *value a at a point* x_0, iff, for every regular sequence $f_n = f*\delta_n$, we have

$$\lim_{n\to\infty} f_n(x_0) = a.$$

It is easily checked that if a distribution f is a continuous function at a point x_0, then the above definition coincides with the value in the ordinary sense. As an alternative to saying that f takes a value at x_0, we also say that the value $f(x_0)$ exists.

From Sections 5.1 and 7.8 it follows that the value of $f*\delta_n$ at x_0 is equal to $(f(x_0-x), \delta_n(x))$. Hence we can equally define the value $f(x_0)$ by

$$f(x_0) = \lim_{n\to\infty} (f(x_0-x), \delta_n(x)).$$

12.2.1. THEOREM. *If a distribution f has a value at* x_0, *then*

$$\lim_{\alpha\to 0} f(x_0+\alpha x) = f(x_0). \tag{1}$$

PROOF. Let Ω be a smooth function of bounded carrier such that $\int\Omega = 1$. Then for any sequence of positive numbers $\alpha_n \neq 0$ tending to 0, the sequence

$$\delta_n(x) = \alpha_n^{-q}\Omega(\alpha_n^{-1}x)$$

is a delta-sequence (see Section 2.1). Assume that $f(x_0) = 0$, i.e., that

$$(f(x_0-x), \delta_n(x)) \to 0.$$

Since the substitution $x = \alpha_n t$ and the inner product with a smooth function of bounded carrier are both regular operations and the equality

$$(f(x_0-\alpha_n t), \Omega(t)) = (f(x_0-x), \delta_n(x))$$

holds for every smooth function f, it also holds for every distribution f. We may therefore write

$$(f(x_0-\alpha_n x), \Omega(x)) \to 0. \tag{2}$$

Now, if φ is an arbitrary smooth function of bounded carrier we can write $\varphi = \beta\Omega$, where Ω is a smooth functions of bounded carrier such that $\int\Omega = 1$ and β is a real number. By (1) we now obtain

$$\big(f(x_0-\alpha_n x),\, \varphi(x)\big) \to 0.$$

Thus $f(x_0-\alpha_n x)$ converges to 0, as $n \to \infty$. Since α_n can be any sequence tending to 0, we in fact have $f(x_0+\alpha x) \to 0$, as $\alpha \to 0$. The case $f(x_0) \neq 0$ reduces to the case discussed above on considering the distribution $f(x)-f(x_0)$ in place of $f(x)$.

REMARK. S. Łojasiewicz has used formula (1) to define the value of f at x_0. Theorem 12.2.1 thus states that the Ł-value (value in the sense of Łojasiewicz) exists, whenever $f(x_0)$ exists in the previous sense, and that the two values are equal. It can be proved, conversely, that if the Ł-value of f exists at x_0 then the value $f(x_0)$ also exists.

We prove it in the case of a single variable. If f has a Ł-value at 0 equal to 0, there exists a continuous function F such that $F^{(k)} = f$ in the neighbourhood of 0 and

$$\lim_{x\to 0} \frac{F(x)}{x^k} = 0$$

(see Section 3.5, Part I). For each number $\varepsilon > 0$, there thus exists a number $\eta > 0$ such that

$$\left|\frac{F(x)}{x^k}\right| < \varepsilon \quad \text{for} \quad |x| < \eta.$$

If δ_n is a delta-sequence, we have, for sufficiently large n

$$\big|\big(f(-x),\, \delta_n(x)\big)\big| = \left|\int_{-\alpha_n}^{\alpha_n} F(-x)\delta_n^{(k)}(x)\right| \leqslant \varepsilon \int_{-\alpha_n}^{\alpha_n} |x^k\delta_n^{(k)}| \leqslant \varepsilon M_k.$$

Hence $\big(f(-x),\, \delta_n\big) \to 0$, which means that $f(0) = 0$.

If f has a Ł-value at x_0 equal to a, the distribution $g(x) = f(x_0+x)-a$ has the Ł-value 0 at $x = 0$ and the proof reduces to the case just discussed.

A similar proof can be given for any number of dimensions: we omit the details.

Following S. Łojasiewicz, the value of a distribution at a point can be used to sharpen the concept of the carrier of a distribution. For any distribution f in O, let $Ł_f$ be the set of all points of O at which the value of f either does not exist or is different from 0. It can be proved that the closure of $Ł_f$ is the carrier in Schwartz's sense (see Section 6.4) of f. Clearly, the set $Ł_f$ gives us more information about the distribution than does its carrier.

12.3. PROPERTIES OF THE DELTA-DISTRIBUTION

The delta-distribution δ has been defined as the (common) limit of all delta-sequences (see Section 3.1, Part II).

12.3.1. THEOREM. *For every smooth function* φ, $(\delta, \varphi) = \varphi(0)$.

PROOF. For any delta-sequence δ_n we have

$$|(\delta_n, \varphi)-\varphi(0)| \leqslant \int_{-\alpha_n}^{\alpha_n} |\delta_n(x)| \cdot |\varphi(x)-\varphi(0)|\, dx \to 0,$$

where α_n is any sequence of positive numbers, convergent to 0, such that $\delta_n(x$ $= 0$ for $|x| \geqslant \alpha_n$.

12.3.2. THEOREM. *If* $(f, \varphi) = \varphi(0)$ *for each* $\varphi \in \mathscr{D}$, *then* $f = \delta$.

PROOF. Let $f_n = f * \delta_n$. Then $(f_n, \varphi) \to \varphi(0)$ and so the interlaced sequence $(f_1, \varphi), (\delta_1, \varphi), (f_2, \varphi), (\delta_2, \varphi), \ldots$ converges to $\varphi(0)$ for each $\varphi \in \mathscr{D}$. The sequence $f_1, \delta_1, f_2, \delta_2, \ldots$ is therefore fundamental, which implies that $f = \delta$.

REMARK. It is plain that $T(\varphi) = \varphi(0)$ is a continuous linear functional on $\mathscr{D}$. Theorem 12.3.2 thus tells us that, in the language of functionals, the delta-distribution should be defined by $\delta(\varphi) = \varphi(0)$. Note that in the functional notation $\delta(\varphi)$, the argument φ is an element of $\mathscr{D}$, this turns out to be awkward to handle in applications.

12.4. PRODUCT OF TWO DISTRIBUTIONS

We say that product of the distributions f and g exists in an open set O, iff the sequence

$$(f * \delta_n) \cdot (g * \delta_n) \tag{1}$$

converges in O for every delta-sequence δ_n. Its limit is then the product of f and g. (The consistency of this definition follows from the fact that the interlaced sequence of two delta-sequences in another delta-sequence.)

If g is a smooth function in an open set O, the limit (1) exists for every distribution defined in O and is equal to the product fg, understood as a regular operation (see Section 2.5, Part II). In fact, $(g * \delta_n)^{(k)}$ then converges to $g^{(k)}$ almost uniformly in O for every fixed $k \in \boldsymbol{P}^q$. Since $f * \delta_n$ converges to f in O, the product (1) converges to fg, by 3.8.1, Part II. We may therefore use the symbol fg for the product in the general case, when g is not necessarily a smooth function.

12.4.2. THEOREM. *If a distribution f takes the value $f(0)$ at the origin, then $f\delta = f(0)\delta$.*

PROOF. Let φ be any smooth function of bounded carrier such that $\varphi(0) \neq 0$. Since $(\delta_n, \varphi) \to \varphi(0)$, we have

$$\gamma_n = (\delta_n, \varphi) \neq 0 \quad \text{for} \quad n \geqslant n_0.$$

It is easily checked that the sequence

$$\sigma_n = \frac{1}{\gamma_n}\, \delta_n \varphi$$

is a delta-sequence for $n \geqslant n_0$. Using the notation $\bar{\sigma}_n(x) = \sigma_n(-x)$, we can write, for $n > n_0$,

$$\big((f*\delta_n)\,\delta_n, \varphi\big) = \gamma_n(f*\delta_n, \sigma_n) = \gamma_n[(f*\delta_n)*\bar{\sigma}_n]\,(0) = \gamma_n[f*(\delta_n*\bar{\sigma}_n)]\,(0),$$

by 5.1(1) and 5.1(3) and Corollary 6.4.7. Since the convolution $\delta_n * \bar{\sigma}_n$ is another delta-sequence, these equalities imply

$$\big((f*\delta_n)\,\delta_n, \varphi\big) \to \varphi(0)f(0). \tag{2}$$

If $\varphi(0) = 0$, we can write $\varphi = \varphi_1 + \varphi_2$, where φ_1 and φ_2 are smooth functions of bounded carrier such that $\varphi_1(0) \neq 0$ and $\varphi_2(0) \neq 0$. Applying the preceding result to φ_1 and φ_2 we have

$$\big((f*\delta_n)\,\delta_n, \varphi_1\big) \to \varphi_1(0)f(0),$$

$$\big((f*\delta_n)\,\delta_n, \varphi_2\big) \to \varphi_2(0)f(0).$$

Adding these together, (2) now follows for every smooth function φ of bounded carrier. Since, by 12.3.1, $\big(f(0)\,\delta, \varphi\big) = f(0)\varphi(0)$, we have $(f*\delta_n)\,\delta_n \to f(0)\,\delta$, which proves our assertion.

12.5. NON EXISTENCE OF δ^2

By the square of the delta-distribution, δ^2, we mean the product $\delta \cdot \delta$. We shall show that this product does not exist, so that the symbol δ^2 is, in fact, meaningless.

In fact, according to the definition of the product, we write

$$\delta \cdot \delta = \lim_{n\to\infty} (\delta*\delta_n)\,(\delta*\delta_n), \quad \text{i.e.,} \quad \delta \cdot \delta = \lim_{n\to\infty} \delta_n^2.$$

There is a smooth function φ, of bounded carrier, such that $\varphi(x) = 1$ for $-\frac{1}{4} \leqslant x \leqslant \frac{1}{4}$ and $\int \varphi = 1$. The sequence

$$\delta_n(x) = n^q \varphi(n^q x)$$

is then a delta-sequence, and furthermore

$$(\delta_n(x))^2 = n^{2q} \quad \text{for} \quad -\frac{1}{4n} \leqslant x \leqslant \frac{1}{4n}.$$

Hence

$$(\delta_n^2, \varphi) \geqslant \int_{I_n} n^{2q} = \left(\frac{n}{2}\right)^q \to \infty,$$

where I_n is the closed interval $-\frac{1}{4n} \leqslant x \leqslant \frac{1}{4n}$ in $\boldsymbol{R}^q$.

This shows that the sequence δ_n^2 does not converge, i.e., that the square δ^2 does not exist.

In spite of this, the square δ^2 is used by physicists in their calculations. We shall see, in Sections 14.2, 14.3 and 15.1, how it is possible to give a satisfactory mathematical justification for such calculations.

Later on, we shall show that the square $\left(\frac{1}{x}\right)^2$ of the distribution $\frac{1}{x}$ does not exist either.

12.6. THE PRODUCT $x\frac{1}{x}$

In this section we treat the one dimensional case, when x is a real variable. The distribution $\frac{1}{x}$ is then defined as the distributional derivative of $\ln|x|$:

$$\frac{1}{x} = (\ln|x|)'.$$

We show that

$$x\frac{1}{x} = 1. \tag{1}$$

This simple equality requires careful justification, since in the definition of the distribution $\frac{1}{x}$, a distributional derivative is involved.

Note that the function $x\ln|x|$ is a primitive function, in the ordinary sense, of

$$1+\ln|x|. \tag{2}$$

Since this function is a locally integrable function, it is equal to the distributional derivative of $x\ln|x|$. Moreover x is a smooth function and so the product $x \cdot \ln|x|$ can be regarded as a regular operation. Hence the ordinary rule for the product continues to hold when we differentiate in the distributional sense and we thus obtain, $x(\ln|x|)' + \ln|x|$, i.e.,

$$x\frac{1}{x} + \ln|x|.$$

Comparing this with (2) we get (1).

12.7. ON THE ASSOCIATIVITY OF THE PRODUCT

The product of three functions is always associative, i.e., $(fg)h = f(gh)$. It may therefore come as something of a surprise that a similar property does not hold for distributions. The following famous example is due to L. Schwartz:

$$\left(\frac{1}{x}x\right)\delta \neq \frac{1}{x}(x\delta).$$

In fact,

$$\left(\frac{1}{x}x\right)\delta = 1 \cdot \delta = \delta,$$

$$\frac{1}{x}(x\delta) = \frac{1}{x} \cdot 0 = 0.$$

However, it should be noted that associativity does hold, whenever at least two of the factors are smooth functions. Explicitly, if f is any distribution and φ, ψ are smooth functions, then

$$(f\varphi)\psi = f(\varphi\psi).$$

This equality follows from the remark that the products with φ, with ψ and with $\varphi\psi$ are regular operations.

13. The Hilbert transform and its applications

13.1. THE HILBERT TRANSFORM

We first define the *Hilbert transform* for functions $\varphi \in \mathscr{D}$. Let A_n be the set of all real numbers x such that $|x| \geqslant \frac{1}{n}$. By the Hilbert transform Φ of $\varphi \in \mathscr{D}$ we mean the limit

$$\Phi(x) = \lim_{n\to\infty} \int_{A_n} \varphi(x-t)\frac{dt}{t}.$$

We show that this limit exists and is equal to the convolution $\varphi * \frac{1}{x}$. In fact, integrating by parts, we find

$$\int_{A_n} \varphi(x-t)\frac{dt}{t} = \ln n\left[\varphi\left(x-\frac{1}{n}\right)-\varphi\left(x+\frac{1}{n}\right)\right]+\int_{A_n}\varphi'(x-t)\ln|t|\,dt.$$

But $\varphi\left(x+\frac{1}{n}\right)-\varphi\left(x-\frac{1}{n}\right) = \frac{2}{n}\varphi'(\xi_n)$, where $|\xi_n| < \frac{1}{n}$. We hence obtain, letting $n \to \infty$,

$$\Phi = \varphi' * \ln|x|. \tag{1}$$

We now show that Φ is a square integrable function. In fact, there is a number x_0 such that $\varphi(x) = 0$ for $|x| > x_0$. This implies that, for $|x| > x_0$,

$$\left|\int_{A_n} \varphi(x-t)\frac{dt}{t}\right| \leqslant \int_{-\infty}^{\infty}\left|\frac{\varphi(x-t)}{t}\right|dt = \int_{-\infty}^{\infty}\left|\frac{\varphi(t)}{x-t}\right|dt \leqslant \frac{M}{|x|-x_0},$$

with $M = \int|\varphi|$. Thus

$$|\Phi| \leqslant \frac{M}{|x|-x_0} \quad \text{for} \quad |x| > x_0.$$

Since the function Φ is also smooth, it is thus square integrable.

Using the distribution $\frac{1}{x}$, formula (1) can be rewritten as

$$\Phi = \varphi * \frac{1}{x}.$$

Hence we can alternatively define the Hilbert transform to be the convolution $\varphi * \frac{1}{x}$. This suggests a generalization of the Hilbert transform to all distributions f for which the convolution $f * \frac{1}{x}$ exists. For instance, the Hilbert transform of δ is $\frac{1}{x}$.

13.2. NON EXISTENCE OF $\left(\frac{1}{x}\right)^2$

By $\left(\frac{1}{x}\right)^2$ we mean the product

$$\left(\frac{1}{x}\right)^2 = \frac{1}{x}\frac{1}{x} = \lim_{n\to\infty}\left(\delta_n * \frac{1}{x}\right)^2. \tag{1}$$

We show that this limit does not exist and so the symbol $\left(\frac{1}{x}\right)^2$ should be regarded as meaningless. In fact, let ψ be a non-negative smooth function of bounded carrier. There is a number x_0 such that $\psi(x) = 0$ for $|x| > x_0$. This implies that for every x satisfying $|x| > x_0$ there is an index n_0 such that

$$\left|\int_{A_n} \psi(x-t)\frac{dt}{t}\right| = \int_{-\infty}^{\infty} \frac{\psi(x-t)}{|t|}\,dt = \int_{-\infty}^{\infty} \frac{\psi(t)}{|x-t|}\,dt \geqslant \frac{\int\psi}{|x_0|+|x|} \qquad \text{for} \quad n > n_0.$$

Hence

$$\left|\psi * \frac{1}{x}\right| \geqslant \frac{\int\psi}{|x_0|+|x|} \qquad \text{for} \quad |x| > x_0.$$

Now let φ be a non-negative function of bounded carrier such that $\varphi(x) \geqslant \frac{1}{2}$ for $|x| \leqslant \frac{1}{2}$, $\int\varphi = 1$ and $\varphi(x) = 0$ for $|x| \geqslant 1$. Then the sequence

$$\delta_n(x) = n\varphi(nx)$$

is a delta sequence and we have

$$\left|\delta_n * \frac{1}{x}\right| \geqslant \frac{1}{\frac{1}{n}+|x|} \qquad \text{for} \quad |x| > \frac{1}{n}.$$

Hence

$$\left(\left(\delta_n * \frac{1}{x}\right)^2, \varphi\right) \geqslant \frac{1}{2}\int_{1/n}^{1/2} \frac{dx}{\left(\frac{1}{n}+x\right)^2} = \frac{n(n-2)}{4(n+2)} \to \infty .$$

This shows that the product (1) does not exist.

It should be noted that we have, by definition,

$$\frac{1}{x^2} = \left(-\frac{1}{x}\right)' = (-\ln|x|)''.$$

The symbol $\frac{1}{x^2}$ therefore represents a distribution and should not confused, in the theory of distributions, with the square $\left(\frac{1}{x}\right)^2$.

13.3. SOME FORMULAE FOR THE HILBERT TRANSFORM

For $t \neq 0$, we have the identity

$$x^k \frac{\varphi(x-t)}{t} - \frac{(x-t)^k \varphi(x-t)}{t} = \sum_{m=1}^{k} (-1)^{m-1} \binom{k}{m} x^{k-m} t^{m-1} \varphi(x-t)$$

for every $k = 1, 2, \ldots$ This identity also holds for $k = 0$, provided we adopt the convention that any sum from "1 to 0" is equal to 0.

If $\varphi \in \mathscr{D}$, we get, integrating over A_n and letting $n \to \infty$,

$$x^k \cdot \left(\varphi * \frac{1}{x}\right) - (x^k \cdot \varphi) * \frac{1}{x} = \sum_{m=1}^{k} (-1)^{m-1} \binom{k}{m} x^{k-m} (x^{m-1} * \varphi). \tag{1}$$

We note that

$$\left(x^k \cdot \left(\varphi * \frac{1}{x}\right), \varphi\right) = \left(\varphi * \frac{1}{x}, x^k \varphi\right),$$

$$\left((x^k \cdot \varphi) * \frac{1}{x}, \varphi\right) = \left(x^k \varphi, -\frac{1}{x} * \varphi\right) = -\left(\varphi * \frac{1}{x}, x^k \varphi\right).$$

Taking the inner product of each side of (1) with φ, and dividing by 2, we therefore obtain

$$\left(\varphi * \frac{1}{x}, x^k \varphi\right) = \frac{1}{2} \sum_{m=1}^{k} (-1)^{m-1} \binom{k}{m} (x^{m-1} * \varphi, x^{k-m} \varphi). \tag{2}$$

In particular, for $k = 0$ and $k = 1$, we obtain the following formulae of Gonzalez-Dominguez and Scarfiello:

$$\left(\varphi * \frac{1}{x}, \varphi\right) = 0, \tag{3}$$

$$\left(\varphi * \frac{1}{x}, x\varphi\right) = \frac{1}{2}\left(\int \varphi\right)^{2}, \tag{4}$$

where $\varphi \in \mathscr{D}$.

13.4. THE PRODUCT $\frac{1}{x}\delta$

This product is defined by the formula

$$\frac{1}{x}\delta = \lim_{n\to\infty}\left(\delta_n * \frac{1}{x}\right)\delta_n .$$

We shall show that the limit exists and is equal to $-\frac{1}{2}\delta'$ so that we have the formula

$$\frac{1}{x}\delta = -\frac{1}{2}\delta'. \tag{1}$$

We are going to give two different proofs of this important equality, both of which depend on the equalities of Gonzalez-Dominguez and Scarfiello.

PROOF 1. Let $f_n = \left(\delta_n * \frac{1}{x}\right)\delta_n$ and

$$F_n(x) = \frac{1}{2}\int_{-\infty}^{x}(x-t)^2 f_n(t)\,dt;$$

then $F_n''' = f_n$. We show that F_n converges uniformly in $(-\infty, \infty)$. Since $f_n = 0$ for $x \leqslant -\alpha_n$, we have

$$F_n(x) = 0 \quad \text{for} \quad x \leqslant -\alpha_n. \tag{2}$$

Since $f_n(x) = 0$ for $x \geqslant \alpha_n$, we have, for $x \geqslant \alpha_n$,

$$F_n'(x) = \int_{-\infty}^{x}(x-t)f_n(t)\,dt = x\int_{-\infty}^{\infty} f_n(t)\,dt - \int_{-\infty}^{\infty} t f_n(t)\,dt$$

$$= x\left(\delta_n * \frac{1}{x}, \delta_n\right) - \left(\delta_n * \frac{1}{x}, x\delta_n\right) = -\tfrac{1}{2}$$

by 13.3 (3) and 13.3 (4). Hence

$$F_n(x) = F_n(\alpha_n) - \frac{x}{2} + \frac{\alpha_n}{2} \quad \text{for} \quad x \geqslant \alpha_n. \tag{3}$$

Finally, if $|x| \leqslant \alpha_n$, we write

$$F_n(x) = \frac{1}{2} \int_{-\alpha_n}^{x} (x-t)^2 \left(\int_{-\alpha_n}^{\alpha_n} \delta_n''(u) L(t-u)\, du \right) \delta_n(t)\, dt$$

where

$$L(x) = x \ln|x| - x.$$

Hence

$$|F_n(x)| \leqslant \frac{1}{2} \int_{-\alpha_n}^{\alpha_n} (2\alpha_n)^2 \left(\int_{-\alpha_n}^{\alpha_n} |\delta_n''(u) L(2\alpha_n)|\, du \right) |\delta_n(t)|\, dt$$

$$= 2L(2\alpha_n)\alpha_n^2 \int |\delta_n''| \cdot \int |\delta_n| \leqslant 2L(2\alpha_n) \cdot M_2 M_0 = \varepsilon_n.$$

Since the function $L(x)$ is continuous and $L(0) = 0$, it follows that

$$|F_n(x)| \leqslant \varepsilon_n \to 0 \quad \text{as} \quad n \to \infty \quad \text{for} \quad |x| \leqslant \alpha_n. \tag{4}$$

Formulae (2), (3) and (4) together imply that the sequence F_n converges uniformly in $-\infty < x < \infty$ to the function F given by

$$F(x) = \begin{cases} 0 & \text{for} \quad x \leqslant 0, \\ -x/2 & \text{for} \quad x \geqslant 0. \end{cases}$$

Hence $f_n \to F''' = -\frac{1}{2}\delta'$, which proves (1).

PROOF 2. If $\varphi \in \mathscr{D}$, we can write

$$\varphi(x) = \varphi(0) + x\varphi'(0) + x^2\psi(x),$$

where ψ is a smooth function. Hence

$$(f_n, \varphi) = \varphi(0)\left(\delta_n * \frac{1}{x}, \delta_n\right) + \varphi'(0)\left(\delta_n * \frac{1}{x}, x\delta_n\right) + \left(\delta_n * \frac{1}{x}, x^2\delta_n\psi\right)$$

and so, by 13.3 (3) and 13.3 (4),

$$(f_n, \varphi) = \tfrac{1}{2}\varphi'(0) + \left(\delta_n * \frac{1}{x}, x^2\delta_n\psi\right). \tag{5}$$

Now

$$\left(\delta_n * \frac{1}{x}, x^2\delta_n\psi\right) = (\delta_n'' * L, x^2\delta_n\psi) = \int_{-\alpha_n}^{\alpha_n} \left(\int_{-\alpha_n}^{\alpha_n} \delta_n''(t) L(x-t)\, dt \right) x^2 \delta_n(x)\psi(x)\, dx$$

and consequently,

$$\left|\left(\delta_n * \frac{1}{x}, x^2 \delta_n \psi\right)\right| \leqslant \int_{-\alpha_n}^{\alpha_n} \left(\int_{-\alpha_n}^{\alpha_n} |\delta_n''(t) L(2\alpha_n)| dt \right) \alpha_n^2 |\delta_n(x)| M dx$$

$$\leqslant ML(2\alpha_n)\alpha_n^2 \int |\delta_n''| \cdot \int |\delta_n| \leqslant MM_2 M_0 L(2\alpha_n),$$

where M is the maximum of $|\psi(x)|$ in the interval $(-\alpha, \alpha)$ and $\alpha = \max_n \alpha_n$. Hence $\left(\delta_n * \frac{1}{x}, x^2 \delta_n \psi\right) \to 0$, and by (5)

$$(f_n, \varphi) \to -\tfrac{1}{2}\varphi'(0) = (-\tfrac{1}{2}\delta', \varphi), \quad \text{as} \quad n \to \infty.$$

Since φ is an arbitrary function in $\mathcal{D}$, this implies that $f_n \to -\frac{1}{2}\delta'$, which proves (1).

13.5. ON THE EQUATION $xf = \delta$

We seek a distribution $f(x)$ such that

$$xf(x) = \delta(x). \tag{1}$$

Looking at this equation for $x \neq 0$, we see that $xf(x) = 0$ for $x \neq 0$ and multiplying by the function $\frac{1}{x}$, which is smooth for $x \neq 0$, we find $f(x) = 0$ for $x \neq 0$. Therefore, every distribution f which satisfies (1) must vanish for $x \neq 0$. We know that the only distributions with this property are of the form

$$f = \gamma_0 \delta + \ldots + \gamma_n \delta^{(n)}$$

where $\gamma_0, \ldots, \gamma_n$ are numbers. Substituting this in (1) and using the equalities $x\delta = 0$, $x\delta^{(k)} = -k\delta^{(k-1)}$, we find

$$-\gamma_1 \delta - \ldots - n\gamma_n \delta^{(n-1)} = \delta$$

and hence $\gamma_1 = -1$, $\gamma_2 = \ldots = \gamma_n = 0$. Every solution of (1) must therefore be of the form

$$f = \gamma_0 \delta - \delta'. \tag{2}$$

On the other hand, it is easy to see that every distribution of the form (2) satisfies (1). Hence (2) is the general solution of (1). It is rather surprising to find that the product $\frac{1}{x}\,\delta$ does not satisfy (1); this is because it is equal to $-\frac{1}{2}\delta'$, and so $x\left(\frac{1}{x}\delta\right) \neq \delta$. Since $\delta = \left(x\frac{1}{x}\right)\delta$, we thus have

$$x\left(\frac{1}{x}\delta\right) \neq \left(x\frac{1}{x}\right)\delta,$$

which is another example of non-associativity. It involves the same factors as the example of Section 12.7, but in a different order.

13.6. GENERALIZATION TO SEVERAL VARIABLES

If $x = (\xi_1, \ldots, \xi_q) \in \boldsymbol{R}^q$, we define $\dfrac{1}{x}$ to be the product

$$\frac{1}{x} = \frac{1}{\xi_1} \cdots \frac{1}{\xi_q}.$$

Since all the variables in this product are separated, it can be regarded as a regular operation and we may therefore apply the usual rules in calculation. We thus easily find

$$\frac{1}{x}\delta(x) = \frac{1}{\xi_1}\delta_1(\xi_1) \cdots \frac{1}{\xi_q}\delta_q(\xi_q) = \left(-\tfrac{1}{2}\,\delta_1'(\xi_1)\right) \cdots \left(-\tfrac{1}{2}\,\delta_q'(\xi_q)\right)$$

$$= (-\tfrac{1}{2})^q \delta_1'(\xi_1) \cdots \delta_q'(\xi_q)$$

and finally

$$\frac{1}{x}\delta = (-\tfrac{1}{2})^q \delta',$$

where δ' denotes the derivative of order $(1, \ldots, 1)$.

14. Applications of the Fourier transform

14.1. THE CONVOLUTION $\frac{1}{x} * \frac{1}{x}$

In this section, we show that the above convolution exists and that the following formula holds

$$\frac{1}{x} * \frac{1}{x} = -\pi^2 \delta, \tag{1}$$

where $x \in \boldsymbol{R}^1$.

In fact, we have by definition

$$\frac{1}{x} * \frac{1}{x} = \lim_{n\to\infty} \left(\delta_n * \frac{1}{x}\right) * \left(\delta_n * \frac{1}{x}\right)$$

provided the limit exists for every delta-sequence δ_n. Since the functions $f_n^{(k)} = (-1)^k k!(\delta_n * x^{-k-1})$ are square integrable, $\frac{1}{x}$ is a tempered distribution, $\delta_n \in \mathscr{S}$ and $\mathscr{F}\left(\frac{1}{x}\right) = i\sqrt{\pi}2^{-1}\operatorname{sgn} x$, the convolutions $f_n * f_n$ exist smoothly, by 8.8.3 and 8.8.4, we obtain the equality

$$\mathscr{F}(f_n * f_n) = -\pi^2 \mathscr{F}(\tilde{\delta}_n),$$

where $\tilde{\delta}_n = \delta_n * \delta_n$. Since $\tilde{\delta}_n$ is tempered to δ, the sequence $\mathscr{F}(\tilde{\delta}_n)$ is tempered to $\mathscr{F}(\delta) = (2\sqrt{\pi})^{-1}$, and consequently, the sequence $\mathscr{F}(f_n * f_n)$ is tempered to $-\pi^{3/2}/2$. It follows that the sequence $f_n * f_n$ is tempered to the distribution whose Fourier transform is $-\pi^{3/2}/2$, i.e., to $-\pi^2\delta$. This establishes the existence of the convolution $\frac{1}{x} * \frac{1}{x}$ and the validity of formula (1).

14.2. THE SQUARE OF $\delta + \frac{1}{\pi^2}\frac{1}{x}$

Following the general definition we write

$$\left(\delta + \frac{1}{\pi i}\frac{1}{x}\right)^2 = \lim_{n\to\infty} B_n^2,$$

where $B_n = \delta_n + \frac{1}{\pi i}\,\delta_n * \frac{1}{x}$. We now show that the limit exists. We have

$$\hat{B}_n = \hat{\delta}_n + \hat{\delta}_n \operatorname{sgn} x = 2H\hat{\delta}_n,$$

where $\hat{B}_n$, $\hat{\delta}_n$ are the Fourier transforms of B_n, δ_n respectively, and H is the Heaviside function. Thus

$$|\hat{\delta}_n(x)| \leqslant \frac{1}{2\sqrt{\pi}} \int_{-\infty}^{\infty} |\delta_n(t)|\,dt \leqslant \frac{M_0}{2\sqrt{\pi}}.$$

On the other hand, for each fixed x, we have

$$\hat{\delta}_n(x) - \frac{1}{2\sqrt{\pi}} = \frac{1}{2\sqrt{\pi}} \int_{-\infty}^{\infty} (e^{ixt/2} - 1)\,\delta_n(t)\,dt \to 0, \quad \text{as} \quad n \to \infty.$$

Hence we see that the sequence $\hat{B}_n$ is bounded and tends pointwise to $H/\sqrt{\pi}$. Moreover $\hat{B}_n(x) = 0$ for $x < 0$. This implies that the sequence of convolutions $\hat{B}_n * \hat{B}_n$ is tempered to $\frac{x}{\pi} H$. It follows that the sequence B_n^2 is tempered to the distribution whose Fourier transform is $\frac{x}{2\pi^{3/2}} H$, i.e., to $-\frac{1}{\pi i}\,\delta' - \frac{1}{\pi^2}\frac{1}{x^2}$. We therefore have

$$\left(\delta + \frac{1}{\pi i}\frac{1}{x}\right)^2 = -\frac{1}{\pi i}\,\delta' - \frac{1}{\pi^2}\frac{1}{x^2}.$$

14.3. THE FORMULA $\delta^2 - \frac{1}{\pi^2}\left(\frac{1}{x}\right)^2 = -\frac{1}{\pi^2}\frac{1}{x^2}$

Since the expressions δ^2 and $\left(\frac{1}{x}\right)^2$ have no meaning, the left hand side of the above formula is apparently meaningless. However, we can write, formally,

$$\delta^2 - \frac{1}{\pi^2}\left(\frac{1}{x}\right)^2 = \left(\delta + \frac{1}{\pi}\frac{1}{x}\right)\left(\delta - \frac{1}{\pi}\frac{1}{x}\right),$$

and then study the product on the right hand side. It turns out that this product does exist. In fact, it is equal to the limit of the sequence

$$\Phi_n = \left(\delta_n + \frac{1}{\pi}\,\delta_n * \frac{1}{x}\right)\left(\delta_n - \frac{1}{\pi}\,\delta_n * \frac{1}{x}\right).$$

The above product can also be written in the form

$$B_n^2 - \frac{2}{\pi i}\,\delta_n\left(\frac{1}{x} * \delta_n\right),$$

where B_n is as in Section 14.2. Hence, the limit exists and is equal to

$$\left(-\frac{1}{\pi i}\,\delta' - \frac{1}{\pi^2}\,\frac{1}{x^2}\right) + \frac{1}{\pi i}\,\delta'.$$

Consequently we have

$$\left(\delta + \frac{1}{\pi}\,\frac{1}{x}\right)\left(\delta - \frac{1}{\pi}\,\frac{1}{x}\right) = -\frac{1}{\pi^2}\,\frac{1}{x^2}.$$

15. Final remarks

15.1. GENERALIZED OPERATIONS

Convolution, inner product, value at a point and the product of two distributions can all be regarded as particular instances of the more general concept of a *generalized operation.*

We recall that an operation $A(\varphi, \psi, \ldots)$ performed on a finite number of imooth functions $\varphi, \psi, \ldots$ defined in open sets $\mathscr{P}, \mathscr{Q}, \ldots$ respectively, is regular sff, for any fundamental sequences in $\mathscr{P}, \mathscr{Q}, \ldots$ of smooth functions $\varphi_n, \psi_n, \ldots,$ the sequence $\Phi_n = A(\varphi_n, \psi_n, \ldots)$ is fundamental in some open set O. If $f, g, \ldots$ are distributions corresponding to fundamental sequences $\varphi_n, \psi_n, \ldots,$ then the symbol $A(f, g, \ldots)$ denotes the distribution defined by Φ_n. In this way, a regular operation can be extended to distributions.

Now, let $A(\varphi, \psi, \ldots)$ be an operation acting on smooth functions $\varphi, \psi, \ldots$ defined in open sets $\mathscr{P}, \mathscr{Q}, \ldots,$ respectively, the result of which is a distribution defined in an open set O. Assume that for certain distributions $f, g, \ldots$ defined in open subsets of $\mathscr{P}, \mathscr{Q}, \ldots$ respectively, the sequence $\Phi_n = A(f * \delta_n, g * \delta_n, \ldots)$ converges distributionally in an open set O for every delta-sequence δ_n. In these circumstances $A(f, g, \ldots)$ denotes the limit of Φ_n, and say that $A(f, g, \ldots)$ exists as a generalized operation. The result of a generalized operation is always unique, i.e., it does not depend on the choice of delta-sequence; this follows from the fact that the interlaced sequence of two delta-sequences is again a delta-sequence.

It is important to note that every regular operation can be regarded as a generalized operation yielding the same result.

The main difference between a regular operation and a generalized operation is that the regular operation can be performed on all distributions, while the generalized operation exists only for certain distributions $f, g, \ldots$ Moreover, calculations with regular operations can be performed on distributions just as for smooth functions, but this is not the case with generalized operations. For instance, associativity of the product docs not hold in general.

In particular, if we put in turn

$$A(\varphi, \psi) = \varphi * \psi, \quad A(\varphi, \psi) = (\varphi, \psi), \quad A(\varphi) = \varphi(x_0), \quad A(\varphi, \psi) = \varphi\psi,$$

then the operation A reduces to convolution, inner product, value at a point and the product, respectively.

There are naturally many other generalized operations, e.g.,

$$A(\varphi, \psi) = \varphi^2 - \frac{1}{\pi^2}\psi^2.$$

This operation can be performed on the distributions δ and $\frac{1}{x}$, because

$$\Phi_n = A\left(\delta * \delta_n, \frac{1}{x} * \delta_n\right) = \delta_n^2 - \frac{1}{\pi^2}\left(\delta_n * \frac{1}{x}\right)^2,$$

and as we have proved in Section 14.3, Φ_n converges to $-\frac{1}{\pi^2}\frac{1}{x^2}$. Hence

$$\delta^2 - \frac{1}{\pi^2}\left(\frac{1}{x}\right)^2 = -\frac{1}{\pi^2}\frac{1}{x^2}.$$

Thanks to the concept of a generalized operation, this formula now has a direct interpretation. Note that the left hand side cannot be interpreted as the difference of two sequences: it must be thought of as a single operation on δ and $\frac{1}{x}$.

15.2. A SYSTEM OF DIFFERENTIAL EQUATIONS

We say that a polynomial is of degree less than $k = (\varkappa_1, \ldots, \varkappa_q) \in \boldsymbol{P}^q$, iff the variables ξ_j $(j = 1, \ldots, q)$ appear in it only to powers smaller than $\varkappa_j$.

15.2.1. THEOREM. *Let $1 \leqslant k \in \boldsymbol{P}^q$. The only distributions f in an interval $I \subset \boldsymbol{R}^q$, bounded or unbounded, satisfying the system of q equations*

$$f^{(\varkappa_j e_j)} = 0 \qquad (j = 1, \ldots, q) \tag{1}$$

are polynomials of degree $< k$.

PROOF. It is obvious that every polynomial f of degree $< k$ satisfies (1). To prove the converse, first note that it is true for $k = (1, \ldots, 1)$. In fact, f is then constant with respect to each of the variables $\xi_1, \ldots, \xi_q$. We now proceed by induction with respect to k. Assume that the assertion is true for all k satisfying $1 \leqslant k \leqslant m$ for some $m = (\mu_1, \ldots, \mu_q) \in \boldsymbol{P}^q$. We have to prove that, given any index i $(1 \leqslant i \leqslant q)$, the assertion is true for $k = m + e_i$. In fact, if f satisfies (1) with $k = m+e_i$, then $f^{(e_i)}$ satisfies (1) with $k = m$. Hence by the induction hypothesis, $f^{(e_i)}$ is a polynomial of degree $< k$, and so we can write

$$f^{(e_i)} = \sum_{0 \leqslant s \leqslant k-1} \alpha_s x^s,$$

where the α_s are numbers. Let

$$g = \sum_{0 \leqslant s \leqslant k-1} \frac{\alpha_s}{\sigma_i+1} x^{s+e_i},$$

where σ_i is the ith coordinate of s. It is easy to see that $g^{(e_i)} = f^{(e_i)}$ and that $g^{(\mu_j e_j)} = 0$ for $j \neq i$. Also $f^{(\mu_j e_j)} = 0$ for $j \neq i$. Setting $h = f-g$, we thus have $h^{(\mu_j e_j)} = 0$ for $j \neq i$ and, moreover, $h^{(e_i)} = 0$. Applying the induction hypothesis to h, we conclude that h is a polynomial of degree $< m-\mu_i e_i+e_i$, and therefore less than $m+e_i$. Similarly, g is a polynomial of degree $< m+e_i$. This implies that the distribution $f = g+h$ is a polynomial of degree $< m+e_i$. The theorem now follows by induction.

15.3. SOME REMARKS ON INTEGRALS OF DISTRIBUTIONS

In Section 9.1, the concept of a smooth integral $\int_{\omega}^{x} f(t)\,dt^k$ where ω is a smooth function, was introduced. This type of integral is a regular operation, in contrast to the ordinary integral $\int_{x_0}^{x} f(t)\,dt^k$, where x_0 is a point, which cannot be defined for arbitrary distributions. However, if the distribution f vanishes in some neighbourhood of x_0 and the carrier of ω is inside that neighbourhood, then the integral $\int_{x_0}^{x} f(t)\,dt^k$ does exist as a generalized operation (see Section 15.1) and is equal to the smooth integral.

In fact, let f be a distribution in a given bounded or unbounded interval $I \subset \boldsymbol{R}^q$. We assume that $f = 0$ in some subinterval $A \subset I$. We assume, moreover, that ω is a smooth function whose carrier is inside A, such that $\int_I \omega = 1$. We recall that, for $k \in \boldsymbol{P}^q$,

$$\int_{\omega}^{x} f(t)\,dt^k = \int_I \omega(y)\,dy \int_y^x f(t)\,dt^k \quad \text{and} \quad \int_{x_0}^{x} f(t)\,dt^k = \lim_{n\to\infty} \int_{x_0}^{x} f_n(t)\,dt^k,$$

where $f_n = f * \delta_n$ and δ_n is a delta-sequence. But, if $x_0 \in A$, we can write

$$\int_y^x f(t)\,dt^k = \lim_{n\to\infty} \int_y^x f_n(t)\,dt^k = \lim_{n\to\infty} \int_{x_0}^{x} f_n(t)\,dt^k$$

for $y \in A$, since $f = 0$ in A. Hence in the present situation, $\int_y^x f(t)\,dt$ does not depend on y. We therefore have, as $\int_I \omega = 1$, the equality

$$\int_\omega^x f(t)\,dt^k = 1 \cdot \lim_{n\to\infty} \int_{x_0}^x f_n(t)\,dt^k.$$

This establishes the existence of $\int_{x_0}^x f$ and the equality of the two integrals. Note that in the formula

$$\Big(\int_{x_0}^x f(t)\,dt^k\Big)^{(m)} = \int_{x_0}^x f(t)\,dt^{k-m} \quad \text{for} \quad m \leqslant k,$$

if the jth coordinate of k vanishes, the value of the integral $\int_{x_0}^x f(t)\,dt^k$ is independent of the jth coordinate of x_0.

In particular, if f is either a continuous or an integrable function in the interval I, then the integrals defined above coincide with the integral $\int_{x_0}^x f(t)\,dt^k$ in the ordinary sense, so that the same symbol can be used for all of them without ambiguity.

15.4. DISTRIBUTIONS WITH A ONE-POINT CARRIER

According to the definition of a carrier (see Section 6.4), a point $x_0 \in \boldsymbol{R}^q$ is the carrier of a distribution f in $\boldsymbol{R}^q$, iff $f(x) = 0$ for $x \neq x_0$, but f is not equal to 0 troughout the whole space $\boldsymbol{R}^q$. In other words, iff f is a null distribution in the set $\boldsymbol{R}^q \setminus x_0(x_0)$, i.e., in the space $\boldsymbol{R}^q$ from which the point x_0 has been removed, but is not a null distribution in the whole space $\boldsymbol{R}^q$.

15.4.1. THEOREM. *If the carrier of a distribution f in $\boldsymbol{R}^q$ is the origin, then f is a linear combination of the delta-distribution and its derivatives. In other words, there exist an index $k \in \boldsymbol{P}^q$ and numbers α_m $(0 \leqslant m \leqslant k)$ such that*

$$f = \sum_{0 \leqslant m \leqslant k} \alpha_m \delta^{(m)}.$$

PROOF. Since f is tempered, there exist a continuous function F and an order $k = (\varkappa_1, \ldots, \varkappa_q)$ such that $f = F^{(k)}$ in $\boldsymbol{R}^q$. We may assume further that $k \geqslant 1$. By 9.1.2, the integral

$$G(x) = \int_{x_0}^x f(t)\,dt^k \qquad (x_0 \neq 0)$$

is also a continuous function in $\boldsymbol{R}^q$. Assume that $x_0 < 0$. Then $G(x) = 0$ in the open set O defined by the inequality $x \ngeq 0$. In fact, if $x_1 \in O$, there is an open interval I containing x_0 and x_1, such that $f = 0$ in I. This implies that $G(x_1) = 0$. Since we can choose x_1 arbitrarily within O, it follows that $G=0$ in O. We have

$$G^{(\varkappa_j e_j)}(x) = \int_{x_0}^{x} f(t)\, dt^{k-\varkappa_j e_j} \quad (j = 1, \ldots, q).$$

We now show that $G^{(\varkappa_j e_j)} = 0$ in the open set O defined by the inequality $x > 0$. In fact, since the jth coordinate of $k - \varkappa_j e_j$ is 0, it follows that $G^{(\varkappa_j e_j)}$ is independent of the jth coordinate of x_0. We may therefore write

$$G^{(\varkappa_j e_j)}(x) = \int_{x_j}^{x} f(t)\, dt^{k-\varkappa_j e_j},$$

where all the coordinates of x_j are negative, except for the jth coordinate ξ_j, which is positive. Now $f = 0$ in the half space U_j defined by the inequality $\xi_j > 0$. This implies that $G^{(\varkappa_j e_j)} = 0$ in U_j. But $O \subset U_j$, and so $G^{(\varkappa_j e_j)} = 0$ in O. By 15.2.1, G is, in O, a polynomial of degree less than k. Hence

$$G(x) = \sum_{0 \leqslant m \leqslant k-1} \beta_m \frac{x^m}{m!}.$$

Since G is continuous in $\boldsymbol{R}^q$ and vanishes in O and, consequently everywhere outside O, we can write

$$G = \sum_{0 \leqslant m \leqslant k-1} \beta_m G_m,$$

where $G_m(x) = x^m/m!$ for $x \geqslant 0$ and $G_m = 0$ otherwise. Hence $G_m^{(k)} = \delta^{(k-1-m)}$ and, consequently,

$$f = G^{(k)} = \sum_{0 \leqslant m \leqslant k-1} \beta_m \delta^{(k-1-m)} = \sum_{0 \leqslant m \leqslant k-1} \alpha_m \delta^{(m)},$$

where $\alpha_m = \beta_{k-1-m}$. Our assertion is thus proved. (The fact that we have $k-1$ instead of k in the summation sign is of no consequence).

16. Appendix

16.1. INDUCTION

As in Section 2.1, N denotes the set of all positive integers. Assume that, for each $n \in N$, we have an assertion A_n. Then the following *Induction Principle* holds:

If 1° *the assertion* A_1 *is true and* 2° A_n *implies* A_{n+1} *for every* $n \in N$, *then the assertion* A_n *is true for every* $n \in N$.

This Induction Principle can be regarded either as a theorem or as an axiom: the discussion of this point belongs in the Foundations of Mathematics. Here, we simply assume that the statement is true and we deduce some of its consequences.

We first recall some definitions and notation. Following Section 2.1, by a *non-negative integer point* p of $\boldsymbol{R}^q$ we mean any system $(\pi_1, \ldots, \pi_q)$ of non-negative integers π_j. The numbers $\pi_1, \ldots, \pi_q$ are the coordinates of the point p. The set of all non-negative integer points of $\boldsymbol{R}^q$ will be denoted by $\boldsymbol{P}^q$. Clearly $\boldsymbol{P}^q \subset \boldsymbol{R}^q$. We denote by e_j $(j = 1, \ldots, q)$ that point of $\boldsymbol{P}^q$ all of whose coordinates are 0, except for the jth coordinate which is 1. We denote by 0 the point all of whose coordinates are 0; this should not give rise to any confusion. By $p+e_j$ we mean the point whose coordinates are the same as those of p, except for the jth coordinate which is π_j+1.

Assume that for each index $p \in \boldsymbol{P}^q$, we have an assertion A_p. Taking the ordinary Induction Principle above as a starting point, we are able to prove the following *Multidimensional Induction Principle*:

If 1* *the assertion* A_0 *is true and* 2* A_p *implies* A_{p+e_j} *for every* $p \in \boldsymbol{P}^q$ *and* $j = 1, \ldots, q$, *then the assertion* A_p *is true for every* $p \in \boldsymbol{P}^q$.

In fact, having first introduced the symbol $\bar{p}$ to denote the sum of the coordinates of p, let B_n $(n \in N)$ be the assertion that all the A_p for which $\bar{p} \leqslant n-1$ are true. Then B_1 is true, since A_0 is true by 1*. Assume that B_n is true for some n. Then by 2*, A_{p+e_j} is true for every p with $\bar{p} \leqslant n-1$ and $j = 1, \ldots, q$. Hence B_{n+1} is true. By the ordinary Induction Principle, it follows that B_n is true for every $n \in N$. In other words that A_p is true for every $p \in \boldsymbol{P}^q$.

16.2. RECURSIVE DEFINITION

We now discuss something closely related to the Induction Principle, namely *recursive definition*. Let X be any non empty set of elements. We assume that, for every $m \in \boldsymbol{P}^q$ and every $j = 1, \ldots, q$, there is defined an operation which assigns to each $x \in X$, an element $F_{mj}(x) \in X$.

16.2.1. THEOREM. *If*

$$F_{m+e_i,j}(F_{mi}(x)) = F_{m+e_j,i}(F_{mj}(x)) \tag{1}$$

for every $m \in \boldsymbol{P}^q$, $i, j = 1, \ldots, q$, and every $x \in X$, then there exists an operation G_m for each $m \in \boldsymbol{P}^q$ such that $G_m(x) \in X$ for $x \in X$ and, moreover, 1′ $G_0(x) = x$, 2′ $G_{m+e_j}(x) = F_{mj}(G_m(x))$ for every $m \in \boldsymbol{P}^q$ and $j = 1, \ldots, q$. The operation G_m is uniquely defined for every $m \in \boldsymbol{P}^q$.

PROOF. Let B_n $(n \in N)$ be the assertion that all the G_m for which $\overline{m} \leqslant n-1$ are defined uniquely. Then B_1 is true, since the operation G_0 is defined uniquely; this follows from 1′ and from the fact that $m+e_j$ in 2′ is different from 0. Assume now that B_n is true for some n, so that given any m with $\overline{m} = n-1$ we can use 2′ to define G_{m+e_j}. To prove that this definition is consistent, we have to show that if $m+e_j = p+e_i$, then

$$F_{mj}(G_m(x)) = F_{pi}(G_p(x)).$$

This is trivially true if $i = j$. If $i \neq j$, we still have $m-e_i = p-e_j$, and so $G_{m-e_i}(x) = G_{p-e_j}(x)$. Hence, by (1)

$$F_{mj}(F_{m-e_i,i}(G_{m-e_i}(x))) = F_{pi}(F_{p-e_j,j}(G_{p-e_j}(x)))$$

and, by 2′,

$$F_{mj}(G_m(x)) = F_{pi}(G_p(x)).$$

Hence B_{n+1} is true. By ordinary induction it follows that B_n is true for every $n \in N$. This means that G_m is defined uniquely for every $m \in \boldsymbol{P}^q$, and the theorem is proved.

The preceding theorem shows under what conditions a recursive definition of G_{m+e_j} by means of G_m is consistent. In particular, if $q = 1$, we always have $i = j$ so that formula (1) holds automatically. Hence is the one-dimensional case, the conditions 1′ and 2′ alone suffice to define $G_0, G_1, \ldots$ and no further conditions are required. Condition 2′ is often called a *recursive formula*.

It should be noted that the proof of the consistency of a recursive definition (or definition by induction) in the multidimensional case does not depend on multidimensional induction. It is based on the ordinary Induction Principle.

16.3. EXAMPLES

EXAMPLE 1. The class $C^\infty(O)$ of smooth functions in an open subset O of $\boldsymbol{R}^q$ can be defined as the intersection of all classes K of continuous functions in O such that $f \in K$ implies $\frac{\partial}{\partial \xi_j} f \in K$ for every $j = 1, \ldots, q$. An advantage of such a definition is that it involves only the first order derivatives. We can then legitimately define the derivative of any given order $m \in \boldsymbol{P}^q$, by means of recursive definition.

Clearly

$$\frac{\partial}{\partial \xi_i}\left(\frac{\partial}{\partial \xi_j} f\right) = \frac{\partial}{\partial \xi_j}\left(\frac{\partial}{\partial \xi_i} f\right)$$

for $f \in C^\infty(O)$. To define $f^{(m)}$ ($m \in \boldsymbol{P}^q$) we take $C^\infty(O)$ to be the set X and we put $F_{mj}(f) = \frac{\partial}{\partial \xi_j} f$, so that F_{mj} does not in fact depend on m in this case. We then set $f^{(m)} = G_m(f)$ by definition. By 1′, we have $f^{(0)} = f$. Condition 2′ implies that $\frac{\partial}{\partial \xi_j} f^{(m)} = f^{(m+e_j)}$. By Theorem 16.2.1 the derivative $f^{(m)}$ is now defined for every $m \in \boldsymbol{P}^q$.

We have still to prove the formula

$$(f^{(k)})^{(m)} = f^{(k+m)} \quad \text{for} \quad k, m \in \boldsymbol{P}^q, \tag{1}$$

which we do by means of q-dimensional induction. The formula involves two indices, k and m, but the induction is carried out only with respect to m. We take for A_m the assertion that formula (1) holds for the given m and all $k \in \boldsymbol{P}^q$. Assertion A_m is clearly true for $m = 0$. Condition 1° is therefore satisfied. Assume that assertion A_m is true for some $m \in \boldsymbol{P}^q$. Then

$$(f^{(k)})^{(m+e_j)} = \frac{\partial}{\partial \xi_j}(f^{(k)})^{(m)} = \frac{\partial}{\partial \xi_j} f^{(k+m)} = f^{(k+m+e_j)}$$

so that condition 2° is satisfied. Formula (1) therefore holds for all m and all k.

From (1) it follows immediately that

$$(f^{(k)})^{(m)} = (f^{(m)})^{(k)} \quad \text{for} \quad k, m \in \boldsymbol{P}^q.$$

EXAMPLE 2. If $m = (\mu_1, \ldots, \mu_q) \in \boldsymbol{P}^q$, we write

$$m! = \mu_1! \ldots \mu_q!.$$

This equality camouflages a recursive definition with respect to q. Moreover, the symbol $\mu_j!$ also requires a recursive definition. We show how it is possible to define $m!$ using a single q-dimensional recursive definition.

We take $\boldsymbol{P}^q$ for X and we put $F_{mj}(x) = (\mu_j+1)x$. It is easy to see that F_{mj} satisfies condition 16.2 (1). By 16.2.1, there is thus a unique operation G_m for each $m \in \boldsymbol{P}^q$, such that 1′ $G_0(x) = x$ and 2′ $G_{m+e_j}(x) = (\mu_j+1)G_m(x)$. We put $m! = G_m(1)$ by definition. From 1′ and 2′ we then obtain the formulae

$$0! = 1, \tag{2}$$

$$(m+e_j)! = (\mu_j+1)m!. \tag{3}$$

16.4. FINITE INDUCTION

We assume that for every index $m \in \boldsymbol{P}^q$ satisfying $m_1 \leqslant m \leqslant m_2$ there is given an assertion B_m. The following *Finite Induction Principle* then holds:

If 1″ *the assertion* B_{m_1} *is true and* 2″ B_m *implies* B_{m+e_j} *for* $m_1 \leqslant m \leqslant m_2-e_j$ *and* $j = 1, \ldots, q$, *then the assertion* B_m *is true for every* $m \in \boldsymbol{P}^q$ *satisfying* $m_1 \leqslant m \leqslant m_2$.

This can be proved as follows. Let $A_p = B_{m_1+p}$ for $0 \leqslant p \leqslant m_2-m_1$ and $A_p = B_{m_2}$ for all the other $p \in \boldsymbol{P}^q$. It is now easy to see that the assertions A_n satisfy conditions 1* and 2* of Section 16.1, so that all the A_n are true. Hence all the original assertions B_m are also true.

16.5. NEWTON'S SYMBOL IN THE MULTIDIMENSIONAL CASE

Let $m = (\mu_1, \ldots, \mu_q) \in \boldsymbol{P}^q$ and $k = (\varkappa_1, \ldots, \varkappa_q) \in \boldsymbol{P}^q$. If $0 \leqslant m \leqslant k$ (i.e., if $0 \leqslant \mu_j \leqslant \varkappa_j$ for $j = 1, \ldots, q$), we write, by definition,

$$\binom{k}{m} = \frac{k!}{m!(k-m)!}. \tag{1}$$

From 16.3 (2), it follows that

$$\binom{k}{0} = \binom{k}{k} = 1, \tag{2}$$

and that

$$\binom{k}{m} = \binom{k}{k-m}. \tag{3}$$

Moreover, using property 16.3 (3) of the factorial, we get

$$(\mu_j+1)\binom{k}{m+e_j} = (\varkappa_j-\mu_j)\binom{k}{m} \tag{4}$$

and

$$(\mu_j+1)\binom{k+e_j}{m+e_j} = (\varkappa_j+1)\binom{k}{m}, \tag{5}$$

provided $0 \leqslant m \leqslant k$.

We recall that $\boldsymbol{B}^q$ denotes the set of all integer points of $\boldsymbol{R}^q$, i.e., of those points all of whose coordinates are integers. Clearly, $\boldsymbol{P}^q \subset \boldsymbol{B}^q \subset \boldsymbol{R}^q$. It will be convenient in the sequel to have defined $\binom{k}{m}$ for $k \in \boldsymbol{P}^q$ and for all $m \in \boldsymbol{B}^q$. We extend the definition (1) by setting $\binom{k}{m} = 0$, whenever m does not satisfy $0 \leqslant m \leqslant k$. We show that formulae (3), (4) and (5) then hold for all $m \in \boldsymbol{B}^q$ and $k \in \boldsymbol{P}^q$. In fact, this is trivial for (3). If $m \nleqslant k$, then also $m+e_j \nleqslant k$ and $m+e_j \nleqslant k+e_j$, and so all the factorials in (4) and (5) are equal to 0. If $0 \nleqslant m$ and $\mu_j+1 = 0$, all the members in (4) and (5) are again equal to 0. It thus only remains to consider the case when $0 \nleqslant m$ and $\mu_j+1 \neq 0$. However it is not difficult to see that then $0 \nleqslant m+e_j$ which together with $0 \nleqslant m$ again implies that all the factorials in (4) and (5) are equal to 0. Formulae (4) and (5) thus hold for all $m \in \boldsymbol{B}^q$ and $k \in \boldsymbol{P}^q$.

We have still to prove the equality

$$\binom{k}{m} + \binom{k}{m+e_j} = \binom{k+e_j}{m+e_j} \tag{6}$$

for all $k \in \boldsymbol{P}^q$ and $m \in \boldsymbol{B}^q$. In fact, if $\mu_j \neq -1$, we have by (4) and (5)

$$\binom{k}{m} + \binom{k}{m+e_j} = \frac{\varkappa_j+1}{\mu_j+1}\binom{k}{m} = \binom{k+e_j}{m+e_j}.$$

If $\mu_j = -1$, then $\binom{k}{m} = 0$ and, by 16.3(3),

$$\binom{k+e_j}{m+e_j} = \frac{k!(\varkappa_j+1)}{(m+e_j)!(k-m-e_j)!(\varkappa_j-\mu_j)} = \binom{k}{m+e_j},$$

which proves (6).

Note finally that, by suitably rearranging the factors in (1), we obtain the formula

$$\binom{k}{m} = \binom{\varkappa_1}{\mu_1}\cdots\binom{\varkappa_q}{\mu_q},$$

which, however, will not be used in the sequel.

16.6. THE FORMULAE OF LEIBNIZ AND OF SCHWARTZ

We prove the following formulae

$$(fg)^{(k)} = \sum_m \binom{k}{m} f^{(m)} g^{(k-m)}, \tag{1}$$

$$f^{(k)}g = \sum_m (-1)^m \binom{k}{m} (fg^{(m)})^{(k-m)}, \tag{2}$$

where f, g are arbitrary smooth functions and $(-1)^m = (-1)^{\mu_1+\cdots+\mu_q}$. Formula (1) is usually attributed to Leibniz; formula (2) is frequently used in Schwartz's famous book on Distributions.

The sums $\sum_m$ in (1) and (2) are usually taken to range over all $m \in \boldsymbol{P}^q$ satisfying $0 \leqslant m \leqslant k \in \boldsymbol{P}^q$. However, according to our definition of $\binom{k}{m}$ given in the preceding section, we can just as easily allow m to range over all $m \in \boldsymbol{B}^q$, without changing the sense of the formulae.

More generally, if we have a sum $\sum_m a_m$ of a finite number of elements a_m, we may always interpret it as a sum over all $m \in \boldsymbol{B}^q$, provided the additional elements are all equal to 0. This interpretation is advantageous because we can then write

$$\sum_m a_m + \sum_m b_m = \sum_m (a_m + b_m), \tag{3}$$

which could lead to confusion, if the sums on the left hand side ranged over different finite sets of indices. We may also write, in general

$$\sum_m a_{m+p} = \sum_m a_m \tag{4}$$

for any fixed $p \in \boldsymbol{B}^q$. All that we require, when writing down any of the formulae 16.3 (1), (3) and (4), is that addition should be associative and commutative.

We prove (1) and (2) by q-dimensional induction. Firstly when $k = 0$, both the formulae reduce to $fg = fg$, and so they are trivially true in this case. Assume now that they hold for some k. Then for (1) we have, by 16.3 (1)

$$\begin{aligned}(fg)^{(k+e_j)} &= ((fg)^{(k)})^{(e_j)} = \left(\sum_m \binom{k}{m} f^{(m)} g^{(k-m)}\right)^{(e_j)} \\ &= \sum_m \binom{k}{m} \left(f^{(m+e_j)} g^{(k-m)} + f^{(m)} g^{(k-m+e_j)}\right) \\ &= \sum_m \binom{k}{m} f^{(m+e_j)} g^{(k-m)} + \sum_m \binom{k}{m} f^{(m)} g^{(k-m+e_j)}.\end{aligned}$$

Applying formula (4) with $p = -e_j$ to the first sum in the last line above, we get, by (3) and (6),

$$(fg)^{(k+e_j)} = \sum_m \left(\binom{k}{m-e_j} + \binom{k}{m}\right) f^{(m)} g^{(k+e_j-m)} = \sum_m \binom{k+e_j}{m} f^{(m)} g^{(k+e_j-m)}.$$

Hence (1) follows by induction.

For formula (2) we get, by 16.3 (1) and 16.6 (3)

$$f^{(k+e_j)}g = (f^{(e_j)})^{(k)}g = \sum_m (-1)^{(m)} \binom{k}{m} (f^{(e_j)}g^{(m)})^{(k-m)}$$

$$= \sum_m (-1)^{(m)} \binom{k}{m} ((fg^{(m)})^{(e_j)} - fg^{(m+e_j)})^{(k-m)}$$

$$= \sum_m (-1)^{(m)} \binom{k}{m} (fg^{(m)})^{(k-m+e_j)} - \sum_m (-1)^m \binom{k}{m} (fg^{(m+e_j)})^{(k-m)}.$$

Applying formula (4) with $p = -e_j$ to the last sum in the last line above, we get, by (3) and (6),

$$f^{(k+e_j)}g = \sum_m (-1)^{(m)} \left(\binom{k}{m-e_j} + \binom{k}{m} \right) (fg^{(m)})^{(k-m+e_j)}$$

$$= \sum_m (-1)^{(m)} \binom{k+e_j}{m} (fg^{(m)})^{(k-m+e_j)}.$$

Hence (2) follows by induction.

Bibliography

1. P. Antosik, On the Mikusiński Diagonal Theorem. *Bull. Acad. Pol. Sci., Sér. sci. math., astronom. et phys.*, 15 (4) (1971), 305–310.
2. P. Antosik, On the Modulus of a Distribution. *Bull. Acad. Pol. Sci., Sér. sci. math., astronom. et phys.*, 15 (10) (1967), 717–722.
3. P. Antosik, The Commutativity of the Limit with Regular Operations. *Bull. Acad. Pol. Sci., Sér. sci. math., astronom. et phys.*, 18(6) (1970), 325–327.
4. P. Antosik (П. Антосик), Порядок относительно меры и его применение к исследованию произведения обобщенных функции. *Studia Math.* 26 (1966), 247–262.
5. P. Antosik and J. Mikusiński, On Hermite Expansions. *Bull. Acad. Pol. Sci., Sér. sci. math., astronom. et phys.*, 16(10) (1968), 787–791.
6. J. M. Gelfand, G. E. Silov (И. М. Гельфанд, Г. Е. Шилов). *Обобщенные функции*, I, II. Moscow 1958.
7. A. Gonzalez-Dominguez and R. Scarfiello, Nota sobre la formula v.p. $\frac{1}{x}\delta = -\frac{1}{2}\delta'$. *Rev. de la Union Matem. Argen.* 1 (1956), 53–67.
8. I. Halperin, *Introduction to the Theory of Distribution.* Toronto 1952.
9. H. König, Neue Begründung der Theorie der Distribution. *Math. Nachr.* 9 (1953), 129–148.
10. J. Korevaar, Distributions Defined from the Point of View of Applied Mathematics. *Proceedings Kon. Nederl. Akad. Wetenschappen*, Series A, No. 2 and *Indag. Math.* 17.2 (1955), 368–383; ibidem Series A, 58.4 and 17.4 (1955), 463–503; ibidem Series A, 58.5 and 17.5 (1955), 563–674.
11. G. Köthe, *Topologische Lineare Räume, I.* Springer, 1966.
12. N. N. Lebiediew (Н. Н. Лебедев), *Специальные функции и их приложения.* Moscow 1953.
13. S. Łojasiewicz, Sur la valeur d'une distribution dans un point. *Bull. Pol. Acad. Sc. Cl. III*, 4 (1956), 239–242.
14. S. Łojasiewicz, Sur la valeur et le limite d'une distribution dans un point. *Studia Math.* 16 (1957), 1–36.
15. S. Łojasiewicz, J. Wloka und Z. Zieleźny, Über eine Definition des Wertes einer Distribution. *Bull. Acad. Pol. Sc. Cl. III*, 3 (1955), 479–481.
16. J. Mikusiński, Sur la méthode de généralisation de M. Laurent Schwartz et sur la convergence faible. *Fund. Math.* 35 (1948), 235–239.
17. J. Mikusiński, Une définition de distribution. *Bull. Acad. Pol. Sc. Cl. III*, 3 (1955), 589–591.
18. J. Mikusiński, A Constructive Theory of Tempered Distributions. *Bull. Acad. Pol. Sci., Sér. sci. math., astronom. et phys.*, 16 (9), (1968), 727–732.
19. J. Mikusiński, A Representation of Tempered Distributions. *Bull. Acad. Pol. Sci., Sér sci. math., astronom et phys.*, 15 (2) (1967), 103–104.

20. J. Mikusiński, Criteria of the Existence and of the Associativity of the Product of Distributions. *Studia Math.* 21 (1962), 253–259.

21. J. Mikusiński, Irregular Operations on Distributions. *Studia Math.* 20 (1961), 163 169.

22. J. Mikusiński, *Lectures on the Constructive Theory of Distributions*. Department of Mathematics, University of Florida, Gainesville 1969.

23. J. Mikusiński, On Convergence of Sequences of Periodic Distributions. *Studia Math.* 31 (1968), 1–14.

24. J. Mikusiński, On Functions and Distributions with a Vanishing Derivative. *Studia Math.* 32 (1969), 9–16.

25. J. Mikusiński, On Spaces of Sequences. *Bull. Acad. Pol. Sci., Sér. sci. math., astronom. et phys.*, 17 (1) (1969), 17–20.

26. J. Mikusiński, On the Square of the Dirac delta-distribution. *Bull. Acad. Pol. Sci., Sér. sci. math., astronom. et phys.*, 14 (9) (1966), 511–513.

27. J. Mikusiński, On the Value of a Distribution at a Point. *Bull. Acad. Pol. Sci., Sér. sci. math., astronom. et phys.*, 8 (10) (1960), 681–683.

28. J. Mikusiński, Sequential Theory of the Convolutions of Distributions. *Studia Math.* 29 (1968), 151–160.

29. J. Mikusiński, Une introduction élémentaire de la transformation de Fourier dans la théorie des distributions. *Mathematica* 8 (31), 1, (1966), 83–90.

30. J. Mikusiński and R. Sikorski, The Elementary Theory of Distributions I. *Rozprawy Mat.* 12 (1957).

31. J. Mikusiński and R. Sikorski, The Elementary Theory of Distributions II. *Rozprawy Mat.* 25 (1961).

32. L. Schwartz, Généralisation de la notion de fonction, de dérivation, de transformation de Fourier, et applications mathématiques et physiques. *Annales Univ. Grenoble* 21 (1945), 57–74.

33. L. Schwartz, *Théorie des distributions I.* Paris 1950.

34. L. Schwartz, *Théorie des distributions II.* Paris 1951.

35. J. Sebastiao e Silva, Sur une construction axiomatique de la théorie des distributions. *Univ. Lisboa, Revista Fac. Ci* (2) 4 (1955), 79–186.

36. R. Sikorski, On Substitution in the Dirac delta-distribution. *Bull. Acad. Pol. Sc.* (1960), 685–689.

37. R. Sikorski, A Definition of the Notion of Distribution. *Bull. Acad. Pol. Sc. Cl. III*, 2 (1954), 207–211.

38. R. Sikorski, Integrals of Distributions. *Studia Math.* 20 (1961), 119–139.

39. K. Skórnik, An Estimation of Fourier Coefficients of Periodic Distributions. *Bull. Acad. Pol. Sci., Sér. sci. math., astronom. et phys.*, 16 (7) (1968), 581–585.

40. K. Skórnik, Postać funkcji lokalnie całkowalnej, której *m*-ta pochodna lokalna znika prawie wszędzie. *Zeszyty Naukowe Wyższej Szkoły Pedagogicznej w Katowicach, Sekcja Matematyki*, Zeszyt nr 5, 1966, 127–152. (The Form of Locally Integrable Function whose *m*-th Derivative Vanishes Almost Everywhere).

41. K. Skórnik, Hereditarily Periodic Distributions. *Studia Math.* 43(1972), 245–272.

42. W. Słowikowski, A. Generalization of the Theory of Distributions. *Bull. Acad. Pol. Sc. Cl. III*, 3 (1955), 3–6.

43. W. Słowikowski, On the Theory of Operator Systems. *Bull. Acad. Pol. Sc. Cl. III*, 3 (1955), 137–142.

44. S. Soboleff, Méthode nouvelle á résoudre le probleme de Cauchy pour les équations hyperboliques normales. *Recueil Math.* (Math. Sbornik) 1 (1936), 39–71.
45. G. Temple, Theories and Applications of Generalized Functions. *Journ. London Math. Soc.* 28 (1953), 134–148.
46. Z. Zieleźny, Sur la définition de Łojasiewicz de la valeur d'une distribution dans un point. *Bull. Acad. Pol. Sc. Cl. III*, 3 (1955), 519–520.

Index of Authors and Terminology